I0043928

Arthur Cayley

An Elementary Treatise on Elliptic Functions

Arthur Cayley

An Elementary Treatise on Elliptic Functions

ISBN/EAN: 9783742807397

Manufactured in Europe, USA, Canada, Australia, Japa

Cover: Foto ©Thomas Meinert / pixelio.de

Manufactured and distributed by brebook publishing software
(www.brebook.com)

Arthur Cayley

An Elementary Treatise on Elliptic Functions

AN ELEMENTARY TREATISE

ON

ELLIPTIC FUNCTIONS

BY

ARTHUR CAYLEY,

SADLERIAN PROFESSOR OF PURE MATHEMATICS IN THE UNIVERSITY OF CAMBRIDGE.

CA

DEIGHT CO.

LO

PREFACE.

THE present treatise is founded upon Legendre's *Traité des Fonctions Elliptiques*, and upon Jacobi's *Fundamenta Nova*, and Memoirs by him in *Crelle's Journal*: comparatively very little use is made of the investigations of Abel or of those of later authors. I show how the transition is made from Legendre's Elliptic Integrals of the three kinds to Jacobi's Amplitude, which is the argument of the Elliptic Functions (the sine, cosine, and delta of the amplitude, or as with Gudermann I write them, sn, cn, dn), and also of Jacobi's functions Z, Π, which replace the integrals of the second and third kinds, and of the functions Θ, H, which he was thence led to. It may be remarked as regards the *Fundamenta Nova*, that in the first part Jacobi (so to speak) hurries on to the problem of transformation without any sufficient development of the theory of the elliptic functions themselves; and that in the concluding part, starting with the developments furnished by the transformation-formulæ, he connects with these, introducing them as the occasion arises, his new functions Z, Π, Θ, H: there are thus various points which require to be more fully discussed. Not included in the *Fundamenta Nova* we have the important theory of the partial differential equation satisfied by the functions Θ, H, and, deduced therefrom, the partial differential equations satisfied by the numerators and denominator in the theories of the multiplication and transformation of the elliptic functions: these I regard as essential parts of Jacobi's theory, and they are here considered accordingly. For further explanation of the range and plan of the present treatise the table of contents, and the first chapter entitled "General Outline," may be consulted. I am greatly indebted to Mr J. W. L. Glaisher of Trinity College for his kind assistance in the revision of the proof-sheets, and for many valuable suggestions.

Cambridge, 1876.

CONTENTS.

CHAPTER I.

General Outline.

CHAPTER II.

The Addition-Equation. Landen's Theorem.

C. *b*

CHAPTER III.

MISCELLANEOUS INVESTIGATIONS.

CHAPTER IV.

ON THE ELLIPTIC FUNCTIONS sn, cn, dn.

CHAPTER V.

THE THREE KINDS OF ELLIPTIC INTEGRALS.

CHAPTER VI.

THE FUNCTIONS $\Pi(u, a)$, Zu, Θu, Hu.

CHAPTER VII.

TRANSFORMATION. GENERAL OUTLINE.

CHAPTER VIII.

THE QUADRIC TRANSFORMATION $n = 2$; AND THE ODD-PRIME TRANSFORMATIONS $n = 3, 5, 7$. PROPERTIES OF THE MODULAR EQUATION AND THE MULTIPLIER.

CHAPTER IX.

JACOBI'S PARTIAL DIFFERENTIAL EQUATIONS FOR THE FUNCTIONS H, Θ, AND FOR THE NUMERATORS AND DENOMINATORS IN THE MULTIPLICATION AND TRANSFORMATION OF THE ELLIPTIC FUNCTIONS sn, cn, dn.

CHAPTER X.

TRANSFORMATION FOR AN ODD AND IN PARTICULAR AN ODD-PRIME ORDER: DEVELOPMENT OF THE THEORY BY MEANS OF THE n-DIVISION OF THE COMPLETE FUNCTIONS.

CHAPTER XI.

THE q-FUNCTIONS: FURTHER THEORY OF THE FUNCTIONS H, Θ.

CHAPTER XII.

REDUCTION OF A DIFFERENTIAL EXPRESSION $\frac{Rdx}{\sqrt{X}}$.

CHAPTER XIII.

QUADRIC TRANSFORMATION OF THE ELLIPTIC INTEGRALS OF THE FIRST AND SECOND KINDS: THE ARITHMETICO-GEOMETRICAL MEAN.

CHAPTER XIV.

THE GENERAL DIFFERENTIAL EQUATION $\frac{dx}{\sqrt{X}} = \frac{dy}{\sqrt{Y}}$.

CHAPTER XV.

ON THE DETERMINATION OF CERTAIN CURVES THE ARC OF WHICH IS REPRESENTED BY AN ELLIPTIC INTEGRAL OF THE FIRST KIND.

CHAPTER XVI.

ON TWO INTEGRALS REDUCIBLE TO ELLIPTIC INTEGRALS.

ADDITION.

FURTHER THEORY OF THE LINEAR AND QUADRIC TRANSFORMATIONS.

A student to whom the subject is new should peruse Chapter I., not dwelling upon it, but returning to it as he finds occasion; and he may afterwards in the first instance confine himself to Chapters II., III., IV., XII. and XIII.

ERRATA.

p. 4, lines 7 and 14, *for* argument *read* amplitude.
p. 64, line 14, *for* these *read* there.
p. 65, second formula, *for* $\operatorname{sn}(u+v)+\operatorname{sn}(u-v)$, *read* $\operatorname{cn}(u+v)+\operatorname{cn}(u-v)$.
p. 71, line 12, *for* $2K'$ *read* $2iK'$, and line 17, *for* $4K'$ *read* $4iK'$.
p. 107, line 8 from bottom, *for* $\frac{du}{1+n\operatorname{sn}^2u}$ *read* $\int_0 \frac{du}{1+n\operatorname{sn}^2u}$.
p. 113, line 4 from bottom, *for* $\Pi(u,a)=\Pi(a,u)$ *read* $\Pi(u,a)-\Pi(a,u)$.
p. 116, bottom line, *for* $\theta_1=\frac{\omega-1}{\omega+1}$ *read* $\sqrt{\theta_1}=\frac{\omega-1}{\omega+1}$.
p. 119, line 9, *for* $\frac{1}{2}$ *read* $\frac{1}{2\sqrt{-\rho}}$.
p. 226, top line, *dele* of the three functions of am.
p. 281, No. 371. It might have been proper to state explicitly that the square-brackets denote infinite products obtained by giving to m the values 0, 1, 2...to infinity.
p. 285, line 8 from bottom, *for* No. 372 *read* No. 378.
p. 298, bottom line, *before* p. 93, *insert* t. I.
p. 308, bottom line, *for* No. 306 *read* No. 307.
p. 328, line 3 from bottom, *for* $-b_1^2F(a,b)$ *read* $-b_1^2F(a_1,b_1)$.

Legendre's earliest systematic work on Elliptic Integrals is the *Exercices de Calcul Intégral sur divers ordres de Transcendantes, et sur les Quadratures*, Paris, t. I., 1811; t. III., 1816, and t. II., 1817: the later work is the *Traité des Fonctions Elliptiques et des Intégrales Euleriennes*, Paris, t. I., 1825; t. II., 1826, and t. III., 1828—32; the greater part of Legendre's own results on the theory of Elliptic Integrals, contained in the first volume of the *Fonctions Elliptiques*, had been already given in the first volume of the *Exercices*. Jacobi's work is the *Fundamenta Nova Theoriæ Functionum Ellipticarum*, Königsberg, 1829: the Memoirs in *Crelle's Journal* extend from 1828 to 1858, some of them, in connexion with the *Fundamenta Nova*, being published shortly after the date of that work. The Memoirs by Abel, published for the most part in the earlier volumes of *Crelle's Journal*, 1826 to 1829, are collected in the *Œuvres Complètes de N. H. Abel*, par B. Holmboe, Christiania, t. I. and II., 1839, except the great memoir on Transcendent Functions, presented to the French Academy, and published, *Mémoires des Savans Etrangers*, t. VII., 1841. A new edition of the works is in preparation.

CHAPTER I.

GENERAL OUTLINE.

Origin of the Elliptic Integrals. Art. Nos. 1 to 11.

1. We consider the integration of a differential expression

$$\frac{R\,dx}{\sqrt{X}},$$

where R is a rational function of x; X a rational and integral quartic function of x, with real coefficients*: the values of the variable x are real, and such that X is positive, or $\sqrt{X}$ real.

2. This can be by a real substitution $\dfrac{p+qx}{1+x}$ in place of x (that is a substitution where p and q are real) reduced to the form

$$\frac{R\,dx}{\sqrt{\pm(1 \pm mx^2)(1 \pm nx^2)}},$$

where R is a rational function of the new x; m and n are real and positive; and the signs are such that the function under the square root is not $-(1+mx^2)(1+nx^2)$.

* The references here and elsewhere to *reality*, and any references to sign or numerical limits, are regarded as in general holding good: it will be understood, however, that imaginary values might be admitted throughout, and the various theorems presented in a more general but less definite form: and there will be occasion to refer to and employ such extensions of the original real theory.

3. The rational function R is the sum of an even function and an odd function of x: the differential expression is thus separated into two parts; that depending on the odd function may be integrated by circular and logarithmic functions (as appears by making therein the substitution $\sqrt{x}$ in place of x); and there remains for consideration only the part depending on the even function of x: or, what is the same thing, we may take R to be an even function of x (that is a rational function of x^2).

4. This being so, we can by a real transformation $\frac{a+bx^2}{c+dx^2}$ in place of x^2 transform the differential expression into the form

$$\frac{R\,dx}{\sqrt{1-x^2 . 1-k^2x^2}},$$

where R is a rational function of the new x^2; k^2 is real, positive and less than 1 (and therefore also k, assumed to be the positive root of k^2, is real, positive and less than 1).

5. In the last-mentioned expression x^2 may be included between the limits 0, 1, or it may be $> \frac{1}{k^2}$; but in the latter case, we can by the substitution $\frac{1}{kx}$ in place of x transform the expression into one of the like form in which the new x^2 lies between the limits 0 and 1: we therefore assume that x^2 lies within these limits.

6. By decomposing R into an integral and fractional part, and the fractional part into simple fractions, and by integrating by parts, the integration is made to depend upon that of the three terms

$$\frac{dx}{\sqrt{1-x^2 . 1-k^2x^2}}, \quad \frac{x^2\,dx}{\sqrt{1-x^2 . 1-k^2x^2}}, \quad \frac{dx}{(1+nx^2)\sqrt{1-x^2 . 1-k^2x^2}},$$

where n is real or imaginary.

Or, what is the same thing, the three terms may be taken to be

$$\frac{dx}{\sqrt{1-x^2 \,.\, 1-k^2x^2}},\quad \frac{(1-k^2x^2)\,dx}{\sqrt{1-x^2 \,.\, 1-k^2x^2}},\quad \frac{dx}{(1+nx^2)\sqrt{1-x^2 \,.\, 1-k^2x^2}},$$

that is

$$\frac{dx}{\sqrt{1-x^2 \,.\, 1-k^2x^2}},\quad \frac{\sqrt{1-k^2x^2}\,dx}{\sqrt{1-x^2}},\quad \frac{dx}{(1+nx^2)\sqrt{1-x^2 \,.\, 1-k^2x^2}}.$$

7. Writing herein $x = \sin\phi$, and putting for shortness

$$\sqrt{1-k^2\sin^2\phi} = \Delta(k, \phi),$$

these are $\dfrac{d\phi}{\Delta(k,\phi)}$, $\Delta(k,\phi)\,d\phi$, and $\dfrac{d\phi}{(1+n\sin^2\phi)\,\Delta(k,\phi)}$,

and we have thus the three kinds of Elliptic *Integrals*: viz. these are

first kind $\quad F(k,\phi) \quad = \displaystyle\int \frac{d\phi}{\Delta(k,\phi)}$,

second kind $\quad E(k,\phi) \quad = \displaystyle\int \Delta(k,\phi)\,d\phi$,

third kind $\quad \Pi(n,k,\phi) = \displaystyle\int \frac{d\phi}{(1+n\sin^2\phi)\,\Delta(k,\phi)}$,

the integral being in each case taken from $\phi = 0$ up to the arbitrary value ϕ. It would of course be allowable under the integral sign to write for ϕ any other letter θ, taking the integral from $\theta = 0$ to $\theta = \phi$.

8. ϕ is the amplitude, k the modulus, n the parameter. The amplitude is a real angle; as already mentioned, the modulus k is positive and less than 1; whence also k', $= \sqrt{1-k^2}$, called the complementary modulus, is real, positive and less than 1. Moreover $\Delta(k,\phi)$, $= \sqrt{1-k^2\sin^2\phi}$, does not become $= 0$, nor consequently change its sign, and it is taken to be always positive. The parameter n, as already mentioned, may be real or imaginary: it is in the first instance taken to be real; and it will appear that the case where it is imaginary can be made

to depend upon that in which it is real. Supposing it to be real, there is a distinction according as it is negative and greater than 1 (viz. in this case the denominator $1+n\sin^2\phi$ becomes $=0$ for a real value of ϕ); or else as it is negative and less than 1, or positive.

9. Instead of the complete notation $\Delta(k,\phi)$, we frequently express only the argument, and write simply $\Delta\phi$; and similarly $F\phi$, $E\phi$, $\Pi\phi$ for $F(k,\phi)$, &c. respectively: viz. in these cases it is assumed to be understood what the unexpressed letters k, or k and n, are. We may in like manner express only the modulus, or the parameter, and write Δk, $\Pi(n,k)$, Πn, or Πk &c., but there is less frequent occasion for this, and the notations when used will be explained.

10. The integrals, taken up to the value $\frac{\pi}{2}$ of the argument, are said to have their complete values, and these are frequently denoted by means of a subscript unity; thus $F\left(k, \frac{\pi}{2}\right)=F_1k$, or simply F_1; and so E_1k, E_1, &c.

11. The three elliptic integrals are not on a par with each other; but they depend, the second and third kinds upon the first kind; or we may say* that they all three depend on the differential expression $\frac{d\phi}{\Delta(k,\phi)}$: thus there is for each of them an addition-theory depending on the integration of the differential equation

$$\frac{d\phi}{\Delta(k,\phi)}+\frac{d\psi}{\Delta(k,\psi)}=0,$$

not for the first kind a theory depending on this equation and for the other two kinds like theories depending on the equations

$$\Delta(k,\phi)\,d\phi+\Delta(k,\psi)\,d\psi=0,$$

$$\frac{d\phi}{(1+n\sin^2\phi)\,\Delta(k,\phi)}+\frac{d\psi}{(1+n\sin^2\psi)\,\Delta(k,\psi)}=0,$$

respectively: these last are equations not admitting of algebraic integration, and which do not present themselves in the theory. And the like as regards multiplication and transformation.

* The statement is made provisionally: the three kinds, as will appear, depend each of them on the functions sn u, cn u, dn u.

The Addition-Theory. Art. Nos. 12 to 14.

12. The differential equation

$$\frac{dx}{\sqrt{X}} + \frac{dy}{\sqrt{Y}} = 0,$$

where Y is the same quartic function of y that X is of x, admits of algebraic integration: and in particular this is the case with the equation

$$\frac{dx}{\sqrt{1-x^2 . 1-k^2x^2}} + \frac{dy}{\sqrt{1-y^2 . 1-k^2y^2}} = 0,$$

and in this last equation we may take the constant of integration, say m, to be the value of either of the variables x, y, when the other of them is put $= 0$.

Writing $x = \sin\phi$, $y = \sin\psi$, $m = \sin\mu$, we obtain for the differential equation

$$\frac{d\phi}{\Delta\phi} + \frac{d\psi}{\Delta\psi} = 0,$$

an algebraic integral such that the constant of integration μ is the value of either of the variables ϕ, ψ when the other of them is $= 0$; viz. this is an integral involving the sines and cosines of ϕ, ψ (and μ), but which (as being algebraic in regard to these sines and cosines) is spoken of as an algebraic integral.

13. The integral in question, say the addition-equation, may be expressed in (among others) the various forms

$$\cos\mu = \cos\phi\cos\psi - \sin\phi\sin\psi\Delta\mu,$$

$$\cos\phi = \cos\psi\cos\mu + \sin\psi\sin\mu\Delta\phi,$$

$$\cos\psi = \cos\phi\cos\mu + \sin\phi\sin\mu\Delta\psi,$$

$$1 - \cos^2\phi - \cos^2\psi - \cos^2\mu + 2\cos\phi\cos\psi\cos\mu \\ - k^2\sin^2\phi\sin^2\psi\sin^2\mu = 0,$$

$$\sin\mu = \sin\phi\cos\psi\Delta\psi + \sin\psi\cos\phi\Delta\phi \quad (+)^*,$$

$$\cos\mu = \cos\phi\cos\psi - \sin\phi\sin\psi\Delta\phi\Delta\psi \quad (+),$$

$$\Delta\mu = \Delta\phi\Delta\psi - k^2\sin\phi\sin\psi\cos\phi\cos\psi \quad (+),$$

* The notation hardly requires explanation; the (+) shows that the function is a fraction the numerator of which is written down, and the denominator of

where in each case there is a denominator

$$= 1 - k^2 \sin^2\phi \sin^2\psi,$$

and

$$\begin{aligned}\sin\phi &= \sin\mu \cos\psi \Delta\psi - \sin\psi \cos\mu\Delta\mu \qquad (+),\\ \cos\phi &= \cos\mu \cos\psi + \sin\mu \sin\psi\Delta\mu\Delta\psi \qquad (+),\\ \Delta\phi &= \Delta\mu\,\Delta\psi + k^2 \sin\mu \sin\psi \cos\mu \cos\psi \qquad (+),\end{aligned}$$

where there is a denominator

$$= 1 - k^2 \sin^2\mu \sin^2\psi,$$

and in these last formulæ we may interchange ϕ, ψ.

14. It is to be remarked that considering μ as variable we have

$$\frac{d\phi}{\Delta\phi} + \frac{d\psi}{\Delta\psi} = \frac{d\mu}{\Delta\mu},$$

viz. the addition-equation is (not the general, but) a particular integral of this differential equation. Writing this equation under the forms

$$\frac{d\phi}{\Delta\phi} = \frac{d\mu}{\Delta\mu} - \frac{d\psi}{\Delta\psi}, \quad \frac{d\psi}{\Delta\psi} = \frac{d\mu}{\Delta\mu} - \frac{d\phi}{\Delta\phi},$$

we naturally regard the integral equation, any form of it which gives μ in terms of ϕ, ψ, as an addition-equation: and any form of it which gives ϕ or ψ in terms of μ, and ψ or ϕ, as a subtraction-equation. The resulting notion of subtraction may be regarded as included in that of addition, and it will hardly be necessary again to refer to it.

The Addition of the three kinds of Elliptic Integrals.
Art. Nos. 15 to 17.

15. We assume throughout ϕ, ψ, μ to be connected by the foregoing addition-equation: recollecting that this is an integral (taken with the constant determined as above) of the differential equation $\frac{d\phi}{\Delta\phi} + \frac{d\psi}{\Delta\psi} = 0$, and reverting to the definition of

which is afterwards stated: it is, I think, a very useful one generally, but there is in Elliptic Functions an especial need of it, from the frequent occurrence therein of groups of complicated algebraical fractions having the same denominator.

the function $F\phi$, it at once appears that for the first kind of elliptic function we have

$$F\phi + F\psi - F\mu = 0$$

(viz. ϕ, ψ, μ being connected as above, the integrals $F\phi$, $F\psi$, $F\mu$ satisfy this relation): this is the addition-theorem for the first kind of elliptic integrals.

16. It can be shown that for the second kind

$$E\phi + E\psi - E\mu = k^2 \sin\phi \sin\psi \sin\mu;$$

and that for the third kind

$$\Pi\phi + \Pi\psi - \Pi\mu = \frac{1}{\sqrt{a}} \tan^{-1} \cdot \frac{n\sqrt{a} \sin\mu \sin\phi \sin\psi}{1 + n - n\cos\mu\cos\phi\cos\psi},$$

$$= \frac{1}{2\sqrt{-a}} \log \frac{1 + n - n\cos\mu\cos\phi\cos\psi + n\sqrt{-a}\sin\mu\sin\phi\sin\psi}{1 + n - n\cos\mu\cos\phi\cos\psi - n\sqrt{-a}\sin\mu\sin\phi\sin\psi},$$

where $a = (1+n)\left(1 + \frac{k^2}{n}\right)$, and n being real, the first or second form is real according as a is positive or negative.

17. The mode of verification is obvious; in fact, representing either of the last-mentioned equations by $U = 0$, and considering U as a function of the variables ϕ, ψ, we have

$$dU = \frac{dU}{d\phi} d\phi + \frac{dU}{d\psi} d\psi$$

$$= \left(\frac{dU}{d\phi}\Delta\phi - \frac{dU}{d\psi}\Delta\psi\right)\frac{d\phi}{\Delta\phi},$$

so that to sustain the assumed equation $U = 0$, we must in virtue of the addition-equation have identically

$$\frac{dU}{d\phi}\Delta\phi - \frac{dU}{d\psi}\Delta\psi = 0,$$

viz. this equation, if true at all, can be nothing else than a form of the addition-equation: or what is the same thing, the addition-equation will be reducible into the last-mentioned form: which being so, it gives $dU = 0$, and thence by integration $U =$ const., and then determining the constant by the condition that for $\psi=0$ the value of ϕ is $=\mu$, the value of the constant must come $=0$; and in this manner we must from the addition-equation arrive at the required equation $U=0$.

The Elliptic Functions am; sinam, cosam, Δ am; *or* sn, cn, dn.
Art. Nos. 18 to 27.

18. We have spoken of ϕ as the amplitude of $F\phi$; or writing $F\phi=u$, then ϕ is the amplitude of u; say $\phi=\text{am}\,u$, and then $\sin\phi$, $\cos\phi$, $\Delta\phi$ are the sine, cosine, and delta of am u, say these are

$$\sin.\text{am}\,u,\ \cos.\text{am}\,u,\ \Delta.\text{am}\,u,$$

which may also be written

$$\text{sinam}\,u,\ \text{cosam}\,u,\ \Delta\,\text{am}\,u,$$

or as an abbreviation

$$\text{sn}\,u,\ \text{cn}\,u,\ \text{dn}\,u.$$

19. But in adopting the last-mentioned forms we introduce a new mode of looking at the functions; viz. sn u is a sort of sine-function, and cn u, dn u are sorts of cosine-functions of u; these are called Elliptic *Functions;* and we may develope the theory from this point of view. Observe that the fundamental equation is $u=F\phi$ or $d\phi=\Delta\phi\,du$: this may be written

$$d\sin\phi=\cos\phi\,\Delta\phi\,du, \text{ or since } \sin\phi=\text{sn}\,u, \text{ this is}$$

$$d\,\text{sn}\,u=\text{cn}\,u\,\text{dn}\,u\,du: \text{ say } \text{sn}'\,u=\text{cn}\,u\,\text{dn}\,u,$$

moreover

$$\text{cn}^2\,u=1-\ \text{sn}^2\,u,$$

$$\text{dn}^2\,u=1-k^2\,\text{sn}^2\,u,$$

and differentiating and substituting for sn$'u$ its value, we find

$$\text{cn}'\,u=-\ \ \text{sn}\,u\,\text{dn}\,u,$$

$$\text{dn}'\,u=-k^2\,\text{sn}\,u\,\text{cn}\,u,$$

and as above

$$\text{sn}'\,u=\ \ \text{cn}\,u\,\text{dn}\,u,$$

which five equations constitute a foundation of the theory. Observe also that $\text{sn}\,0=0$, $\text{cn}\,0=1$, $\text{dn}\,0=1$, $\text{sn}\,(-u)=-\text{sn}\,u$, $\text{cn}\,(-u)=\text{cn}\,u$, $\text{dn}\,(-u)=\text{dn}\,u$.

20. But this theory is already furnished by the addition-equation; viz. starting from the equation $F\phi + F\psi = F\mu$, then writing $F\phi = u$, $F\psi = v$ (and therefore $\phi = \text{am}\, u$, $\psi = \text{am}\, v$) we have $F\mu = u + v$ or $\mu = \text{am}\,(u+v)$: the equations which determine $\sin\mu$, $\cos\mu$, $\Delta\mu$ in terms of the sin, cos, and Δ of ϕ and ψ give the sn, cn and dn of $u+v$ in terms of those of u and v: viz. these equations are

$$\text{sn}\,(u+v) = \text{sn}\,u\;\text{cn}\,v\;\text{dn}\,v + \text{sn}\,v\;\text{cn}\,u\;\text{dn}\,u \qquad (\div),$$

$$\text{cn}\,(u+v) = \text{cn}\,u\;\text{cn}\,v - \text{sn}\,u\;\text{dn}\,u\;\text{sn}\,v\;\text{dn}\,v \qquad (\div),$$

$$\text{dn}\,(u+v) = \text{dn}\,u\;\text{dn}\,v - k^2\,\text{sn}\,u\;\text{cn}\,u\;\text{sn}\,v\;\text{cn}\,v \qquad (\div),$$

where the denominator is

$$= 1 - k^2\,\text{sn}^2 u\;\text{sn}^2 v,$$

and we may on the left-hand sides write $u - v$ instead of $u + v$, changing in each of the three numerators the sign of the second term.

21. These equations may be obtained independently: viz. in any one of them differentiating the right-hand side in regard to u and substituting for $\text{sn}'u$, $\text{cn}'u$, $\text{dn}'u$, their values, we obtain a symmetrical function of u, v; hence the same result as would have been obtained by differentiating in regard to v: the expression in question is thus a function of $u + v$; and writing therein $v = 0$, we find it to be the sn, cn or dn (as the case may be) of $u + v$; which proves the equations.

22. We thus see that F is an inverse function, the direct function being sn; and that cn, dn are connected therewith as the cosine with the sine. It may be remarked that there are six quotients, sn ÷ cn, sn ÷ dn; cn ÷ sn, dn ÷ sn; dn ÷ cn, cn ÷ dn, which are in some sort analogous to the functions tan, cot: if all these functions had to be considered, appropriate notations would be $\frac{\text{sn}}{\text{cn}}$, &c. $\left(\text{viz. } \frac{\text{sn}\,u}{\text{cn}\,u} = \frac{\text{sn}}{\text{cn}}\,u, \text{\&c.}\right)$. These are not required: it is however in some of the formulæ convenient to have a symbol for the single quotient sn ÷ cn: and considering this

as standing for sin . am ÷ cos . am, it is = tan . am, and we accordingly write it as tn: viz. we have $\frac{\text{sn}\, u}{\text{cn}\, u}$, = tan . am u, = tn u.

23. In further illustration suppose that the theory of the circular functions sine, cosine, was unknown, and that we had defined Fx to be the function

$$\int_0 \frac{dx}{\sqrt{1-x^2}}.$$

Then taking the variables x, y to be connected by the differential equation

$$\frac{dx}{\sqrt{1-x^2}} + \frac{dy}{\sqrt{1-y^2}} = 0,$$

and supposing that z is the value of y answering to $x = 0$, we have

$$Fx + Fy = Fz.$$

But the differential equation admits of algebraic integration: and determining in each case the constant by the condition that for $x = 0$, y shall be $= z$, the algebraic integral may be expressed in the two forms

$$x\sqrt{1-y^2} + y\sqrt{1-x^2} \qquad = z,$$

$$xy - \sqrt{1-x^2}\sqrt{1-y^2} = \sqrt{1-z^2},$$

so that either of these equations represents the above-mentioned transcendental integral; and we have thus a circular theory precisely analogous to the elliptic theory in its original form. But here the function Fx is the inverse function $\sin^{-1}x$, and the last-mentioned two equations are the equivalents of the equation

$$\sin^{-1}x + \sin^{-1}y = \sin^{-1}z,$$

whence writing $\sin^{-1}x = u$, $\sin^{-1}y = v$, and therefore $x = \sin u$, $y = \sin v$, $z = \sin(u+v)$: also assuming $\sqrt{1-\sin^2 u} = \cos u$, and therefore $\sqrt{1-\sin^2 v} = \cos v$, and $\sqrt{1-\sin^2(u+v)} = \cos(u+v)$, the equations in question become

$$\sin(u+v) = \sin u \cos v + \sin v \cos u,$$

$$\cos(u+v) = \cos u \cos v - \sin u \sin v,$$

and it is clearly convenient to use these functions sin, cos in place of F, denoting as above $\sin^{-1}$.

24. In the theory of the circular functions we have an addition-theory, which gives rise to and may be considered as including a subtraction-theory: and this leads to a multiplication- and division-theory: viz. we find from $\sin u$, $\cos u$, the functions $\sin$ or $\cos nu$, $\sin$ or $\cos \frac{m}{n}u$; we have similarly for the elliptic functions sn, cn, dn a multiplication- and division-theory. These will be considered in detail; they are referred to here only for the sake of the remark that there is for the elliptic functions a "transformation-theory" which has no analogue in the circular functions, viz. we determine in terms of the functions of u the like functions with an argument $\frac{u}{M}$, and a new modulus λ in place of the original k: the transformation is of any integer order n, and there is, for each value of n, a relation called the modular equation between k and the new modulus λ. And it is convenient to notice that in the multiplication-theory the sn, cn and dn of nu, and in the transformation-theory the same functions of $\left(\frac{u}{M}, \lambda\right)$, are fractions having a common denominator, so that in each case there are three numerators and a denominator which come into consideration.

25. The circular theory gives rise to a numerical transcendant π, viz. $\frac{\pi}{2} = \frac{1}{2}3{\cdot}14159\ldots$ is a quantity such that $\sin\frac{\pi}{2} = 1$, $\cos\frac{\pi}{2} = 0$, $\frac{\pi}{2}$ being the smallest positive value of the argument for which the two functions have these values: and in developing the theory from the integral $\int\frac{dx}{\sqrt{1-x^2}}$, $\frac{\pi}{2}$ would be the complete function defined from the equation

$$\frac{\pi}{2} = \int_0^1 \frac{dx}{\sqrt{1-x^2}}.$$

Moreover the circular functions are periodic, having for their common period four times this quantity, $=2\pi$: viz. we have

$$\frac{\sin}{\cos}(u+2\pi)=\frac{\sin}{\cos}u.$$

26. Corresponding to $\frac{\pi}{2}$ we have in elliptic functions in the first instance the complete function F_1k, also denoted by K, viz. K is a real positive quantity defined by the equation

$$K=\int_0^{\frac{\pi}{2}}\frac{d\phi}{\sqrt{1-k^2\sin^2\phi}},$$

or, what is the same thing,

$$K=\int_0^1\frac{dx}{\sqrt{1-x^2\,.\,1-k^2x^2}},$$

where K is of course not a mere numerical transcendant, but a function of k: K is such that we have $\operatorname{sn}K=1$, $\operatorname{cn}K=0$, $\operatorname{dn}K=k'$. Writing $v=K$, we obtain simple expressions for the sn, cn, dn of $u+K$, and thence for those of $u+2K$ and $u+4K$; viz. it ultimately appears that the sn, cn and dn of $u+4K$ are the same as the sn, cn and dn of u respectively: or the functions have a real period $4K$.

27. But the form of the integral suggests the consideration of another quantity

$$\int_0^{\frac{1}{k}}\frac{dx}{\sqrt{1-x^2\,.\,1-k^2x^2}};$$

this is a complex quantity transformable into the form

$$\int_0^1\frac{dx}{\sqrt{1-x^2\,.\,1-k^2x^2}}+i\int_0\frac{dx}{\sqrt{1-x^2\,.\,1-k'^2x^2}},$$

viz. K' being the same function of the complementary modulus k' that K is of k, the value is $=K+iK'$.

We have

$$\operatorname{sn}(K+iK')=\frac{1}{k},\ \operatorname{cn}(K+iK')=\frac{ik'}{k},\ \operatorname{dn}(K+iK')=0,$$

and then forming the sn, cn and dn of $u+K+iK'$, &c. it ultimately appears that the functions of $u+4(K+iK')$ are equal to those of u respectively: viz. there is a second period $4(K+iK')$. But as above seen $4K$ was a period, and thus the periods may be taken to be $4K$, $4iK'$ respectively—only it must be borne in mind that K, $K+iK'$ have, K, iK' have *not*, analogous relations to the elliptic functions. *This is the theorem of the double periodicity of the elliptic functions.*

Further theory in regard to the third kind of Elliptic Integrals: Addition of Parameters, and Interchange of Amplitude and Parameter. Art. Nos. 28 to 31.

28. We may differentiate an algebraic function

$$\sin\phi\cos\phi\, R(\sin^2\phi)\,\Delta(k,\phi),$$

where $R(\sin^2\phi)$ denotes a rational function of $\sin^2\phi$; and thereby obtain an expression involving two or more terms of the form $\dfrac{d\phi}{(1+n\sin^2\phi)\,\Delta(k,\phi)}$ with different values of n. Conversely, integrating such expression we obtain an equation containing two or more terms of the form $\Pi(n,k,\phi)$, that is elliptic integrals of the third kind with different parameters. In particular there may be two parameters only; viz. these being n, n', then we have either $nn'=k^2$ or $(1+n)(1+n')=k'^2$: the resulting formulæ are useful for the reduction of an integral of the third kind to a like integral where the parameter is of one of the standard forms $\cot^2\theta$, $-1+k'^2\sin^2\theta$, $-k^2\sin^2\theta$.

29. There may be three parameters; the theorem is in this case a theorem for the "addition of the parameters." To explain this, suppose that two of the parameters are $-k^2\sin^2 p$, $-k^2\sin^2 q$ (this, if p, q are taken to be real, is a particular assumption, limiting the generality of the result; but allowing them to be imaginary, it is no restriction): then the third parameter is $-k^2\sin^2 r$, where the angles p, q, r are connected together by that very relation which is the addition-equation for the integrals of the first kind, $Fp+Fq-Fr=0$ (rather it is, in the first instance, $Fp+Fq+Fr=$ const., reducible to the last-mentioned particular form): the theorem then gives $\Pi(-k^2\sin^2 r,k,\phi)$ in terms of $\Pi(-k^2\sin^2 p,k,\phi)$ and $\Pi(-k^2\sin^2 q,k,\phi)$; and it is in this sense a theorem for the addition of parameters.

30. The theorem leads to an expression for an integral of given imaginary parameter in terms of two integrals of real parameter, one of them of the form $-k^2\sin^2\theta$, the other of the form $\cot^2\lambda$ or $-1+k'^2\sin^2\lambda$.

31. There is a further theory of the "interchange of amplitude and parameter:" differentiating the two sides of the equation

$$\Pi(n,k,\phi)=\int\frac{d\phi}{(1+n\sin^2\phi)\,\Delta(k,\phi)},$$

in regard to n, and, after multiplication by a factor, conversely integrating in regard to this variable, we obtain

$$\sqrt{(1+n)\left(1+\frac{k^2}{n}\right)}\,\Pi(n,k,\phi)$$

expressed as a sum of certain integrals in respect to n. Expressing this parameter in one of the standard forms, for instance $n=-k^2\sin^2\theta$, the integrals in regard to n become integrals in regard to θ, viz. these are the elliptic integrals $F(k,\theta)$, $E(k,\theta)$ and an integral of the third kind $\Pi(n',k,\theta)$, where the parameter n' is $=-k^2\sin^2\phi$: that is $n=-k^2\sin^2\theta$, $n'=-k^2\sin^2\phi$. We have a relation between the integrals $\Pi(n,k,\phi)$, $\Pi(n',k,\theta)$; this relation [involving also $F(k,\phi)$, $E(k,\phi)$, $F(k,\theta)$, $E(k,\theta)$] is a form of the so-called theorem for the interchange of amplitude and parameter: those belonging to the other two forms of parameter $n=\cot^2\theta$ and $n=-1+k'^2\sin^2\theta$ are less elegant, inasmuch as in the θ functions the modulus is k' instead of k.

The second and third kinds of Elliptic Integrals expressed in terms of the argument u; new Notations. Art. Nos. 32 to 34.

32. The introduction of u, $=F\phi$, as the argument in place of u, in fact supersedes the consideration of the elliptic integral of the first kind: by introducing u as the argument in the integrals of the second and third kinds, we obtain

$$E(k,\phi)=\int_0(1-k^2\,\mathrm{sn}^2u)\,du,\quad \Pi(n,k,\phi)=\int_0\frac{du}{1+n\,\mathrm{sn}^2u},$$

which functions *changing the notation* might be called $E(k, u)$ and $\Pi(n, k, u)$ respectively. But it is found convenient to consider somewhat different functions; viz. in place of the integral of the second kind Jacobi considers

$$Zu = u\left(1 - \frac{E}{K}\right) - k^2 \int_0 \operatorname{sn}^2 u\, du,$$

where E, K denote the complete functions $E_1 k$, $F_1 k$ respectively: Zu is of course a function of k, so that its complete expression is $Z(k, u)$: it is $= -\frac{E}{K} u + E(k, u)$, differing from $E(k, u)$ by a multiple of u.

33. As regards the third kind, the parameter is taken to be $= -k^2 \operatorname{sn}^2 a$ (to meet every case a must not be restricted to real values) and the function considered is

$$\Pi(u, a) = \int_0 \frac{k^2 \operatorname{sn} a \operatorname{cn} a \operatorname{dn} a \operatorname{sn}^2 u\, du}{1 - k^2 \operatorname{sn}^2 a \operatorname{sn}^2 u},$$

[being of course a function also of k, so that its complete expression would be $\Pi(u, a, k)$]: viz. writing $n = -k^2 \operatorname{sn}^2 a$, this is in fact a multiple of

$$\int \frac{\sin^2\phi\, d\phi}{(1 + n \sin^2\phi)\, \Delta(k, \phi)}, = \frac{1}{n}\left[F(k, \phi) - \Pi(n, k, \phi)\right].$$

34. The advantage of the new forms is very great: thus the addition-theorem for the second kind of integral is

$$Zu + Zv - Z(u + v) = k^2 \operatorname{sn} u \operatorname{sn} v \operatorname{sn}(u + v)$$

and that for the third kind gives in like manner the value of

$$\Pi(u, a) + \Pi(v, a) - \Pi(u + v, a)$$

in terms of the functions of $u, v, u + v$: the theorem for the addition of the parameters gives a very similar expression for

$$\Pi(u, a) + \Pi(u, b) - \Pi(u, a + b),$$

and the theorem for the interchange of amplitude and parameter is in fact a relation between $\Pi(u, a)$ and $\Pi(a, u)$.

The Functions Θu, H*u*. *The q-formulæ.* Art. Nos. 35 to 42.

35. From the function sn*u* we derive a new function Θu by the equation

$$\Theta u = \sqrt{\frac{2Kk'}{\pi}}\, e^{\frac{1}{2}u^2\left(1-\frac{E}{K}\right) - k^2\int du\int du\, \mathrm{sn}^2 u}$$

(K, E denoting as before the complete functions $F_1 k$, $E_1 k$): this may be regarded as one of a system of four functions, Θu, $\Theta(u+K)$, $\Theta(u+iK')$, $\Theta(u+K+iK')$; or writing

$$Hu = -ie^{-\frac{\pi(K'-2iu)}{4K}}\,\Theta(u+iK'),$$

the functions may be taken to be Θu, $\Theta(u+K)$, Hu, $H(u+K)$.

36. The function Zu is at once expressed in terms of Θu and its derived function $\Theta' u$; viz. we have $Zu = \frac{\Theta' u}{\Theta u}$.

The function $\Pi(u, a)$ has a simple expression in terms of Θ, viz. we have

$$\Pi(u, a) = u\frac{\Theta' a}{\Theta a} + \tfrac{1}{2}\log\frac{\Theta(u-a)}{\Theta(u+a)}.$$

37. Writing herein $u+a$ for u, we have

$$\Pi(u+a, a) = (u+a)\frac{\Theta' a}{\Theta a} - \tfrac{1}{2}\log\frac{\Theta(u+2a)}{\Theta u};$$

and for the values $a = \tfrac{1}{2}K$, $\tfrac{1}{2}iK'$, $\tfrac{1}{2}K + \tfrac{1}{2}iK'$ the function $\Pi(u+a, a)$ is expressible in finite terms by means of the functions log sn*u*, log cn*u*, log dn*u* respectively: the resulting equations give, after all reductions, the formulæ next referred to*.

38. The functions sn*u*, cn*u*, dn*u* are found to be fractional functions, the three numerators and the denominator being the four functions above spoken of; viz. we have

$$\mathrm{sn}\,u = \frac{1}{\sqrt{k}}Hu \div,\quad \mathrm{cn}\,u = \sqrt{\frac{k'}{k}}H(u+K) \div,\quad \mathrm{dn}\,u = \sqrt{k'}\,\Theta(u+K) \div,$$

where denom. $= \Theta u$.

* This is not Jacobi's method nor perhaps the most direct or natural way of obtaining the formulæ in question; but the connexion of the formulæ with the expression for $\Pi(u, a)$ is very noticeable.

39. These functions H, Θ are in fact doubly infinite products; viz. writing for shortness

$$\begin{aligned}
(m, m') &= \quad 2mK + \quad 2m'iK',\\
(\overline{m}, m') &= (2m+1)K + \quad 2m'iK',\\
(m, \overline{m}') &= \quad 2mK + (2m'+1)iK',\\
(\overline{m}, \overline{m}') &= (2m+1)K + (2m'+1)iK';
\end{aligned}$$

then, disregarding certain constant factors, we have

$$\begin{aligned}
Hu \quad &= u\Pi\left\{1 + \frac{u}{(m, m')}\right\},\\
H(u+K) &= \Pi\left\{1 + \frac{u}{(\overline{m}, m')}\right\},\\
\Theta u \quad &= \Pi\left\{1 + \frac{u}{(m, \overline{m}')}\right\},\\
\Theta(u+K) &= \Pi\left\{1 + \frac{u}{(\overline{m}, \overline{m}')}\right\},
\end{aligned}$$

where (except that in the first product the simultaneous values $m = 0$, $m' = 0$ are to be omitted) m, m' have all positive or negative integer values, including zero, but under the following condition, viz. taking μ, μ' to denote each of them an indefinitely large positive integer, μ being also indefinitely large in comparison with μ', so that $\mu' \div \mu = 0$, then

for m the limits are	$m = -\mu$	to	$+\mu$,
" m' " "	$m' = -\mu'$	"	$+\mu'$,
" $\overline{m}$ " "	$m = -\mu - 1$	"	$+\mu$,
" $\overline{m}'$ " "	$m' = -\mu' - 1$	"	$+\mu'$.

40. Giving to m all its values, and reducing by means of the factorial expressions of $\sin x$, $\cos x$, the expressions become singly infinite products of circular functions such as

$$\sin\frac{\pi}{2K}(u + 2m'iK');$$

or writing $\frac{\pi u}{2K} = x$, these are expressible as products or series involving $\cos 2x$, or the multiple sines or cosines of x, with coefficients which are functions of the quantity $q, = e^{-\frac{\pi K'}{K}}$; viz. we have thus the q-formulæ which Jacobi obtains in quite a different manner (from a transformation formula, by writing therein $\frac{u}{n}$ for u, and taking n infinite), and which in fact led him to the functions H and Θ. The formulæ are very remarkable as well in themselves as from their origin, and the connexion which they establish between Elliptic Functions and the theory of Numbers: as a specimen take here the identity

$$[1 + 2q + 2q^4 + 2q^9 + \dots]^4 = 1 + 8\left\{\frac{q}{1-q} + \frac{2q^2}{1+q^2} + \frac{3q^3}{1-q^3} + \dots\right\},$$

which not only shows that every number is the sum of four squares, but affords the means of finding the number of decompositions.

41. The four functions Θu, $\Theta(u+K)$, Hu, $H(u+K)$ considered as functions of $\omega, = \frac{\pi K'}{K}$, and $v = \frac{\pi u}{2K}$, each of them satisfy the partial differential equation

$$\frac{d^2\sigma}{dv^2} - 4\frac{d\sigma}{d\omega} = 0.$$

This equation, not given in the *Fundamenta Nova*, but obtained by Jacobi (*Crelle*, t. III. p. 306, 1828), is, in fact, an immediate consequence from the expressions of the functions as series in terms of q ($= e^{-\omega}$) and u; but it is also obtainable from the finite expressions of these functions.

42. There is no proper addition-equation for the functions H, Θ: the nearest analogue is the system of equations

$$\Theta(u+v)\,\Theta(u-v) = \frac{\Theta^2 u\,\Theta^2 v - H^2 u\,H^2 v}{\Theta^2 0},$$

$$H(u+v)\,H(u-v) = \frac{H^2 u\,\Theta^2 v - \Theta^2 u\,H^2 v}{\Theta^2 0};$$

involving, it will be observed, the H, Θ as well of $u - v$ as of $u + v$. But these formulæ show, what follows also from the double-product expressions, that these functions have a multiplication-theory; and the double-product expressions also show that they have a transformation theory.

The Numerators and Denominator in the multiplication and transformation of the Elliptic Functions. Art. No. 43.

43. We are thus, in the multiplication and in the transformation of the elliptic functions, led to expressions for the three numerators and the denominator of the functions of nu, or of $\left(\frac{u}{M}, \lambda\right)$ (*ante*, No. 24), in terms of the functions H, Θ; and by the aid of the above-mentioned partial differential equation we obtain partial differential equations satisfied by the numerator- and denominator-functions in question: thus, considering the denominator only, and writing for convenience $x = \sqrt{k}\,\mathrm{sn}\,u$, $\alpha = k + \frac{1}{k}$, $\nu = n$, in the case of the transformation of the nth order $\mathrm{sn}\left(\frac{u}{M}, \lambda\right)$, but $= n^2$ in the case of multiplication $\mathrm{sn}\,nu$; then the denominator, considered as a function of x and α, satisfies the partial differential equation

$$(1 - \alpha x^2 + x^4)\frac{d^2 z}{dx^2} + (\nu - 1)(\alpha x - 2x^3)\frac{dz}{dx}$$

$$+ \nu(\nu - 1)x^2 z - 2\nu(\alpha^2 - 4)\frac{dz}{d\alpha}.$$

(Jacobi, *Crelle*, t. IV. p. 185, 1829.) As regards the transformation formula, it is to be observed that λ, quâ function of k, must consequently be considered as a function of α, and the expression of z as a function of x, and of α directly and through λ, is so complex, that not only the equation is practically useless, but it is difficult to verify it even in the simple case of a cubic transformation: but as regards the multiplication formula, the equation is very convenient for the determination of the actual expression of s as a function of x and a, or, what is the same thing, of sn u and k. The equation requires some change of form to adapt it to the three numerators respectively: and the resulting equations are in like manner practically useless for transformation but very convenient for multiplication.

Concluding Remarks. Art. No. 44.

44. The foregoing outline is purposely very brief as to the theory of transformation, and as to the ulterior theory of the third kind of Elliptic Integrals; as to these it is completed by the outlines prefixed to the chapters on these subjects respectively, and generally the outlines or introductory paragraphs to the several chapters may be consulted: as thus extended, the outline is intended to cover the whole of the present treatise up to the end of chapter XI., and also chapter XII., which contains the reduction of the differential expression $Rdx \div \sqrt{X}$ to the like expression with the radical in the standard form $\sqrt{1-x^2 . 1-k^2x^2}$, as mentioned at the beginning of this outline. The remaining chapters, XIII. to XVI., I regard as supplementary; the outlines or introductory paragraphs will explain what the contents of these are; I only remark here that chapter XIII., relating in fact to Landen's transformation, belongs to the elementary part of the subject, and might have been brought in at a much earlier stage; the only reason for deferring it was the convenience of using the form of radical $\sqrt{a^2\cos^2\phi + b^2\sin^2\phi}$, instead of the standard form $\sqrt{1-k^2\sin^2\phi}$; generally whatever relates to the non-standard form of radical is given in these supplementary chapters.

CHAPTER II.

THE ADDITION-EQUATION: LANDEN'S THEOREM.

45. As already mentioned the addition-equation is the integral of the differential equation

$$\frac{d\phi}{\Delta\phi} + \frac{d\psi}{\Delta\psi} = 0,$$

($\Delta\phi = \sqrt{1 - k^2 \sin^2\phi}$, &c.) the constant of integration, μ being the value of either variable when the other is put $= 0$. Of the proofs which are here given several are only verifications of the theorem assumed to be known: but the first one is a direct investigation. The fifth proof (Jacobi's by means of two fixed circles) leads so naturally to Landen's theorem, that, although belonging to a different part of the subject, I have given it in the present chapter.

First Proof (Walton, *Quarterly Math. Journ.* t. XI. pp. 177—178, 1870). Art. No. 46.

46. Rationalising the differential equation, we have

$$d\phi^2 - d\psi^2 = -k^2 (\sin^2\phi\, d\psi^2 - \sin^2\psi\, d\phi^2),$$

or, as this may be written,

$$(d\phi^2 - d\psi^2)(\cos^2\phi - \cos^2\psi)$$
$$= -k^2 (\sin^2\phi\, d\psi^2 - \sin^2\psi\, d\phi^2)(\cos^2\phi - \cos^2\psi).$$

The left-hand side is

$$= -\sin(\phi + \psi)(d\phi + d\psi)\sin(\phi - \psi)(d\phi - d\psi),$$
$$= -d\cos(\phi + \psi) \,.\, d\cos(\phi - \psi),$$

or putting $x = \cos\phi\cos\psi$, $y = \sin\phi\sin\psi$, and therefore

$$\cos(\phi+\psi) = x - y, \quad \cos(\phi-\psi) = x + y,$$

this is $= dy^2 - dx^2$.

The right-hand side, omitting the factor $-k^2$, is

$$(\cos\phi + \cos\psi)(\sin\phi d\psi + \sin\psi d\phi)$$
$$\times(\cos\phi - \cos\psi)(\sin\phi d\psi - \sin\psi d\phi),$$

where the first factor is

$$= \cos\phi\sin\psi d\phi + \cos\psi\sin\phi d\psi + \sin\phi\cos\phi d\psi + \sin\psi\cos\psi d\phi;$$

viz. writing this under the form

$$\cos\phi\sin\psi d\phi + \cos\psi\sin\phi d\psi + \sin\phi\cos\phi(\cos^2\psi + \sin^2\psi)\,d\psi$$
$$+ \sin\psi\cos\psi(\cos^2\phi + \sin^2\phi)\,d\phi,$$

it is $$= dy + xdy - ydx;$$

and similarly the second factor is

$$= -dy + xdy - ydx.$$

Hence, restoring the factor $-k^2$, the right-hand side is

$$= k^2[dy^2 - (xdy - ydx)^2],$$

or the differential equation is

$$dy^2 - dx^2 = k^2[dy^2 - (xdy - ydx)^2];$$

viz. writing herein $\frac{dy}{dx} = p$, this is

$$p^2 - 1 = k^2[p^2 - (y - px)^2];$$

or we have

$$(y - px)^2 = p^2 + \frac{1}{k^2}(1 - p^2), = \frac{1}{k^2}[(k^2 - 1)p^2 + 1],$$

which is an equation of Clairaut's form; or taking γ as the arbitrary constant, the integral is

$$y = \gamma x + \frac{1}{k}\sqrt{(k^2 - 1)\gamma^2 - 1},$$

that is

$$\sin\phi\sin\psi = \gamma\cos\phi\cos\psi + \frac{1}{k}\sqrt{(k^2 - 1)\gamma^2 - 1}.$$

Let μ be the value of ψ corresponding to the value $\phi = 0$, then writing $\phi = 0$, $\psi = \mu$, we have

$$\gamma \cos\mu + \frac{1}{k}\sqrt{(k^2-1)\gamma^2+1} = 0,$$

giving $\gamma^2 k^2 (1-\sin^2\mu) = \gamma^2 k^2 - \gamma^2 + 1$, that is $\gamma^2(1-k^2\sin^2\mu) = 1$, or $\gamma = \frac{1}{\Delta\mu}$, whence $\frac{1}{k}\sqrt{(k^2-1)\gamma^2+1} = \frac{-\cos\mu}{\Delta\mu}$; and substituting, we obtain

$$\cos\mu = \cos\phi\cos\psi - \sin\phi\sin\psi\Delta\mu,$$

the required addition-equation.

Second Proof (Jacobi, *Crelle*, t. VIII. p. 332, 1832). Art. No. 47.

47. Assume $a + b\cos\phi\cos\psi + c\sin\phi\sin\psi = 0$, then differentiating we have

$$(-b\sin\phi\cos\psi + c\cos\phi\sin\psi)\,d\phi$$
$$+ (-b\cos\phi\sin\psi + c\sin\phi\cos\psi)\,d\psi = 0;$$

say this is $$Md\phi + Nd\psi = 0.$$

But we have

$$M^2 + (b\cos\phi\cos\psi + c\sin\phi\sin\psi)^2 = b^2\cos^2\psi + c^2\sin^2\psi,$$
$$N^2 + (b\cos\phi\cos\psi + c\sin\phi\sin\psi)^2 = b^2\cos^2\phi + c^2\sin^2\phi,$$

that is

$$M^2 = -a^2 + b^2\cos^2\psi + c^2\sin^2\psi = b^2 - a^2 - (b^2-c^2)\sin^2\psi,$$
$$N^2 = -a^2 + b^2\cos^2\phi + c^2\sin^2\phi = b^2 - a^2 - (b^2-c^2)\sin^2\phi,$$

and the differential equation thus is

$$\frac{d\phi}{\sqrt{b^2-a^2-(b^2-c^2)\sin^2\phi}} + \frac{d\psi}{\sqrt{b^2-a^2-(b^2-c^2)\sin^2\psi}} = 0;$$

viz. an integral of this equation is

$$a + b\cos\phi\cos\psi + c\sin\phi\sin\psi = 0.$$

But observe that the differential equation contains the single constant $\frac{b^2-c^2}{b^2-a^2}$, the integral equation the two constants $\frac{b}{a}$, $\frac{c}{a}$, which of course cannot be expressed in terms of $\frac{b^2-c^2}{b^2-a^2}$, but only in terms of this and an arbitrary constant, say μ. Hence the assumed equation is the general integral of the differential equation.

To complete the investigation write $\frac{b^2-c^2}{b^2-a^2}=k^2$, and assume $\frac{a}{b}=-\cos\mu$, then the equation $\frac{b^2-c^2}{b^2-a^2}=k^2$, or $c^2=b^2-(b^2-a^2)k^2$ becomes $c^2=b^2(1-k^2\sin^2\mu)$, or say $c=-b\Delta\mu$: substituting these values of a and c, the equation becomes

$$-\cos\mu+\cos\phi\cos\psi-\sin\phi\sin\psi\Delta\mu=0;$$

viz. we have

$$\cos\mu=\cos\phi\cos\psi-\sin\phi\sin\psi\Delta\mu,$$

as the integral of the differential equation $\frac{d\phi}{\Delta\phi}+\frac{d\psi}{\Delta\psi}=0$.

And it is clear that μ is the value of either variable corresponding to the value 0 of the other variable.

Forms of the Addition-Equation. Art. Nos. 48 and 49.

48. We have

$$(\cos\mu-\cos\phi\cos\psi)^2-\sin^2\phi\sin^2\psi\,\Delta^2\mu=0;$$

or expanding and reducing

$$1-\cos^2\phi-\cos^2\psi-\cos^2\mu+2\cos\phi\cos\psi\cos\mu$$
$$-k^2\sin^2\phi\sin^2\psi\sin^2\mu=0,$$

which is symmetrical in regard to the three quantities: hence we have also

$$(\cos\phi-\cos\mu\cos\psi)^2=\sin^2\mu\sin^2\psi\,\Delta^2\phi,$$
$$(\cos\psi-\cos\mu\cos\phi)^2=\sin^2\mu\sin^2\phi\,\Delta^2\psi,$$

and extracting the square roots, it appears that the signs on the right-hand side must be $+$: we thus have

$$\cos\phi - \cos\mu\cos\psi = \quad \sin\mu\sin\psi\,\Delta\phi,$$
$$\cos\psi - \cos\mu\cos\phi = \quad \sin\mu\sin\phi\,\Delta\psi,$$

to which join the original equation

$$\cos\mu - \cos\phi\cos\psi = -\sin\phi\sin\psi\,\Delta\mu.$$

49. From the rationalised equation, writing $\sin^2\mu = 1 - \cos^2\mu$, we obtain

$$(1 - k^2\sin^2\phi\sin^2\psi)\cos^2\mu - 2\cos\phi\cos\psi\cos\mu$$
$$= 1 - \cos^2\phi - \cos^2\psi - k^2\sin^2\phi\sin^2\psi,$$

that is

$$[(1 - k^2\sin^2\phi\sin^2\psi)\cos\mu - \cos\phi\cos\psi]^2$$
$$= (1 - k^2\sin^2\phi\sin^2\psi)(1 - \cos^2\phi - \cos^2\psi - k^2\sin^2\phi\sin^2\psi)$$
$$+ \cos^2\phi\cos^2\psi,$$

which is easily seen to be

$$= \sin^2\phi\sin^2\psi\,\Delta^2\phi\,\Delta^2\psi;$$

and then extracting the square roots, the sign on the right-hand side is $-$, and we have

$$(1 - k^2\sin^2\phi\sin^2\psi)\cos\mu = \cos\phi\cos\psi - \sin\phi\sin\psi\,\Delta\phi\,\Delta\psi,$$

which gives the value of μ in terms of ϕ and ψ.

Combining with this the equation

$$\cos\mu - \cos\phi\cos\psi = -\sin\phi\sin\psi\,\Delta\mu,$$

we have the value of $\Delta\mu$: and if from $\cos\mu$ we proceed to find the value of $\sin^2\mu$, we have

$$(1 - k^2\sin^2\phi\sin^2\psi)^2\sin^2\mu = (1 - k^2\sin^2\phi\sin^2\psi)^2$$
$$- (\cos\phi\cos\psi - \sin\phi\sin\psi\,\Delta\phi\,\Delta\psi)^2,$$

which is readily found to be

$$= (\sin\phi\cos\psi\,\Delta\psi + \sin\psi\cos\phi\,\Delta\phi)^2;$$

and extracting the square roots, the sign on the right-hand

side is +: we have thus the formulæ

$$\sin\mu = \sin\phi\cos\psi\,\Delta\psi + \sin\psi\cos\phi\,\Delta\phi, \quad (\div)$$

$$\cos\mu = \cos\phi\cos\psi - \sin\phi\sin\psi\,\Delta\phi\,\Delta\psi, \quad (\div)$$

$$\Delta\mu = \Delta\phi\,\Delta\psi - k^2\sin\phi\sin\psi\cos\phi\cos\psi, \quad (\div)$$

where the denominator is

$$= 1 - k^2\sin^2\phi\sin^2\psi.$$

And we have in like manner

$$\sin\phi = \sin\mu\cos\psi\,\Delta\psi - \sin\psi\cos\mu\,\Delta\mu, \quad (\div)$$

$$\cos\phi = \cos\mu\cos\psi + \sin\mu\sin\psi\,\Delta\mu\,\Delta\psi, \quad (\div)$$

$$\Delta\phi = \Delta\mu\,\Delta\psi + k^2\sin\mu\sin\psi\cos\mu\cos\psi, \quad (\div)$$

where the denominator is

$$= 1 - k^2\sin^2\mu\sin^2\psi.$$

And we may in these formulæ interchange ϕ, ψ.

Third Proof of the Addition-Equation (a verification).
Art. No. 50.

50. Writing the equation in the form

$$\cos\mu\operatorname{cosec}\phi\operatorname{cosec}\psi - \cot\phi\cot\psi = -\Delta\mu,$$

then differentiating the left-hand side the coefficient of $d\phi$ is

$$-\cos\mu\operatorname{cosec}\phi\cot\phi\operatorname{cosec}\psi + \operatorname{cosec}^2\phi\cot\psi,$$

$$= \frac{1}{\sin^2\phi\sin\psi}(\cos\psi - \cos\mu\cos\phi),$$

which in virtue of the form

$$\cos\psi - \cos\mu\cos\phi = \sin\mu\sin\phi\,\Delta\psi,$$

is
$$= \frac{\sin\mu}{\sin\phi\sin\psi}\Delta\psi:$$

and similarly the coefficient of $d\psi$ is

$$= \frac{\sin\mu}{\sin\phi\sin\psi}\Delta\phi,$$

so that omitting the common factor, the differential equation becomes

$$d\phi\Delta\psi + d\psi\Delta\phi = 0,$$

which is right.

Fourth Proof (Legendre, *Traité des Fonctions Elliptiques*, t. I. p. 20, by a spherical triangle). Art. No. 51.

51. Consider a spherical triangle ABC, obtuse-angled at C, such that the sides CB, CA are $= \phi, \psi$ respectively, and

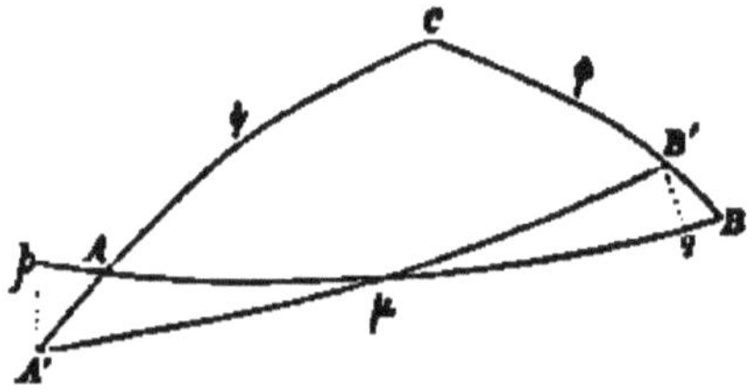

that the cosine of the angle C is $= -\Delta\mu$. This being so, the equation $\cos\mu - \cos\phi\cos\psi = -\sin\phi\sin\psi\,\Delta\mu$ shows that the side AB is $= \mu$ (so that by sliding the constant arc AB, $= \mu$, along the two fixed sides CA, CB, we obtain the different values of ϕ, ψ which satisfy the relation in question). And the other two equations $\cos\phi - \cos\mu\cos\psi = \sin\mu\sin\psi\,\Delta\phi$, and $\cos\psi - \cos\mu\cos\phi = \sin\mu\sin\phi\,\Delta\phi$, show that $\cos A = \Delta\phi$, and $\cos B = \Delta\psi$: so that the sides a, b, c of the spherical triangle are ϕ, ψ, μ respectively, and the cosines of the opposite angles A, B, C are $\Delta\phi, \Delta\psi, -\Delta\mu$ respectively.

Now considering the consecutive position $A'B'$ of the side AB, and letting fall on AB the perpendiculars $A'p$ and $B'q$, the equation $A'B' = AB$ gives $Ap = Bq$, that is $AA'\cos A = BB'\cos B$, or $db\cos A + da\cos B = 0$; viz. this is the differential equation $d\phi\Delta\psi + d\psi\Delta\phi = 0$.

Fifth Proof (Jacobi, *Crelle*, t. III. p. 376, 1828, by two fixed circles). Art. No. 52.

52. Consider two fixed circles as shown in the figure, and suppose that we have

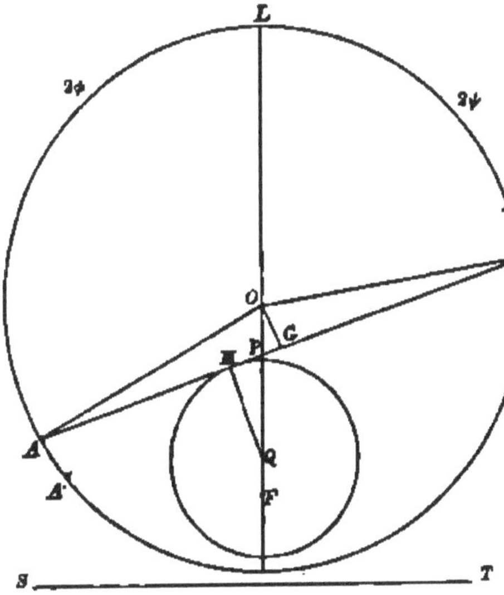

Radius of larger circle = R,

" smaller " = r,

Distance OQ of centres = D.

Write moreover

$$\Delta\mu = \frac{R-D}{R+D}, \quad \cos\mu = \frac{r}{R+D},$$

whence easily

$$k^2 = \frac{4DR}{(R+D)^2 - r^2}.$$

viz. k and μ are given functions of R, r, D: and it may be noticed that

$$(R+D)^2 - r^2 = \frac{4R^2 \sin^2\mu}{(1+\Delta\mu)^2}.$$

Imagine now a variable tangent AB, and assume

$$\angle AOL = 2\phi, \quad \angle BOL = 2\psi,$$

then letting fall on AB the perpendicular OG, we have

$$\angle AOG = \pi - (\phi+\psi), \quad \angle QOG = \phi - \psi;$$

and thence projecting AO, OQ on QM, we have

$$R\cos(\phi+\psi) + D\cos(\phi-\psi) = r;$$

that is, $\quad (R+D)\cos\phi\cos\psi - (R-D)\sin\phi\sin\psi = r,$

or what is the same thing

$$\cos\phi\cos\psi - \sin\phi\sin\psi\,\Delta\mu = \cos\mu,$$

which is the integral equation.

Also
$$\begin{aligned} AM^2 &= AQ^2 - MQ^2 \\ &= R^2 + D^2 + 2DR\cos 2\phi - r^2 \\ &= (R+D)^2 - r^2 - 4DR\sin^2\phi \\ &= [(R+D)^2 - r^2]\,\Delta^2\phi. \end{aligned}$$

And similarly

$$BM^2 = [(R+D)^2 - r^2]\,\Delta^2\psi.$$

Now varying the tangent let the new position be $A'B'$; then clearly $AA' : BB' = AM : BM$; that is

$$d\phi : -d\psi = AM : BM,$$

or
$$\frac{d\phi}{AM} + \frac{d\psi}{BM} = 0;$$

viz. substituting for AM, BM their values, we have the required differential equation

$$\frac{d\phi}{\Delta\phi} + \frac{d\psi}{\Delta\psi} = 0,$$

corresponding to the above integral equation.

Landen's Theorem, from the foregoing geometrical figure.
Art. Nos. 53 to 56.

53. Suppose that the large circle and also k remaining constant, the small circle is varied; that is, let r, D vary subject to the foregoing condition

$$k^2 = \frac{4DR}{(R+D)^2 - r^2},$$

it is readily shown that the radical axis of the two circles remains unaltered. In fact, taking the centre of the larger circle as origin and the axis of x vertically downwards, the equations of the two circles are

$$x^2 + y^2 - R^2 = 0,$$

$$(x - D)^2 + y^2 - r^2 = 0,$$

and thence for the radical axis

$$2Dx - R^2 - D^2 + r^2 = 0, \text{ or } x = \frac{R^2 + D^2 - r^2}{2D}, = \frac{2R}{k^2},$$

which is constant. In particular the smaller circle may reduce itself to the point F' (one of the limit-circles of the original two circles, or what is the same thing an antipoint of their points of intersection, viz. that antipoint which lies within the smaller circle): and then taking the distance $OF = \delta$, we have

$$\frac{4\delta R}{(R+\delta)^2} = \frac{4DR}{(R+D)^2 - r^2},$$

or what is the same thing

$$\delta(R^2 + D^2 - r^2) = D(R^2 + \delta^2).$$

54. Reverting now to the original two circles, if in the figure $\angle AOG = \omega (= \pi - \phi - \psi)$ and $\angle QOG = \chi (= \phi - \psi)$, then obviously $AA' \cos MAO = AMd\chi$, that is $R . 2d\phi \sin \omega = AMd\chi$; or what is the same thing $2AGd\phi = AMd\chi$; hence the equation $\frac{d\phi}{AM} = -\frac{d\psi}{BM}$ may be completed into

$$\frac{d\phi}{AM} = -\frac{d\psi}{BM} = \frac{d\chi}{2AG}.$$

and observing that $AG^2 = AO^2 - OG^2 = R^2 - (D\cos\chi - r)^2$, the equation is

$$\frac{d\phi}{\Delta(k,\phi)} = \frac{-d\psi}{\Delta(k,\psi)} = \frac{d\chi\sqrt{(R+D)^2 - r^2}}{2\sqrt{R^2 - (D\cos\chi - r)^2}}.$$

We have $QP = \dfrac{r}{\cos\chi}$, and thence $OP = D - \dfrac{r}{\cos\chi}$, whence from the triangle OAP, in which the angles A, P are $= \phi + \psi - \frac{1}{2}\pi$ (that is $2\phi - \chi - \frac{1}{2}\pi$) and $\frac{1}{2}\pi + \chi$ respectively, we have

$$D - \frac{r}{\cos\chi} : -\cos(2\phi - \chi) = R : \cos\chi;$$

that is $$D\cos\chi - r = -R\cos(2\phi - \chi),$$

which is an integral equation corresponding to the above differential equation

$$\frac{d\phi}{\Delta(k,\phi)} = \frac{d\chi\sqrt{(R+D)^2 - r^2}}{2\sqrt{R^2 - (D\cos\chi - r)^2}}.$$

Writing now $\angle APO = \theta$, then $\chi = \theta - \frac{1}{2}\pi$, $2\phi - \chi = 2\phi - \theta + \frac{1}{2}\pi$, and the integral and differential equations become respectively

$$D\sin\theta - r = R\sin(2\phi - \theta),$$

and $$\frac{d\phi}{\Delta(k,\phi)} = \frac{d\theta\sqrt{(D+R)^2 - r^2}}{2\sqrt{R^2 - (D\sin\theta - r)^2}}.$$

55. Suppose now that the smaller circle reduces itself to the point F, then retaining θ to denote the angle in this state of the figure, we must in place of D, r write δ, 0; and the equations become

$$\delta\sin\theta = R\sin(2\phi - \theta),$$

$$\frac{d\phi}{\Delta(k,\phi)} = \frac{d\theta\,(R+\delta)}{2\sqrt{R^2 - \delta^2\sin^2\theta}};$$

or writing herein $\lambda = \frac{\delta}{R}$, these are

$$\lambda \sin\theta = \sin(2\phi - \theta),$$

and
$$\frac{d\phi}{\Delta(k, \phi)} = \tfrac{1}{2}(1+\lambda)\frac{d\theta}{\Delta(\lambda, \theta)},$$

where in virtue of the relations $k^2 = \frac{4\delta R}{(R+\delta)^2}$ and $\lambda = \frac{\delta}{R}$, we have $k^2 = \frac{4\lambda}{(1+\lambda)^2}$, and therefore also $k' = \frac{1-\lambda}{1+\lambda}$ and $\lambda = \frac{1-k'}{1+k'}$.

56. The result would have come out more simply by considering *ab initio* the smaller circle as replaced by the point F: viz. the chord AB would then pass through the point F, and the points M, Q each coincide with F: but it was interesting to consider the theory in connexion with the original figure of the two circles.

The theorem gives, it will be observed, a transformation of the differential expression $\frac{d\phi}{\Delta(k, \phi)}$ into an expression $\frac{d\phi}{\Delta(\lambda, \theta)}$, involving a new modulus λ: viz. considering λ as derived from k by the equation $\lambda = \frac{1-k'}{1+k'}$, then we have between the two variable angles ϕ, θ an integral equation $\lambda \sin\theta = \sin(2\phi - \theta)$ answering to the differential relation $\frac{d\phi}{\Delta(k, \phi)} = \frac{\frac{1}{2}(1+\lambda)\,d\theta}{\Delta(k, \theta)}$: or since ϕ, θ vanish together this last is equivalent to

$$F(k, \phi) = \tfrac{1}{2}(1+\lambda)\,F(\lambda, \theta).$$

The integral equation gives $\lambda\tan\theta = \sin 2\phi - \cos 2\phi \tan\theta$, that is

$$\tan\theta = \frac{\sin 2\phi}{\lambda + \cos 2\phi};$$

whence $\sin\theta = \frac{\sin 2\phi}{\sqrt{1 + 2\lambda\cos 2\phi + \lambda^2}}, = \frac{\sin 2\phi}{\sqrt{(1+\lambda)^2 - 4\lambda\sin^2\phi}}$, or observing that $\frac{4\lambda}{(1+\lambda)^2} = k^2$ and $1 + \lambda = \frac{2}{1+k'}$, this is

$$\sin\theta = \tfrac{1}{2}(1+k')\frac{\sin 2\phi}{\sqrt{1 - k^2\sin^2\phi}}.$$

Sixth Proof of the Addition-Equation. Art. No. 57.

57. The rationalised equation in ϕ, ψ, μ may be written

$$\sin^2\mu - \cos^2\phi - \cos^2\psi + 2\cos\phi\cos\psi\cos\mu - k^2\sin^2\mu\,(1-\cos^2\phi)\,(1-\cos^2\psi) = 0;$$

viz. this is

$$k'^2\sin^2\mu - \Delta^2\mu\,(\cos^2\phi + \cos^2\psi) + 2\cos\mu\cos\phi\cos\psi - k^2\sin^2\mu\cos^2\phi\cos^2\psi = 0,$$

or as it may also be written

$$\begin{array}{llll} & k'^2\sin^2\mu & \cdot & -\Delta^2\mu\cos^2\phi \\ +2\,(& \cdot & \cos\mu\cos\phi & \cdot \quad)\cos\psi \\ +\;(& -\Delta^2\mu & \cdot & -k^2\sin^2\mu\cos^2\phi)\cos^2\psi = 0; \end{array}$$

viz. the left-hand side is a quadriquadric function of $\cos\phi$, $\cos\psi$: say this is u, and represent it successively under the forms $A' + 2B'\cos\phi + C'\cos^2\phi$, and $A + 2B\cos\psi + C\cos^2\psi$, where of course A', B', C' are given functions of $\cos\psi$, and A, B, C are the like given functions of $\cos\phi$: we have

$$\frac{du}{d\cos\phi} = 2\,(C'\cos\phi + B'),$$

but the equation $u = 0$ gives $(C'\cos\phi + B')^2 = (B'^2 - A'C')$, whence $\dfrac{du}{d\cos\phi} = 2\sqrt{B'^2 - A'C'}$, or what is the same thing $\dfrac{du}{d\phi} = -2\sin\phi\sqrt{B'^2 - A'C'}$, and similarly $\dfrac{du}{d\psi} = -2\sin\psi\sqrt{B^2 - AC}$: wherefore the differential equation is

$$\sqrt{B'^2 - A'C'}\sin\phi\,d\phi + \sqrt{B^2 - AC}\sin\psi\,d\psi = 0;$$

we have

$$\begin{aligned} B^2 - AC &= \cos^2\mu\cos^2\phi + (k'^2\sin^2\mu - \Delta^2\mu\cos^2\phi)\,(\Delta^2\mu + k^2\sin^2\mu\cos^2\phi) \\ &= \quad k'^2\sin^2\mu\,\Delta^2\mu \qquad \cdot \qquad \cdot \\ &\quad + (\cos^2\mu + k^2k'^2\sin^4\mu - \Delta^4\mu)\cos^2\phi \\ &\quad - k^2\sin^2\mu\,\Delta^2\mu\cos^4\phi, \end{aligned}$$

and coefficient of $\cos^2\phi$ is

$$\begin{aligned} & 1-\sin^2\mu+k^2k'^2\sin^4\mu \\ & -1+2k^2\sin^2\mu-k^4\sin^4\mu, \\ = \; & (k^2-k'^2)\sin^2\mu\,\Delta^2\mu\,; \end{aligned}$$

whence the value of B^2-AC is

$$\begin{aligned} & \sin^2\mu\,\Delta^2\mu\,\{k'^2+(k^2-k'^2)\cos^2\phi-k^2\cos^4\phi\}, \\ = \; & \sin^2\mu\,\Delta^2\mu\,.\,(1-\cos^2\phi)\,(k'^2+k^2\cos^2\phi), \\ = \; & \sin^2\mu\,\Delta^2\mu\,\sin^2\phi\,\Delta^2\phi\,; \end{aligned}$$

that is

$$\sqrt{B^2-AC}=\sin\mu\sin\phi\,\Delta\mu\,\Delta\phi\,;$$

and similarly

$$\sqrt{B'^2-A'C'}=\sin\mu\sin\psi\,\Delta\mu\,\Delta\psi,$$

whence the foregoing result is

$$d\phi\,\Delta\psi+d\psi\,\Delta\phi=0,$$

the required differential equation.

It may be remarked that this, like the third proof, *ante*, No. 50, is a verification, the difference being that we use the rationalised integral equation instead of the original irrational equation: and that they are each of them closely connected with the second proof, *ante*, No. 47, although this is less in the form of a verification.

CHAPTER III.

MISCELLANEOUS INVESTIGATIONS.

THE present chapter contains, in relation to the first and second kinds of elliptic integrals, various matters not very closely connected which it was convenient to give here before going on in the following Chapter IV. with the main theory: the contents will be seen from the headings of the several articles.

Arcs of curves representing or represented by the elliptic integrals $E(k, \phi)$, $F(k, \phi)$. Art. Nos. 58 to 62.

58. The elliptic integral of the second kind occurs naturally as representing the arc of an ellipse: viz. taking the equation of the ellipse to be $\frac{x^2}{a^2}+\frac{y^2}{b^2}=1$, this is satisfied on writing therein $x=a\sin\phi$, $y=b\cos\phi$ (observe that ϕ is the complement of the eccentric anomaly, or say of the parametric angle): we then have $dx=a\cos\phi\,d\phi$, $dy=-b\sin\phi$: and thence

$$ds^2=(a^2\cos^2\phi+b^2\sin^2\phi)\,d\phi^2$$

$$=[a^2-(a^2-b^2)\sin^2\phi]\,d\phi^2,$$

so that taking $k=\frac{\sqrt{a^2-b^2}}{a}$ (= excentricity) we have

$$ds=a\Delta(k, \phi)\,d\phi,$$

and thence $$s=aE(k, \phi),$$

the arc being measured from the extremity of the major axis: the length of the quadrant is $=aE_1k$. In the case of the circle the length is $aE_1 0$, $= a \cdot \frac{1}{2}\pi$; and, as the minor axis diminishes, k increases and $E_1 k$ diminishes, until ultimately for the indefinitely thin ellipse k becomes $= 1$, and $s = aE_1 1$, $= a$.

59. We may also represent the arc of the hyperbola: taking the equation to be $\frac{x^2}{a^2} - \frac{y^2}{b^2} = 1$, and expressing x, y in terms of the parametric angle u, that is writing $x = a \sec u$, $y = b \tan u$, we have $dx = a \sec u \tan u \, du$, $dy = b \sec^2 u \, du$, and thence

$$ds = \frac{du}{\cos^2 u} \sqrt{b^2 + a^2 \sin^2 u},$$

which does not immediately express the arc s by means of an elliptic integral: to obtain an expression of the required form assume

$$k = \frac{a}{\sqrt{a^2 + b^2}} \text{ and therefore } k' = \frac{b}{\sqrt{a^2 + b^2}}$$

(k = reciprocal of the eccentricity); and consider an angle ϕ connected with u by the equation $\tan u = k' \tan \phi$: the expressions of x, y in terms of ϕ are

$$x = \frac{a}{\cos \phi} \Delta\phi, \text{ giving } dx = \frac{ak'^2 \sin \phi \, d\phi}{\cos^2 \phi \, \Delta\phi},$$

$$y = bk' \tan \phi, \quad \text{,,} \quad dy = \frac{bk' \, d\phi}{\cos^2 \phi}$$

[$\Delta\phi$ is written for shortness to denote $\Delta(k, \phi)$ and so presently $F\phi$, $E\phi$ to denote $F(k, \phi)$, $E(k, \phi)$ respectively]: and thence

$$ds = \frac{bk' \, d\phi}{\cos^2 \phi \, \Delta\phi},$$

a value which of course may also be obtained from the foregoing expression of ds in terms of du.

We obtain by differentiation

$$d \,.\, \Delta\phi \tan \phi = \left(\frac{k'^2}{\cos^2 \phi \, \Delta\phi} - \frac{k'^2}{\Delta\phi} + \Delta\phi\right) d\phi,$$

and conversely, integrating from zero, we have

$$\Delta\phi \tan \phi = k'^2 \int \frac{d\phi}{\cos^2 \phi \, \Delta\phi} - k'^2 F\phi + E\phi,$$

whence substituting for the integral and observing that $\frac{b}{k'} = \frac{a}{k}$, we have

$$s = \frac{a}{k}\,[\tan\phi\,\Delta\phi + k'^2 F\phi - E\phi],$$

where (see figure) s denotes the arc AM measured from the vertex A of the hyperbola.

60. As regards the geometric signification, observe that for the point M on the hyperbola, the construction of the angles u, ϕ is as follows, viz. drawing the lines NQ, NR, $= b$ and bk'

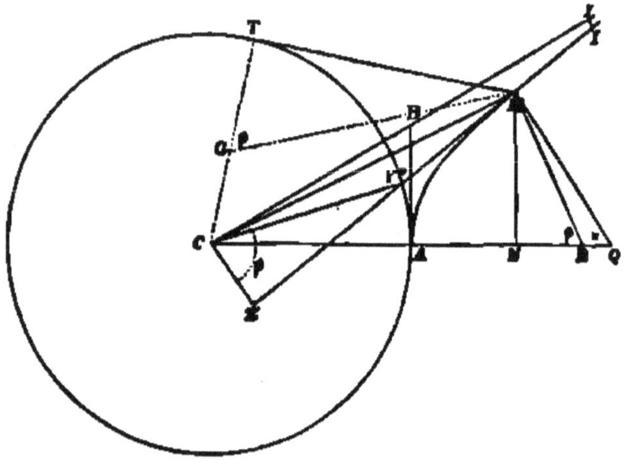

respectively, and joining these with the point M, then $\angle Q = u$, $\angle R = \phi$. To obtain a different construction for the angle ϕ, with centre C and radius CA $(= a)$ describe a circle, and drawing from M the tangent MT, and the radius CT, we have $MT^2 = x^2 + y^2 - a^2 = \left(1 + \frac{a^2}{b^2}\right) y^2$, that is $MT = \frac{y}{k'}$; hence measuring off from T the distance $TG = b$, and joining GM, the angle MGT is $= \phi$. Moreover the perpendicular CZ on the tangent at M is given by $\frac{1}{CZ^2} = \frac{x^2}{a^4} + \frac{y^2}{b^4}$, and substituting for x, y their values in terms of ϕ we find $CZ = a\cos\phi$; hence if Y be the intersection of this tangent with the circle, we have also $\angle YCZ = \phi$. Further $MZ^2 = x^2 + y^2 - a^2\cos^2\phi$; or since $x^2 + y^2 = a^2 + b^2\tan^2\phi$, this is

$$\begin{aligned} MZ^2 &= a^2\sin^2\phi + b^2\tan^2\phi, \\ &= a^2\tan^2\phi\left(\cos^2\phi + \frac{b^2}{a^2}\right), = a^2\tan^2\phi\left(\frac{1}{k^2} - \sin^2\phi\right), \\ &= \frac{a^2\tan^2\phi}{k^2}\Delta^2\phi; \end{aligned}$$

or finally $MZ = \frac{a}{k}\tan\phi\,\Delta\phi$.

Hence the formula is

$$s = MZ + \frac{a}{k}(k'^2 F\phi - E\phi),$$

or what is the same thing

$$MZ - MA = \frac{a}{k}(E\phi - k'^2 F\phi),$$

the quantity on the right-hand side being it is clear positive, viz. it is in fact

$$= ak\int\frac{\cos^2\phi}{\Delta\phi}.$$

As ϕ approaches 90° the point M goes off towards infinity, and the point Z tends to coincide with C: hence writing $\phi = 90°$, we obtain

$$IC - IA = \frac{a}{k}(E_1 - k'^2 F_1)$$

(where I represents the point at infinity on the curve or the asymptote) as the expression for the excess of the length of the asymptote over the arc of the curve.

61. It is less obvious how to find a curve the arc of which shall express the elliptic function of the first kind. Legendre

remarked that in the particular case $k = \frac{1}{\sqrt{2}}$, the solution was afforded by the lemniscate $(x^2+y^2)^2 = a^2(x^2-y^2)$. Observe that the curve is a horizontal figure-of-eight, the extremities being given by $y=0$, $x=\pm a$, and the branches at the origin being inclined to the axis of x at angles $=\pm 45°$. The equation is satisfied on writing therein

$$x = a\cos\phi\sqrt{1-\tfrac{1}{2}\sin^2\phi},$$

$$y = \frac{a}{\sqrt{2}}\sin\phi\cos\phi$$

(in fact these values give $x^2+y^2 = a^2\cos^2\phi$, $x^2-y^2 = a^2\cos^4\phi$): and hence determining the element of arc ds, $=\sqrt{dx^2+dy^2}$, we have

$$dx = \frac{a\sin\phi}{\sqrt{1-\frac{1}{2}\sin^2\phi}}\left(-\tfrac{3}{2}+\sin^2\phi\right)d\phi,\quad dy = \frac{a}{\sqrt{2}}\left(1-2\sin^2\phi\right)d\phi,$$

whence attending to the identity

$$\sin^2\phi\left(-\tfrac{3}{2}+\sin^2\phi\right)^2+\tfrac{1}{2}\left(1-\tfrac{1}{2}\sin^2\phi\right)\left(1-2\sin^2\phi\right)^2 = \tfrac{1}{2},$$

we have
$$ds^2 = \tfrac{1}{2}a^2\frac{d\phi^2}{1-\frac{1}{2}\sin^2\phi},$$

or finally
$$ds = \frac{a}{\sqrt{2}}\frac{d\phi}{\sqrt{1-\frac{1}{2}\sin^2\phi}},$$

whence
$$s = \frac{a}{\sqrt{2}}F\left(\frac{1}{\sqrt{2}},\ \phi\right),$$

s denoting the arc measured from the extremity $x=a$, $y=0$ $(\phi=0)$ to the point belonging to the value ϕ of the parametric angle. The same result may be obtained by means of the polar equation $r^2 = a^2\cos 2\theta$, introducing instead of θ the variable ϕ connected with it by the equation $\sin\phi = \sqrt{2}\sin\theta$. At the origin we have $\phi = 90°$, and the length of the quadrant of the curve is thus $=\frac{a}{\sqrt{2}}F_1\left(\frac{1}{\sqrt{2}}\right)$.

It thus appears that the leminiscate serves to express the function F of modulus $\dfrac{1}{\sqrt{2}}$.

62. For the general representation of the function $F(k, \phi)$ Legendre used the sextic curve

$$x = h \sin\phi \left(1 + \tfrac{1}{3} m \sin^2\phi\right),$$

$$y = bh \cos\phi \left(1 + m - \tfrac{1}{3} m \cos^2\phi\right),$$

where, k being the modulus, the values of h, m are

$$h = \frac{1 - 2k^2}{k'^2}, \quad m = \frac{3k^2}{1 - 2k^2},$$

and it is then easily found that

$$s = F(k, \phi) - \frac{k^2}{k'^2} \sin\phi \cos\phi\, \Delta(k, \phi),$$

where observe it is not the arc s, but the difference of this arc and an algebraic function, which is equal to the function $F(k, \phi)$: and the solution is not an elegant one.

63. A very beautiful solution was obtained by Serret (improved upon by Liouville), *Liouv.* t. X, 1845, pp. 257 and 351: and I have found that the theory admits of further development: I reserve the whole investigation for a subsequent chapter, remarking here that Serret's solution was suggested to him by a different treatment of the leminiscate; viz. the equation of the curve is satisfied by

$$x = a\frac{z + z^3}{1 + z^4}, \quad y = a\frac{z - z^3}{1 + z^4},$$

values which lead to

$$ds^2 = dx^2 + dy^2 = \frac{2a^2}{1 + z^4} dz^2, \text{ or } ds = \frac{a\sqrt{2}\,dz}{\sqrt{1 + z^4}},$$

so that the arc is expressed as a multiple of

$$\int \frac{dz}{\sqrt{1 + z^4}}.$$

which is an expression in the nature of an elliptic integral. To compare with the former solution observe that we have

$$\cos\phi = \frac{z\sqrt{2}}{\sqrt{1+z^4}},$$

$$\sin\phi = \frac{1-z^2}{\sqrt{1+z^4}},$$

$$\sqrt{1-\tfrac{1}{2}\sin^2\phi} = \frac{1}{\sqrt{2}}\frac{1+z^2}{\sqrt{1+z^4}},$$

and thence

$$d\phi = \frac{\sqrt{2}\,(1+z^2)\,dz}{1+z^4} \quad\text{and}\quad \frac{d\phi}{\sqrt{1-\tfrac{1}{2}\sin^2\phi}} = \frac{2dz}{\sqrt{1+z^4}}.$$

March of the Functions $F(k,\phi)$, $E(k,\phi)$. Art. Nos. 64 to 70.

64. To gain some idea of the march of the functions $F\phi$, $E\phi$, we may, taking ϕ as abscissa, trace the curves $y = \frac{1}{\Delta\phi}$, $y = \Delta\phi$: the areas of these curves included between the axis of y and the ordinate corresponding to the abscissa ϕ will of course represent the values of the integrals $F\phi$, $E\phi$.

65. If $k = 0$, then $\Delta\phi = 1$, and the curves $y = \Delta\phi$, $y = \frac{1}{\Delta\phi}$, each reduce themselves to the line $y = 1$. Here of course

$$F\phi = E\phi = \phi.$$

If $k > 0$, < 1, which is the standard case, then the curve $y = \Delta\phi$ is an undulating curve lying wholly below the line $y = 1$, and the curve $y = \frac{1}{\Delta\phi}$ an undulating curve lying wholly above this line. As ϕ increases from zero the functions $F\phi$, $E\phi$ each continually increase from zero, the function $F\phi$ being always the larger; and it is moreover clear that for a given value of ϕ, as k increases the function $F\phi$ increases and

$E\phi$ diminishes, and conversely as k decreases then $F\phi$ diminishes and $E\phi$ increases. In particular $k=0$, $F_1 = K = \frac{1}{2}\pi$,

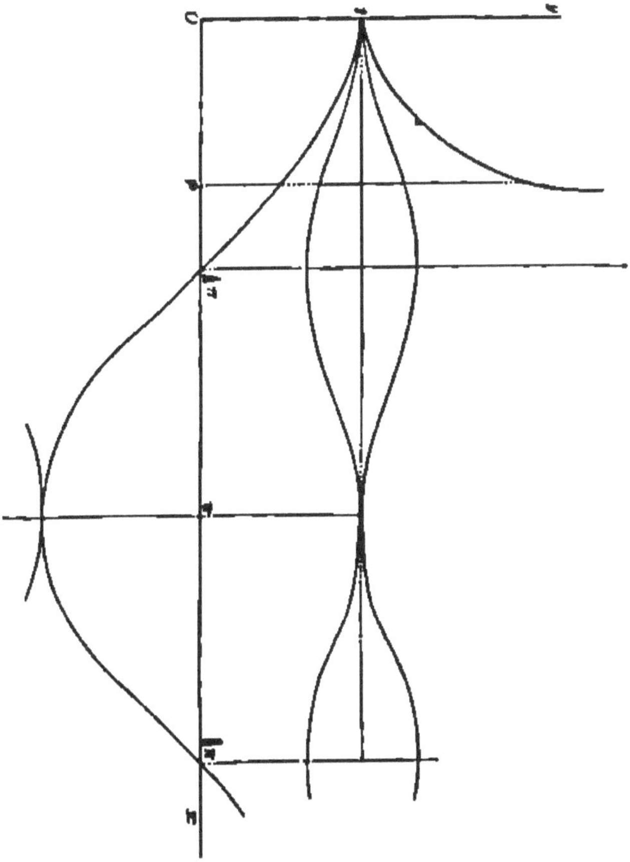

and also $E_1 = \frac{1}{2}\pi$, so that as k increases from zero, F_1 or K increases from $\frac{1}{2}\pi$, and E_1 diminishes from $\frac{1}{2}\pi$.

66. We see moreover that for each of the functions $F\phi$, $E\phi$ (k having a given value) it is sufficient to know the values of the functions for values of ϕ from 0 to $\frac{1}{2}\pi$: in fact we have $F(-a) = -Fa$, $F\pi = 2F_1$, and then

$$Fa = F\pi - F(\pi - a), \quad = 2F_1 - F(\pi - a),$$

giving the values from $a = \pi$ to $\frac{1}{2}\pi$, and

$$Fa = F\pi + F(a - \pi), \quad = 2F_1 + F(a - \pi),$$

giving the values from $a = \pi$ to 2π; and so on. Or what is the same thing we have in general

$$F(m\pi \pm a) = 2mF_1 \pm Fa,$$

and similarly $$E(m\pi \pm a) = 2mE_1 \pm Ea.$$

67. If $k = 1$ there is an entire change in the form of the curves, viz. the curve $y = \Delta\phi$ becomes $y = \cos\phi$, which is a curve lying as before wholly below the line $y = 1$, but which, instead of being included between this and the line $y = 0$, passes below the last-mentioned line, and is in fact included between the lines $y = +1$, $y = -1$. And the curve $y = \dfrac{1}{\Delta\phi}$ becomes $y = \dfrac{1}{\cos\phi}$, where the ordinate becomes infinite for $\phi = \frac{1}{2}\pi$: we have then between the values $\frac{1}{2}\pi$, $\frac{3}{2}\pi$ a branch lying wholly below the line $y = -1$, the ordinates at the limits being $= -\infty$, then from $\frac{3}{2}\pi$ to $\frac{5}{2}\pi$ a like branch lying wholly above the line $y = +1$, the ordinates at the limits being each $= +\infty$; and so on.

Observe that in this case $E\phi = \int \cos\phi \, d\phi = \sin\phi$, so that $E_1 = 1$, and, completing a former statement, we may say that as k increases from 0 to 1, E_1 decreases from $\frac{1}{2}\pi$ to 1.

We have also $F\phi = \displaystyle\int \frac{d\phi}{\cos\phi}$, which admits of finite integration, viz. we have $$F\phi = \log\tan(\tfrac{1}{4}\pi + \tfrac{1}{2}\phi),$$

(observe that log tan is here the hyperbolic logarithm of the tangent,) and in particular $F_1 = \infty$ (a value agreeing with the form of the curve), so that, completing a former statement, we may say that as k increases from 0 to 1, F_1 increases from $\frac{1}{2}\pi$ to ∞.

68. This case (corresponding to the extreme value $k = 1$ of the modulus) is one of great interest: writing

$$u = F\phi = \log \tan (\tfrac{1}{4}\pi + \tfrac{1}{2}\phi),$$

we have $\phi =$ amplitude u (for this particular value $k = 1$), or as it is convenient to write it $\phi = \text{gud}\, u$ (read Gudermannian of u, after Gudermann, by whom the form was specially considered), and then $\sin\phi \qquad = \sin \text{gud}\, u$,

$$\cos\phi = \Delta\phi = \cos \text{gud}\, u,$$

or as we may for shortness write them

$$\sin\phi = \text{sg}\, u, \quad \cos\phi = \Delta\phi = \text{cg}\, u;$$

viz. we have here the two new functions sg, cg, replacing the sn, cn, dn of the general case.

69. We have in a subsequent part of the subject to consider the expressions $\frac{K'}{K}$, and $q = e^{-\frac{\pi K'}{K}}$; and it is convenient to notice here that

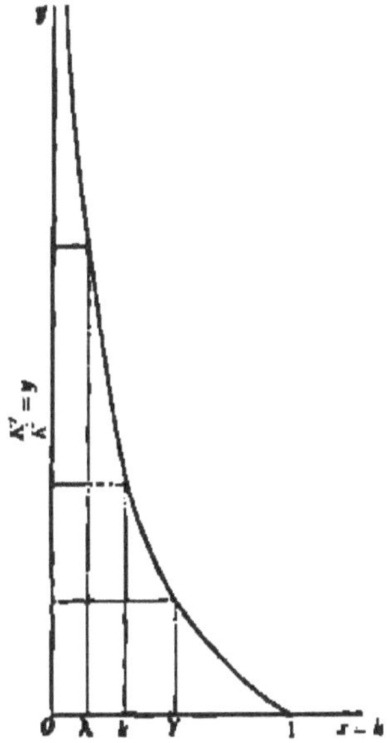

$$k=0;\quad k'=1,\quad K=\tfrac{1}{2}\pi,\quad K'=\infty,\quad \frac{K'}{K}=\infty,\quad q=0:$$

$$k=\frac{1}{\sqrt{2}};\quad k'=\frac{1}{\sqrt{2}},\quad K=K'=F_1\left(\frac{1}{\sqrt{2}}\right),\quad \frac{K'}{K}=1,\quad q=e^{-\pi}:$$

$$k=1;\quad k'=0,\quad K=\infty,\quad K'=\tfrac{1}{2}\pi,\quad \frac{K'}{K}=0,\quad q=1:$$

viz. as k increases from 0 to 1, $\frac{K'}{K}$ diminishes from ∞ to 0 and q increases from 0 to 1. The annexed figure shows the curve $x=k$, $y=\frac{K'}{K}$. It shows also a construction which will present itself in the sequel: viz. considering an abscissa $x=k$, and the abscissæ $x=\lambda$, $x=\gamma$, which belong to the double ordinate and the half-ordinate respectively; then if Γ, Γ' be the complete functions to the modulus γ, and Λ, Λ' the complete functions to the modulus λ, we have it is clear

$$2\frac{\Gamma'}{\Gamma}=\frac{K'}{K},\quad 2\frac{K'}{K}=\frac{\Lambda'}{\Lambda}.$$

70. Conversely if λ be such that $\frac{\Lambda'}{\Lambda}=2\frac{K'}{K}$, then λ is less than k; and similarly if γ be such that $\frac{1}{2}\frac{K'}{K}=\frac{\Gamma'}{\Gamma}$, then γ is greater than k: and not only so, but if, starting from k, we repeat this process of the double ordinate so as to obtain a series of moduli $\lambda, \lambda_1, \lambda_2, \ldots$ then we approximate very rapidly to a modulus $=0$: and similarly if, starting from k, we repeat the process of the half-ordinate so as to obtain a series of moduli $\gamma, \gamma_1, \gamma_2, \ldots$ then we approximate very rapidly to the modulus $=1$. And the like conclusions follow if n denoting any number greater than 1 (say n a positive integer $=$ or >2), we have

$$\frac{\Lambda'}{\Lambda}=n\frac{K'}{K},\quad \frac{K'}{K}=n\frac{\Gamma'}{\Gamma}.$$

Properties of the Functions $F(k, \phi)$, $E(k, \phi)$, *but chiefly the complete functions* $F_1 k$, $E_1 k$. Art. Nos. 71 to 78.

71. Starting from the expressions

$$F(k, \phi) = \int \frac{d\phi}{\Delta(k, \phi)}, \quad E(k, \phi) = \int \Delta(k, \phi)\, d\phi,$$

when k is small we may under the integral sign expand in ascending powers of k, and then integrating from 0 to $\frac{1}{2}\pi$ by the formula

$$\int_0^{\frac{1}{2}\pi} \sin^{2n}\phi\, d\phi = \frac{1.3\ldots 2n-1}{2.4\ldots 2n}\,\frac{\pi}{2},$$

obtain the formulæ

$$F_1 k = \tfrac{1}{2}\pi\left(1 + \frac{1^2}{2^2}k^2 + \frac{1^2.3^2}{2^2.4^2}k^4 + \frac{1^2.3^2.5^2}{2^2.4^2.6^2}k^6 + \ldots\right),$$

$$E_1 k = \tfrac{1}{2}\pi\left(1 - \frac{1}{2^2}k^2 - \frac{1^2.3}{2^2.4^2}k^4 - \frac{1^2.3^2.5}{2^2.4^2.6^2}k^6 - \ldots\right),$$

or what is the same thing, introducing the notation of hypergeometric series

$$F(\alpha, \beta, \gamma, x) = 1 + \frac{\alpha.\beta}{1.\gamma}x + \frac{\alpha.\alpha+1.\beta.\beta+1}{1.2.\gamma.\gamma+1}x^2 + \&c.,$$

these are
$$F_1 k = \tfrac{1}{2}\pi . F(\ \tfrac{1}{2}, \tfrac{1}{2}, 1, k^2),$$
$$E_1 k = \tfrac{1}{2}\pi . F(-\tfrac{1}{2}, \tfrac{1}{2}, 1, k^2).$$

72. Suppose k is very nearly 1, k' is small and we have $k = 1 - \frac{1}{2}k'^2$; to find the value of $F_1 k$ we may write

$$F_1 k = \int_{\frac{1}{2}\pi-\epsilon}^{\frac{1}{2}\pi} \frac{d\phi}{\sqrt{\cos^2\phi + k'^2\sin^2\phi}} + \int_0^{\frac{1}{2}\pi-\epsilon} \frac{d\phi}{\sqrt{\cos^2\phi + k'^2\sin^2\phi}},$$

where ϵ may be taken an indefinitely small quantity which is nevertheless indefinitely large as regards k'. This being so, writing in the first integral $\frac{1}{2}\pi - u$ in place of ϕ, since throughout the integral u is small, the integral becomes $\int_0^\epsilon \frac{du}{\sqrt{k'^2 + k^2u^2}}$, which is $=\frac{1}{k}\log\frac{k\epsilon+\sqrt{k'^2+k^2\epsilon^2}}{k'}$, or neglecting k' in regard to $k\epsilon$, this is $=\frac{1}{k}\log\frac{2k\epsilon}{k'}$, or say $=\log\frac{2\epsilon}{k'}$. In the second integral $k'\sin\phi$ is throughout small as regards $\cos\phi$, and the integral is

$$=\int_\epsilon^{\frac{1}{2}\pi-\epsilon}\frac{d\phi}{\cos\phi},$$

which is $=\log\tan(\frac{1}{2}\pi-\frac{1}{2}\epsilon)$, or what is the same thing $=\log\frac{2}{\epsilon}$. Hence we have

$$F_1k=\log\frac{2\epsilon}{k'}+\log\frac{2}{\epsilon},\quad =\log\frac{4}{k'},$$

as an approximate value of F_1k, k being nearly equal to unity.

73. The functions $F(k,\phi)$, $E(k,\phi)$, considered as functions of k, satisfy certain differential equations.

Write for shortness E, F to denote the functions $E(k,\phi)$, $F(k,\phi)$, and Δ to denote $\Delta(k,\phi)$. Then

$$\frac{dE}{dk}=-\int\frac{k\sin^2\phi\,d\phi}{\Delta},\quad \frac{dF}{dk}=\int\frac{k\sin^2\phi\,d\phi}{\Delta^3},$$

and writing herein $\sin^2\phi=\frac{1}{k^2}(1-\Delta)$, the two expressions depend on the integrals $\int\frac{d\phi}{\Delta}$, $\int\Delta d\phi$, $\int\frac{d\phi}{\Delta^3}$: the two first of these are F, E respectively: as regards the third of them, we have

$$\frac{d}{d\phi}\frac{\sin\phi\cos\phi}{\Delta}=\frac{1-2\sin^2\phi+k^2\sin^4\phi}{\Delta^3},$$

or what is the same thing

$$k^2\frac{d}{d\phi}\frac{\sin\phi\cos\phi}{\Delta}=\frac{\Delta^4-k'^2}{\Delta^3},\quad =\Delta-\frac{k'^2}{\Delta^3},$$

and thence by integration

$$\int\frac{d\phi}{\Delta^3}=\frac{1}{k'^2}E-\frac{k^2\sin\phi\cos\phi}{k'^2\Delta}.$$

The foregoing expressions of $\frac{dE}{dk}$, $\frac{dF}{dk}$, thus become

$$\frac{dE}{dk} = \frac{1}{k}\ (E-F),$$

$$\frac{dF}{dk} = \frac{1}{kk'^2}(E-k'^2F) - \frac{k\sin\phi\cos\phi}{k'^2},$$

whence also

$$F = \quad E - k\frac{dE}{dk},$$

$$E = k'^2\left(F + k\frac{dF}{dk}\right) + \frac{k^2\sin\phi\cos\phi}{\Delta};$$

and in particular if $\phi = \frac{1}{2}\pi$, and E, F now denote the complete functions E_1k, F_1k, then

$$\frac{dE}{dk} = \frac{1}{k}\ (E-F),$$

$$\frac{dF}{dk} = \frac{1}{kk'^2}(E-k'^2F).$$

Let E', F' denote the complementary complete functions E_1k', F_1k'; then observing that $\frac{d}{dk} = -\frac{k}{k'}\frac{d}{dk'}$, we have

$$\frac{dE'}{dk} = -\frac{k}{k'^2}(E'-F'),$$

$$\frac{dF'}{dk} = -\frac{k}{k'^2k}(E'-k^2F').$$

74. If we now consider the expression $EF' + E'F - FF'$, and form its differential coefficient in regard to k, this (substituting therein for $\frac{dE}{dk}$, &c. their values) is found to be $=0$: the expression in question is therefore $=$ a constant; and if to find its value we take k to be indefinitely small, then writing it under the form $(E-F)F' + E'F$, and observing that F' is equal to the indefinitely large quantity $\log\frac{4}{k}$, but that this is multiplied by the indefinitely small quantity $E-F, = -\frac{1}{4}\pi k^2$, and

consequently that the product is $=0$, there remains only the second term $E'F$, which is $=\frac{1}{2}\pi$ (viz. for $k=0$, we have $E'=1$, $F=\frac{1}{2}\pi$); we have therefore

$$EF'+E'F-FF'=\tfrac{1}{2}\pi\,;$$

or writing this at full length,

$$E_1k\,.\,F_1k'+E_1k'\,.\,F_1k-F_1k\,.\,F_1k'=\tfrac{1}{2}\pi,$$

a relation between the original and complementary complete functions E_1k, F_1k, E_1k', F_1k'. [Later on, instead of these quantities we write K, K', E, E', and the equation is

$$EK'+E'K-KK'=\tfrac{1}{2}\pi.]$$

75. The equation in question has recently been proved in a very elegant manner by Mr J. W. L. Glaisher, *Messenger of Mathematics*, t. IV. 1874, p. 95. Writing for convenience $k^2=c$, $k'^2=c'$, and $u=EF'+E'F-FF'$, then from the definitions of the functions,

$$u=\iint\frac{(1-cx^2)+(1-c'y^2)-1}{\sqrt{1-x^2\,.\,1-cx^2\,.\,1-y^2\,.\,1-c'y^2}}\,dx\,dy,$$

where, and in what follows, the integrals in regard to x, y respectively are taken from 0 to 1. Differentiating with regard to c, observing that $dc'=-dc$, and reducing, we have

$$2\frac{du}{dc}=\iint\frac{y^2-x^2+cx^4-c'y^4+c'x^2y^4-cx^4y^2}{(1-x^2\,.\,1-y^2)^{\frac{1}{2}}\,.\,(1-cx^2\,.\,1-c'y^2)^{\frac{3}{2}}}\,dx\,dy,$$

where the numerator is

$$=(1-x^2)\,(1-c'y^4)-(1-y^2)\,(1-cx^4)\,;$$

hence
$$2\frac{du}{dc}=\int\frac{\sqrt{1-x^2}\,dx}{(1-cx^2)^{\frac{3}{2}}}\int\frac{(1-c'y^4)\,dy}{(1-y^2)^{\frac{1}{2}}\,(1-c'y^2)^{\frac{3}{2}}}$$
$$-\int\frac{\sqrt{1-y^2}\,dy}{(1-c'y^2)^{\frac{3}{2}}}\int\frac{(1-cx^4)\,dx}{(1-x^2)^{\frac{1}{2}}\,(1-cx^2)^{\frac{3}{2}}}$$
$$=pq'-p'q \text{ suppose,}$$

where
$$p=\int\frac{\sqrt{1-x^2}\,dx}{(1-cx^2)^{\frac{3}{2}}},\quad q=\int\frac{(1-cx^4)\,dx}{(1-x^2)^{\frac{1}{2}}\,(1-cx^2)^{\frac{3}{2}}},$$

and p', q' are the like functions with c' in place of c.

We have

$$q = \int \frac{[(1-x^2)+(x^2-cx^4)]\,dx}{(1-x^2)^{\frac{1}{2}}(1-cx^2)^{\frac{3}{2}}}$$

$$= p + \int \frac{x^2\,dx}{\sqrt{1-x^2 . 1-cx^2}}$$

$$= p + \left(\frac{-x\sqrt{1-x^2}}{\sqrt{1-cx^2}}\right) + \int \frac{\sqrt{1-x^2}\,dx}{(1-cx^2)^{\frac{3}{2}}},$$

where the second term, taken between the limits, vanishes; and we have therefore $q = p + p, = 2p$. And similarly $q' = 2p'$; hence

$$2\frac{du}{dc} = p.2p' - p'.2p, = 0;$$

hence u is independent of c, and putting $c = 0$, we find that its value is $= \frac{1}{2}\pi$, and the theorem is thus proved.

76. Reverting to the equations

$$F = E - k\frac{dE}{dk}, \quad E = k'^2\left(F + k\frac{dF}{dk}\right) + \frac{k^2 \sin\phi\cos\phi}{\Delta},$$

and from these eliminating successively E and F, we find

$$(1-k^2)\frac{d^2F}{dk^2} + \frac{1-3k^2}{k}\frac{dF}{dk} - F + \frac{\sin\phi\cos\phi}{\Delta^3} = 0,$$

$$(1-k^2)\frac{d^2E}{dk^2} + \frac{1-k^2}{k}\frac{dE}{dk} + E - \frac{\sin\phi\cos\phi}{\Delta} = 0;$$

and in particular if $\phi = \frac{1}{2}\pi$, and E, F now again denote the complete functions E_1k, F_1k, then

$$(1-k^2)\frac{d^2F}{dk^2} + \frac{1-3k^2}{k}\frac{dF}{dk} - F = 0,$$

$$(1-k^2)\frac{d^2E}{dk^2} + \frac{1-k^2}{k}\frac{dE}{dk} + E = 0.$$

We have consequently a particular solution of each of the differential equations

$$(1-k^2)\frac{d^2y}{dk^2}+\frac{1-3k^2}{k}\frac{dy}{dk}-y=0,$$

$$(1-k^2)\frac{d^2z}{dk^2}+\frac{1-k^2}{k}\frac{dz}{dk}+z=0;$$

and we can in terms of the foregoing expressions obtain the complete integrals of these equations; for this purpose, transforming to a new variable k', connected with k by the equation $k^2+k'^2=1$, it is easily found that the transformed equations are

$$(1-k'^2)\frac{d^2y}{dk'^2}+\frac{1-3k'^2}{k'}\frac{dy}{dk'}-y=0,$$

$$(1-k'^2)\frac{d^2z}{dk'^2}-\left(\frac{1+k'^2}{k'}\right)\frac{dz}{dk'}+z=0,$$

where the new equation in y is as regards k' of the same form as the original equation in regard to k: hence, F_1k being a particular solution, another particular solution is F_1k'; and we have the general solution $y=\alpha F_1k+\alpha' F_1k'$. And moreover, observing that the equation in E is satisfied by the value $E=k'^2\left(F+k\frac{dF}{dk}\right)$, it appears that the equation in z must be satisfied by the value $z=k'^2\left(y+k\frac{dy}{dk}\right)$, viz. this is

$$=k'^2(\alpha F_1k+\alpha' F_1k')+k'^2k\left(\alpha\frac{d}{dk}F_1k-\alpha'\frac{k}{k'}\frac{d}{dk'}F_1k'\right):$$

reducing by the formulæ

$$\frac{dF_1k}{dk}=\frac{1}{k'^2k}(E_1k-k'^2F_1k),\quad \frac{dF_1k'}{dk'}=\frac{1}{k'k^2}(E_1k'-k^2F_1k'),$$

this is $z=\alpha E_1k+\alpha'(F_1k'-E_1k')$: where, instead of α, α', we may of course write β, β'; we have thus

$$y=\alpha F_1k+\alpha' F_1k',$$

$$z=\beta E_1k+\beta'(F_1k'-E_1k'),$$

as the complete integrals of the differential equations in y, z. And more generally the equations being

$$(1-k^2)\frac{d^2y}{dk^2}+\frac{1-3k^2}{k}\frac{dy}{dk}-y+\frac{\sin\phi\cos\phi}{\Delta^3}=0,$$

$$(1-k^2)\frac{d^2z}{dk^2}+\frac{1-k^2}{k}\frac{dz}{dk}+z-\frac{\sin\phi\cos\phi}{\Delta}=0;$$

then, to obtain the complete solutions, we must to the expression for y add the term $F(k,\phi)$, and to that for z the term $E(k,\phi)$.

77. To obtain developments for F_1k, E_1k when k is nearly $=1$, or k' is small, observe that F_1k is a solution of

$$(1-k'^2)\frac{d^2y}{dk'^2}+\frac{1-3k'^2}{k'}\frac{dy}{dk'}-y=0,$$

having, when k' is small, the value $\log\frac{4}{k'}$: and conversely, that a solution of the differential equation satisfying the foregoing condition will be the required value of F_1k. Such a solution is $y=P\log\frac{1}{k'}+Q$, where $P-1$ and Q are each a function of the form $Bk'^2+Ck'^4+\ldots$. Substituting in the differential equation, we have first

$$(1-k'^2)\frac{d^2P}{dk'^2}+\frac{1-3k'^2}{k'}\frac{dP}{dk'}-P=0,$$

and then

$$(1-k'^2)\frac{d^2Q}{dk'^2}+\frac{1-3k'^2}{k'}\frac{dQ}{dk'}-Q-2\frac{1-k'^2}{k'}\frac{dP}{dk'}+2P=0;$$

and the first equation then gives

$$P=1+\frac{1^2}{2^2}k'^2+\frac{1^2\,.\,3^2}{2^2\,.\,4^2}k'^4+\ldots,=F(\tfrac{1}{2},\tfrac{1}{2},1,k'^2).$$

Represent this for a moment by

$$1+m_1k'^2+m_2k'^4+\&c.,$$

and assume for convenience

$$Q = -m_1 A_1 k^2 - m_2 A_2 k^4 - m_3 A_3 k^6 - \ldots$$

Substituting these values, the equation to be satisfied is found to be

$$\begin{array}{llllll}
 & k^0 & k^2 & k^4 & k^6 & k^8 \\
0 = & -4m_1 & -8m_2 & -12m_3 & -16m_4 & -20m_5 \;\ldots \\
 & & +4m_1 & +8m_2 & +12m_3 & +16m_4 \;\ldots \\
 & +2 & +2m_1 & +2m_2 & +2m_3 & +2m_4 \;\ldots \\
 & -2m_1A_1 & -12m_2A_2 & -30m_3A_3 & -56m_4A_4 & -90m_5A_5 \;\ldots \\
 & & +2m_1A_1 & +12m_2A_2 & +30m_3A_3 & +56m_4A_4 \;\ldots \\
 & -2m_1A_1 & -4m_2A_2 & -6m_3A_3 & -8m_4A_4 & -10m_5A_5 \;\ldots \\
 & & +6m_1A_1 & +12m_2A_2 & +18m_3A_3 & +24m_4A_4 \;\ldots \\
 & & +m_1A_1 & +m_2A_2 & +m_3A_3 & +m_4A_4 \;\ldots
\end{array}$$

viz. this gives

$$\begin{aligned}
2 - 4m_1 - 4m_1A_1 & = 0, \\
6m_1 - 8m_2 - 16m_2A_2 + 9m_1A_1 & = 0, \\
10m_2 - 12m_3 - 36m_3A_3 + 25m_2A_2 & = 0, \\
14m_3 - 16m_4 - 64m_4A_4 + 49m_3A_3 & = 0, \\
\&c., \qquad & \&c.;
\end{aligned}$$

or, observing that $4m_1 = 1$, $16m_2 = 9m_1$, $36m_3 = 25m_2$, &c., we have

$$\begin{aligned}
2 - 4m_1 & = 4m_1 \,.\, A_1, \\
6m_1 - 8m_2 & = 9m_1 (A_2 - A_1), \\
10m_2 - 12m_3 & = 25m_2 (A_3 - A_2), \\
14m_3 - 16m_4 & = 49m_3 (A_4 - A_3), \\
\&c., \qquad & \&c.;
\end{aligned}$$

that is

$$\begin{aligned}
2 - \frac{1}{1} & = 1A_1 \qquad\qquad , = \frac{1}{1}, \\
6 - \frac{9}{2} & = 9 (A_2 - A_1), = \frac{3}{2}, \\
10 - \frac{25}{3} & = 25 (A_3 - A_2), = \frac{5}{3}, \\
14 - \frac{49}{4} & = 49 (A_4 - A_3), = \frac{7}{4}, \\
\&c., \qquad & \&c.;
\end{aligned}$$

or finally
$$A_1 = \frac{2}{1\,.\,2},$$
$$A_2 = \frac{2}{1\,.\,2} + \frac{2}{3\,.\,4},$$
$$A_3 = \frac{2}{1\,.\,2} + \frac{2}{3\,.\,4} + \frac{2}{5\,.\,6},$$
$$A_4 = \frac{2}{1\,.\,2} + \frac{2}{3\,.\,4} + \frac{2}{5\,.\,6} + \frac{2}{7\,.\,8},$$
&c., &c.,

and thence

$$\begin{aligned} F_1 k = \log \frac{4}{k'} & \\ & + \frac{1^2}{2^2} k'^2 \left(\log \frac{4}{k'} - \frac{2}{1\,.\,2}\right) \\ & + \frac{1^2\,.\,3^2}{2^2\,.\,4^2} k'^4 \left(\log \frac{4}{k'} - \frac{2}{1\,.\,2} - \frac{2}{3\,.\,4}\right) \\ & + \frac{1^2\,.\,3^2\,.\,5^2}{2^2\,.\,4^2\,.\,6^2} k'^6 \left(\log \frac{4}{k'} - \frac{2}{1\,.\,2} - \frac{2}{3\,.\,4} - \frac{2}{5\,.\,6}\right) \\ & + \&c., \end{aligned}$$

where the limit of the subtracted series is $= \log 4$, or 1·38629... From this we obtain $E_1 k$ by the formula

$$E_1 k = k'^2 F_1 k - k'(1 - k'^2) \frac{d}{dk'} F_1 k:$$

leading to

$$\begin{aligned} E_1 k = 1 & \\ & + \frac{1}{2} k'^2 \left(\log \frac{4}{k'} - \frac{1}{1\,.\,2}\right) \\ & + \frac{1^2\,.\,3}{2^2\,.\,4} k'^4 \left(\log \frac{4}{k'} - \frac{2}{1\,.\,2} - \frac{1}{3\,.\,4}\right) \\ & + \frac{1^2\,.\,3^2\,.\,5}{2^2\,.\,4^2\,.\,6} k'^6 \left(\log \frac{4}{k'} - \frac{2}{1\,.\,2} - \frac{2}{3\,.\,4} - \frac{1}{5\,.\,6}\right) \\ & + \&c., \end{aligned}$$

where in the several subtracted series, the numerator of the last fraction is 1, but the other numerators are each 2: the limit of the subtracted series is as in the former case $= \log 4$, or $1{\cdot}38629\ldots$: hence in the two cases respectively the successive partial series converge to $\log \frac{4}{k'} - \log 4, = -\log k'$. We have thus the values of $F_1 k$, $E_1 k$ for k nearly $= 1$, corresponding in a remarkable manner to those previously given for the case of k small.

78. Kummer has given, *Crelle*, t. xv. (1836) p. 83, the following general formulæ in relation to hypergeometric series,

$$2cF(\tfrac{1}{2}\alpha, \tfrac{1}{2}\beta, \tfrac{1}{2}, q^2)$$

$$= F[\alpha, \beta, \tfrac{1}{2}(\alpha+\beta+1), \tfrac{1}{2}(1+q)] + F[\alpha, \beta, \tfrac{1}{2}(\alpha+\beta+1), \tfrac{1}{2}(1-q)];$$

$$2dqF[\tfrac{1}{2}(\alpha+1), \tfrac{1}{2}(\beta+1), \tfrac{3}{2}, q^2]$$

$$= F[\alpha, \beta, \tfrac{1}{2}(\alpha+\beta+1), \tfrac{1}{2}(1+q)] - F[\alpha, \beta, \tfrac{1}{2}(\alpha+\beta+1), \tfrac{1}{2}(1-q)];$$

where c, d are constants to be determined: as regards c, writing $q = 0$, we have at once $c = F[\alpha, \beta, \tfrac{1}{2}(\alpha+\beta+1), \tfrac{1}{2}]$: as regards d, imagining the series on the right-hand side expanded, taking their difference and dividing by q, and then writing $q = 0$, we find $2d = F'[\alpha, \beta, \tfrac{1}{2}(\alpha+\beta+1), \tfrac{1}{2}]$, where in general $F'(\alpha, \beta, \gamma, m)$ denotes $\frac{d}{dx} F(\alpha, \beta, \gamma, x)$, writing therein $x = m$.

Taking now $\alpha = \beta = \tfrac{1}{2}$; and $q = 1 - 2k^2$, whence

$$\tfrac{1}{2}(1+q) = k'^2, \quad \tfrac{1}{2}(1-q) = k^2,$$

we find

$$2cF(\tfrac{1}{4}, \tfrac{1}{4}, \tfrac{1}{2}, q^2) = F(\tfrac{1}{2}, \tfrac{1}{2}, 1, k'^2) + F(\tfrac{1}{2}, \tfrac{1}{2}, 1, k^2), = (F_1 k' + F_1 k) \div \tfrac{1}{2}\pi,$$

$$2dqF(\tfrac{3}{4}, \tfrac{3}{4}, \tfrac{3}{2}, q^2) = F(\tfrac{1}{2}, \tfrac{1}{2}, 1, k'^2) - F(\tfrac{1}{2}, \tfrac{1}{2}, 1, k^2), = (F_1 k' - F_1 k) \div \tfrac{1}{2}\pi,$$

in virtue of the expression for $F_1 k$, $F_1 k'$ obtained ante, No. 71.

Hence, conversely

$$F_1 k = \tfrac{1}{2}\pi\, [cF(\tfrac{1}{4}, \tfrac{1}{4}, \tfrac{1}{2}, q^2) - dqF(\tfrac{3}{4}, \tfrac{3}{4}, \tfrac{3}{2}, q^2)],$$

$$F_1 k' = \tfrac{1}{2}\pi\, [cF(\tfrac{1}{4}, \tfrac{1}{4}, \tfrac{1}{2}, q^2) + dqF(\tfrac{3}{4}, \tfrac{3}{4}, \tfrac{3}{2}, q^2)];$$

viz. the complete functions F_1k, F_1k' are here expressed by means of two series, each proceeding in powers of q, $=1-2k^2$. Writing $k=k'=\frac{1}{\sqrt{2}}$, whence $q=0$, we find

$$\tfrac{1}{4}\pi c = F_1\left(\frac{1}{\sqrt{2}}\right), = \int_0^{\frac{\pi}{2}} \frac{d\phi}{\sqrt{1-\frac{1}{2}\sin^2\phi}},$$

and with a little more difficulty $\frac{1}{4}\pi d = \frac{1}{2}\pi + F_1\left(\frac{1}{\sqrt{2}}\right)$, and we have thus the expressions for F_1k, F_1k' given by Jacobi, *Fund. Nova*, pp. 67 and 68.

The Gudermannian. Art. Nos. 79 to 85.

79. It has been already remarked that, for $k=1$, the function $F(k, \phi)$ becomes $=\log\tan(\frac{1}{4}\pi+\frac{1}{2}\phi)$, and that instead of the general function am u, we have the gudermannian gd u, giving rise to the two functions sin gd u and cos gd u, or say sg u and cg u. We have in regard to these a theory corresponding to that of the functions of am u (sn u, cn u, dn u), discussed in the following Chapter: and it is convenient to consider in the first instance the special case in question, $k=1$.

80. Starting from

$$F\phi = \log\tan(\tfrac{1}{4}\pi+\tfrac{1}{2}\phi) = u,$$

where as a definition $\phi = \mathrm{gd}\,u$, or what is the same thing,

$$u = \log\tan(\tfrac{1}{4}\pi + \tfrac{1}{2}\,\mathrm{gd}\,u);$$

we have

$$e^u = \tan(\tfrac{1}{4}\pi+\tfrac{1}{2}\,\mathrm{gd}\,u) = \frac{1+\tan\frac{1}{2}\mathrm{gd}\,u}{1-\tan\frac{1}{2}\mathrm{gd}\,u} = \frac{\cos\frac{1}{2}\mathrm{gd}\,u+\sin\frac{1}{2}\mathrm{gd}\,u}{\cos\frac{1}{2}\mathrm{gd}\,u-\sin\frac{1}{2}\mathrm{gd}\,u}$$

$$= \frac{1+\sin\mathrm{gd}\,u}{\cos\mathrm{gd}\,u} = \frac{\cos\mathrm{gd}\,u}{1-\sin\mathrm{gd}\,u};$$

and thence

$$\sin \operatorname{gd} u \text{ or } \operatorname{sg} u = \frac{e^u - e^{-u}}{e^u + e^{-u}}, = \frac{-i \sin iu}{\cos iu}, = \frac{\sinh u}{\cosh u}, = \tanh u,$$

and

$$\cos \operatorname{gd} u \text{ or } \operatorname{cg} u = \frac{1}{e^u + e^{-u}}, = \frac{1}{\cos iu}, \quad = \frac{1}{\cosh u}, = \operatorname{sech} u,$$

(where $\sinh u, = \frac{1}{2}(e^u - e^{-u})$, and $\cosh u, = \frac{1}{2}(e^u + e^{-u})$, denote the hyperbolic sine and cosine of u; and similarly $\tanh u$ and $\operatorname{sech} u$ denote the hyperbolic tangent and secant of u).

It may be added that

$$\operatorname{cg}^2 u + \operatorname{sg}^2 u = 1,$$

and further $\quad \operatorname{gd}' u = \operatorname{cg} u, \ \operatorname{sg}' u = \operatorname{cg}^2 u, \ \operatorname{cg}' u = -\operatorname{sg} u \operatorname{cg} u,$

also $\quad \operatorname{sg} iu = i \tan u, \ \operatorname{cg} iu = \sec u.$

81. The equations may also be written

$\operatorname{sg} u = -i \tan iu,$	$\sin iu = i \operatorname{tg} u,$
$\operatorname{cg} u = \dfrac{1}{\cos iu},$	$\cos iu = \dfrac{1}{\operatorname{cg} u},$
$\operatorname{tg} u = -i \sin iu,$	$\tan iu = i \operatorname{sg} u;$

($\operatorname{tg} u$ denoting $\tan \operatorname{gd} u$) which may also be arrived at as follows, viz. considering the angles θ, ϕ connected by the equation $\cos\theta \cos\phi = 1$, or as it may in various forms be written,

$\sin\theta = i \tan\phi,$	$\sin\phi = -i \tan\theta,$
$\cos\theta = \dfrac{1}{\cos\phi},$	$\cos\phi = \dfrac{1}{\cos\theta},$
$\tan\theta = i \sin\phi,$	$\tan\phi = -i \sin\theta,$

then $\cos\theta\, d\theta = i \sec^2\phi\, d\phi$, that is $d\theta = \dfrac{i\,d\phi}{\cos\phi}$, or

$$\theta = i \log \tan\left(\tfrac{1}{4}\pi + \tfrac{1}{2}\phi\right);$$

whence assuming $\phi = \operatorname{gd} u$ we have $\theta = iu$, and thence the foregoing relations.

82. We easily obtain the addition-equations

$$\operatorname{sg}(u+v) = \operatorname{sg} u \operatorname{sg} v, \qquad (+)$$

$$\operatorname{cg}(u+v) = \operatorname{cg} u \operatorname{cg} v, \qquad (+)$$

where $\qquad$ denom. $= 1 + \operatorname{sg} u \operatorname{sg} v$;

viz. if for a moment $e^u = \alpha$, $e^v = \beta$, then

$$\operatorname{sg} u = \frac{\alpha^2-1}{\alpha^2+1},\ \operatorname{cg} u = \frac{2\alpha}{\alpha^2+1},\ \operatorname{sg} v = \frac{\beta^2-1}{\beta^2+1},\ \operatorname{cg} v = \frac{2\beta}{\beta^2+1},$$

and substituting these values, the expressions for $\operatorname{sg}(u+v)$, $\operatorname{cg}(u+v)$ come out $= \frac{\alpha^2\beta^2-1}{\alpha^2\beta^2+1}$ and $\frac{2\alpha\beta}{\alpha^2\beta^2+1}$ respectively; which proves the formulæ.

83. To deduce the equations from the general formulæ for $\operatorname{sn}(u+v)$, $\operatorname{cn}(u+v)$, $\operatorname{dn}(u+v)$, (see next Chapter,) observe that putting $k=1$, and consequently $\operatorname{sn} = \operatorname{sg}$, $\operatorname{cn} = \operatorname{dn} = \operatorname{cg}$, these become

$$\operatorname{sg}(u+v) = \operatorname{sg} u \operatorname{cg}^2 v + \operatorname{sg} v \operatorname{cg}^2 u, \qquad (+)$$

$$\operatorname{cg}(u+v) = \operatorname{cg} u \operatorname{cg} v - \operatorname{cg} u \operatorname{sg} u \operatorname{cg} v \operatorname{sg} v, \qquad (+)$$

where $\qquad$ denom. $= 1 - \operatorname{sg}^2 u \operatorname{sg}^2 v$.

Here in $\operatorname{sg}(u+v)$ the numerator is $\operatorname{sg} u(1-\operatorname{sg}^2 v) + \operatorname{sg} v(1-\operatorname{sg}^2 u)$, which is $= (\operatorname{sg} u + \operatorname{sg} v)(1 - \operatorname{sg} u \operatorname{sg} v)$, and in $\operatorname{cg}(u+v)$ the numerator is $= \operatorname{cg} u \operatorname{cg} v(1 - \operatorname{sg} u \operatorname{sg} v)$, and the denominator is $= (1 + \operatorname{sg} u \operatorname{sg} v)(1 - \operatorname{sg} u \operatorname{sg} v)$; whence, throwing out the factor $(1 - \operatorname{sg} u \operatorname{sg} v)$, we have the formulæ in question.

84. It is easy to derive the formulæ for the sg and cg of the sum of any number of functions. Writing for convenience $\operatorname{sg} u = x$, $\operatorname{cg} u = \sqrt{1-x^2}$, $= x'$, $\operatorname{sg} v = y$, $\operatorname{cg} v = \sqrt{1-y^2}$, $= y'$, the foregoing formulæ may be written

$$\operatorname{sg}(u+v) = x+y, \qquad (+)$$

$$\operatorname{cg}(u+v) = x'y', \qquad (+)$$

where $\qquad$ denom. $= 1 + xy$;

and then introducing a new angle w, and writing $\text{sg}\, w = z$, $\text{cg}\, w = z'$, we find

$$\text{sg}(u+v+w) = x + y + z + xyz, \qquad (\div)$$

$$\text{cg}(u+v+w) = x'y'z', \qquad (\div)$$

where $\qquad \text{denom.} = 1 + xy + xz + yz;$

and so when there is a fourth angle ω, $\text{sg}\,\omega = t$, $\text{cg}\,\omega = t'$, we have

$$\text{sg}(u+v+w+\omega) = x + y + z + t + xyz + xyt + xzt + yzt, \qquad (\div)$$

$$\text{cg}(u+v+w+\omega) = x'y'z't', \qquad (\div)$$

where $\qquad \text{denom.} = 1 + xy + xz + yz + zt + yt + zt + xyzt;$

and so on, the law being obvious.

85. If the angles of all of them $= u$, retaining x to denote $\text{cg}\, u$, and putting for x' its value $= \sqrt{(1+x)(1-x)}$, we have

$$\text{sg}\, nu = \tfrac{1}{2}[(1+x)^n - (1-x)^n], \qquad (\div)$$

$$\text{cg}\, nu = \quad (1+x)^{\frac{1}{2}n}(1-x)^{\frac{1}{2}n}, \qquad (\div)$$

where $\qquad \text{denom.} = \tfrac{1}{2}[(1+x)^n + (1-x)^n];$

and observe that, n being even, the expressions are rational, but n being odd, the numerator of $\text{cg}\, nu$ contains the factor $\sqrt{1-x^2}$.

The formulæ are valuable for their own sake; and they afford very convenient verifications of formulæ relating to the general functions sn, cn, dn: viz. putting in these $k = 1$, they must of course reduce themselves to the far more simple formulæ for sg, cg.

The foregoing values of $\text{sg}\, u$, $\text{cg}\, u$, give

$$e^{i\,\text{gd}\, u} = \frac{2e^u + i(e^{2u} - 1)}{e^{2u} + 1}, = \frac{i(e^u - i)^2}{e^{2u} + 1};$$

that is $\qquad e^{i\,\text{gd}\, u} = \dfrac{i(e^u - i)}{e^u + i};$

and again, $e^{i\,\text{gd}\,u} = \dfrac{2+i(e^u-e^{-u})}{e^u+e^{-u}}, = \dfrac{1+\sin ui}{\cos ui}, = \dfrac{\cos\frac{1}{2}ui+\sin\frac{1}{2}ui}{\cos\frac{1}{2}ui-\sin\frac{1}{2}ui},$

$$= \frac{1+\tan\frac{1}{2}ui}{1-\tan\frac{1}{2}ui};$$

that is
$$e^{i\,\text{gd}\,u} = \tan\left(\tfrac{1}{4}\pi + \tfrac{1}{2}ui\right),$$

or what is the same thing,

$$i\,\text{gd}\,u = \log\tan\left(\tfrac{1}{4}\pi + \tfrac{1}{2}ui\right);$$

with which compare the original equation

$$u = \log\tan\left(\tfrac{1}{4}\pi + \tfrac{1}{2}\,\text{gd}\,u\right).$$

If in the first of these for u we write $\dfrac{1}{i}\,\text{gd}\,u$, it becomes

$$i\,\text{gd}\left(\frac{1}{i}\,\text{gd}\,u\right) = \log\tan\left(\tfrac{1}{4}\pi + \tfrac{1}{2}\,\text{gd}\,u\right),$$

that is
$$i\,\text{gd}\left(\frac{1}{i}\,\text{gd}\,u\right) = u,$$

a remarkable property of the function $\text{gd}\,u$; there is no analogue to this as regards the general function $\text{am}\,u$.

CHAPTER IV.

ON THE ELLIPTIC FUNCTIONS sn, cn, dn.

WE now commence a systematic development of the theory of the elliptic functions properly so called, the functions sn, cn, dn.

Addition and Subtraction Formulæ. Art. Nos. 86 to 93.

86. The formulæ are

$$\operatorname{sn}(u+v) = \operatorname{sn} u \operatorname{cn} v \operatorname{dn} v + \operatorname{sn} v \operatorname{cn} u \operatorname{dn} u, \qquad (\div)$$

$$\operatorname{cn}(u+v) = \operatorname{cn} u \operatorname{cn} v \quad - \operatorname{sn} u \operatorname{dn} u \operatorname{sn} v \operatorname{dn} v, \qquad (\div)$$

$$\operatorname{dn}(u+v) = \operatorname{dn} u \operatorname{dn} v \quad - k^2 \operatorname{sn} u \operatorname{cn} u \operatorname{sn} v \operatorname{cn} v, \qquad (\div)$$

where denominator

$$= 1 - k^2 \operatorname{sn}^2 u \operatorname{sn}^2 v;$$

and

$$\operatorname{sn}(u-v) = \operatorname{sn} u \operatorname{cn} v \operatorname{dn} v - \operatorname{sn} v \operatorname{cn} u \operatorname{dn} u, \qquad (\div)$$

$$\operatorname{cn}(u-v) = \operatorname{cn} u \operatorname{cn} v \quad + \operatorname{sn} u \operatorname{dn} u \operatorname{sn} v \operatorname{dn} v, \qquad (\div)$$

$$\operatorname{dn}(u-v) = \operatorname{dn} u \operatorname{dn} v \quad + k^2 \operatorname{sn} u \operatorname{cn} u \operatorname{sn} v \operatorname{cn} v, \qquad (\div)$$

with same denominator

$$= 1 - k^2 \operatorname{sn}^2 u \operatorname{sn}^2 v.$$

As remarked in Chapter I., these are given by the addition-equation, or they may be deduced from

$$\operatorname{cn}^2 u = 1 - \operatorname{sn}^2 u, \quad \operatorname{dn}^2 u = 1 - k^2 \operatorname{sn}^2 u,$$

$$\operatorname{sn}' u = \quad \operatorname{cn} u \operatorname{dn} u,$$

$$\operatorname{cn}' u = \quad - \operatorname{sn} u \operatorname{dn} u,$$

$$\operatorname{dn}' u = - k^2 \operatorname{sn} u \operatorname{cn} u.$$

87. Putting for shortness $\operatorname{sn} u = x$, $\operatorname{sn} v = y$, these are

$$\operatorname{sn}(u \pm v) = x\sqrt{1-y^2}\sqrt{1-k^2y^2} \pm y\sqrt{1-x^2}\sqrt{1-k^2x^2}, \quad (\div)$$

$$\operatorname{cn}(u \pm v) = \sqrt{1-x^2}\sqrt{1-y^2} \quad \mp xy\sqrt{1-k^2x^2}\sqrt{1-k^2y^2}, \quad (\div)$$

$$\operatorname{dn}(u \pm v) = \sqrt{1-k^2x^2}\sqrt{1-k^2y^2} \mp k^2xy\sqrt{1-x^2}\sqrt{1-y^2}, \quad (\div)$$

where denominator

$$= 1 - k^2x^2y^2.$$

88. Represent for a moment the last-mentioned numerators and denominator by $A \pm A'$, $B \pm B'$, $C \pm C'$, and D, viz.

$$A = x\sqrt{1-y^2}\sqrt{1-k^2y^2}, \quad A' = y\sqrt{1-x^2}\sqrt{1-k^2x^2},$$

$$B = \sqrt{1-x^2}\sqrt{1-y^2}, \qquad B' = -xy\sqrt{1-k^2x^2}\sqrt{1-k^2y^2},$$

$$C = \sqrt{1-k^2x^2}\sqrt{1-k^2y^2}, \quad C' = -k^2xy\sqrt{1-x^2}\sqrt{1-y^2},$$

$$D = 1 - k^2x^2y^2,$$

then we have evidently

$$\operatorname{sn}(u+v) + \operatorname{sn}(u-v) = 2A, \qquad (\div)$$

$$\operatorname{sn}(u+v) - \operatorname{sn}(u-v) = 2A', \qquad (\div)$$

$$\operatorname{cn}(u+v) + \operatorname{cn}(u-v) = 2B, \qquad (\div)$$

$$\operatorname{cn}(u+v) - \operatorname{cn}(u-v) = 2B', \qquad (\div)$$

$$\operatorname{dn}(u-v) + \operatorname{dn}(u+v) = 2C, \qquad (\div)$$

$$\operatorname{dn}(u+v) - \operatorname{dn}(u-v) = 2C', \qquad (\div)$$

where throughout

$$\text{denominator} = D.$$

89. But there are other formulæ depending on the property that the rational functions $A^2 - A'^2$, $B^2 - B'^2$, $C^2 - C'^2$ contain D as a factor. In fact writing

$$P = x^2 - y^2,$$

$$Q = 1 - x^2 - y^2 + k^2x^2y^2,$$

$$R = 1 - k^2x^2 - k^2y^2 + k^2x^2y^2,$$

we have
$$A^2 - A'^2 = PD,$$
$$B^2 - B'^2 = QD,$$
$$C^2 - C'^2 = RD;$$
and thence
$$\text{sn}(u+v)\,\text{sn}(u-v) = P, \qquad (\div)$$
$$\text{cn}(u+v)\,\text{cn}(u-v) = Q, \qquad (\div)$$
$$\text{dn}(u+v)\,\text{dn}(u-v) = R, \qquad (\div)$$
where denominator $= D$.

I write down at full length the first of these as it is a formula of frequent occurrence,

$$\text{sn}(u+v)\,\text{sn}(u-v) = (\text{sn}^2 u - \text{sn}^2 v) \div (1 - k^2\,\text{sn}^2 u\,\text{sn}^2 v).$$

90. We may deduce a variety of other formulæ, for instance

$$[1 + \text{sn}(u+v)]\,[1 + \text{sn}(u-v)]$$
$$= (D + 2A + P) \div D;$$

where the numerator is

$$1 - k^2x^2y^2 + 2x\sqrt{1-y^2}\sqrt{1-k^2y^2} + x^2 - y^2,$$
$$= (\sqrt{1-y^2} + x\sqrt{1-k^2y^2})^2.$$

91. To complete the theory we consider the expressions

$$S = \sqrt{1-x^2}\sqrt{1-k^2x^2}\sqrt{1-y^2}\sqrt{1-k^2y^2}, \quad S' = -k'^2xy, \quad (k'^2 = 1-k^2).$$
$$T = x\sqrt{1-k^2x^2}\sqrt{1-y^2}, \qquad T' = -y\sqrt{1-k^2y^2}\sqrt{1-x^2},$$
$$U = x\sqrt{1-x^2}\sqrt{1-k^2y^2}, \qquad U' = \ \ y\sqrt{1-y^2}\sqrt{1-k^2x^2}.$$

It then appears that each of the functions sn, cn, dn of $u \pm v$ can be expressed in a fourfold form as follows:

$$\text{sn}(u+v) = \frac{A+A'}{D} = \frac{P}{A-A'} = \frac{U+U'}{B-B'} = \frac{T-T'}{C-C'},$$
$$\text{cn}(u+v) = \frac{B+B'}{D} = \frac{U-U'}{A-A'} = \frac{Q}{B-B'} = \frac{S+S'}{C-C'},$$
$$\text{dn}(u+v) = \frac{C+C'}{D} = \frac{T+T'}{A-A'} = \frac{S-S'}{B-B'} = \frac{R}{C-C'};$$

with the like formulæ changing simultaneously the signs of v, A', B', C', S', T', U'. Any equations obtained by comparing different values of the same function are of course identities: thus we have $\frac{A+A'}{D} = \frac{P}{A-A'}$, that is, $A^2 - A'^2 = PD$ as above, &c. Again

$$(B+B')(C-C') = D(S+S'), \quad (B-B')(C+C') = D(S-S'),$$

or what is the same thing,

$$BC - B'C' = DS, \quad -BC' + B'C = DS', \text{ \&c.},$$

with various other identities.

92. By selecting the proper expressions we obtain at once formulæ involving different functions of $u+v$ and $u-v$ respectively: thus let it be required to find the product

$$\operatorname{sn}(u+v)\operatorname{cn}(u-v);$$

these are expressions for the factors involving $B-B'$ in the denominator and the numerator respectively, viz. we have

$$\operatorname{sn}(u+v)\operatorname{cn}(u-v) = \frac{U+U'}{B-B'} \cdot \frac{B-B'}{D},$$

$$= \frac{U+U'}{D},$$

or what is the same thing, the value is

$$= \frac{\operatorname{sn} u \operatorname{cn} u \operatorname{dn} v + \operatorname{sn} v \operatorname{cn} v \operatorname{dn} u}{1 - k^2 \operatorname{sn}^2 u \operatorname{sn}^2 v}.$$

Similarly,

$$\operatorname{cn}(u+v)\operatorname{sn}(u-v) = \frac{U-U'}{D},$$

and by combination of these formulæ, we obtain

$$\sin[\operatorname{am}(u+v) + \operatorname{am}(u-v)] = \frac{2U}{D},$$

$$\sin[\operatorname{am}(u+v) - \operatorname{am}(u-v)] = \frac{2U'}{D}.$$

93. Although the formulæ are so numerous that they cannot be remembered, and in the manner just explained any one of them may be obtained with extreme facility, yet for convenience of reference I reproduce the whole series of 33 formulæ given *Fund. Nova*, pp. 32—34. We have throughout

$$\text{denom.} = 1 - k^2\,\mathrm{sn}^2u\,\mathrm{sn}^2v.$$

(1) to (21).

$$\mathrm{sn}(u+v)+\mathrm{sn}(u-v) = 2\,\mathrm{sn}\,u\,\mathrm{cn}\,v\,\mathrm{dn}\,v, \quad (\div)$$

$$\mathrm{cn}(u+v)+\mathrm{cn}(u-v) = 2\,\mathrm{cn}\,u\,\mathrm{cn}\,v, \quad (\div)$$

$$\mathrm{dn}(u+v)+\mathrm{dn}(u-v) = 2\,\mathrm{dn}\,u\,\mathrm{dn}\,v, \quad (\div)$$

$$\mathrm{sn}(u+v)-\mathrm{sn}(u-v) = 2\,\mathrm{sn}\,v\,\mathrm{cn}\,u\,\mathrm{dn}\,u, \quad (\div)$$

$$\mathrm{cn}(u-v)-\mathrm{cn}(u+v) = 2\,\mathrm{sn}\,u\,\mathrm{sn}\,v\,\mathrm{dn}\,u\,\mathrm{dn}\,v, \quad (\div)$$

$$\mathrm{dn}(u-v)-\mathrm{dn}(u+v) = 2k^2\,\mathrm{sn}\,u\,\mathrm{sn}\,v\,\mathrm{cn}\,u\,\mathrm{cn}\,v, \quad (\div)$$

$$\mathrm{sn}(u+v)\,\mathrm{sn}(u-v) = \mathrm{sn}^2u-\mathrm{sn}^2v, \quad (\div)$$

$$1+k^2\,\mathrm{sn}(u+v)\,\mathrm{sn}(u-v) = \mathrm{dn}^2v+k^2\,\mathrm{sn}^2u\,\mathrm{cn}^2v, \quad (\div)$$

$$1+\mathrm{sn}(u+v)\,\mathrm{sn}(u-v) = \mathrm{cn}^2v+\mathrm{sn}^2u\,\mathrm{dn}^2v, \quad (\div)$$

$$1+\mathrm{cn}(u+v)\,\mathrm{cn}(u-v) = \mathrm{cn}^2u+\mathrm{cn}^2v, \quad (\div)$$

$$1+\mathrm{dn}(u+v)\,\mathrm{dn}(u-v) = \mathrm{dn}^2u+\mathrm{dn}^2v, \quad (\div)$$

$$1-k^2\,\mathrm{sn}(u+v)\,\mathrm{sn}(u-v) = \mathrm{dn}^2u+k^2\,\mathrm{sn}^2v\,\mathrm{cn}^2u, \quad (\div)$$

$$1-\mathrm{sn}(u+v)\,\mathrm{sn}(u-v) = \mathrm{cn}^2u+\mathrm{sn}^2v\,\mathrm{dn}^2u, \quad (\div)$$

$$1-\mathrm{cn}(u+v)\,\mathrm{cn}(u-v) = \mathrm{sn}^2u\,\mathrm{dn}^2v+\mathrm{sn}^2v\,\mathrm{dn}^2u, \quad (\div)$$

$$1-\mathrm{dn}(u+v)\,\mathrm{dn}(u-v) = k^2(\mathrm{sn}^2u\,\mathrm{cn}^2v+\mathrm{sn}^2v\,\mathrm{cn}^2u), \quad (\div)$$

$$[1\pm\mathrm{sn}(u+v)]\,[1\pm\mathrm{sn}(u-v)] = (\mathrm{cn}\,v\pm\mathrm{sn}\,u\,\mathrm{dn}\,v)^2, \quad (\div)$$

$$[1\pm\mathrm{sn}(u+v)]\,[1\mp\mathrm{sn}(u-v)] = (\mathrm{cn}\,u\pm\mathrm{sn}\,v\,\mathrm{dn}\,u)^2, \quad (\div)$$

$$[1\pm k\,\mathrm{sn}(u+v)]\,[1\pm k\,\mathrm{sn}(u-v)] = (\mathrm{dn}\,v\pm k\,\mathrm{sn}\,u\,\mathrm{cn}\,v)^2, \quad (\div)$$

$$[1\pm k\,\mathrm{sn}(u+v)]\,[1\mp k\,\mathrm{sn}(u-v)] = (\mathrm{dn}\,u\pm k\,\mathrm{sn}\,v\,\mathrm{cn}\,u)^2, \quad (\div)$$

$$[1\pm\mathrm{cn}(u+v)]\,[1\pm\mathrm{cn}(u-v)] = (\mathrm{cn}\,u\pm\mathrm{cn}\,v)^2, \quad (\div)$$

$$[1\pm\mathrm{cn}(u+v)]\,[1\mp\mathrm{cn}(u-v)] = (\mathrm{sn}\,u\,\mathrm{dn}\,v\mp\mathrm{sn}\,v\,\mathrm{dn}\,u)^2, \quad (\div)$$

$$[1 \pm \operatorname{dn}(u+v)][1 \pm \operatorname{dn}(u-v)] = (\operatorname{dn} u \pm \operatorname{dn} v)^2, \quad (+)$$

$$[1 \pm \operatorname{dn}(u+v)][1 \mp \operatorname{dn}(u-v)] = k^2 (\operatorname{sn} u \operatorname{cn} v \mp \operatorname{sn} v \operatorname{cn} u)^2, \quad (+)$$

$$\operatorname{sn}(u+v)\operatorname{cn}(u-v) = \operatorname{sn} u \operatorname{cn} u \operatorname{dn} v + \operatorname{sn} v \operatorname{cn} v \operatorname{dn} u, \quad (+)$$

$$\operatorname{sn}(u-v)\operatorname{cn}(u+v) = \operatorname{sn} u \operatorname{cn} u \operatorname{dn} v - \operatorname{sn} v \operatorname{cn} v \operatorname{dn} u, \quad (+)$$

$$\operatorname{sn}(u+v)\operatorname{dn}(u-v) = \operatorname{sn} u \operatorname{dn} u \operatorname{cn} v + \operatorname{sn} v \operatorname{dn} v \operatorname{cn} u, \quad (+)$$

$$\operatorname{sn}(u-v)\operatorname{dn}(u+v) = \operatorname{sn} u \operatorname{dn} u \operatorname{cn} v - \operatorname{sn} v \operatorname{dn} v \operatorname{cn} u, \quad (+)$$

$$\operatorname{cn}(u+v)\operatorname{dn}(u-v) = \operatorname{cn} u \operatorname{cn} v \operatorname{dn} u \operatorname{dn} v - k'^2 \operatorname{sn} u \operatorname{sn} v, \quad (+)$$

$$\operatorname{cn}(u-v)\operatorname{dn}(u+v) = \operatorname{cn} u \operatorname{cn} v \operatorname{dn} u \operatorname{dn} v + k'^2 \operatorname{sn} u \operatorname{sn} v, \quad (+)$$

$$\sin[\operatorname{am}(u+v) + \operatorname{am}(u-v)] = 2 \operatorname{sn} u \operatorname{cn} u \operatorname{dn} v, \quad (+)$$

$$\sin[\operatorname{am}(u+v) - \operatorname{am}(u-v)] = 2 \operatorname{sn} v \operatorname{cn} v \operatorname{dn} u, \quad (+)$$

$$\cos[\operatorname{am}(u+v) + \operatorname{am}(u-v)] = \operatorname{cn}^2 u - \operatorname{sn}^2 u \operatorname{dn}^2 v, \quad (+)$$

$$\cos[\operatorname{am}(u+v) - \operatorname{am}(u-v)] = \operatorname{cn}^2 v - \operatorname{sn}^2 v \operatorname{dn}^2 u, \quad (+)$$

(22) to (33).

The Periods $4K$, $4iK'$. Art. No. 94.

94. The theory of the periods depends on the equations

$$\begin{array}{l|l|l} \operatorname{sn} 0 = 0, & \operatorname{sn} K = 1, & \operatorname{sn}(K+iK') = \dfrac{1}{k}, \\ \operatorname{cn} 0 = 1, & \operatorname{cn} K = 0, & \operatorname{cn}(K+iK') = \dfrac{-ik'}{k}, \\ \operatorname{dn} 0 = 1, & \operatorname{dn} K = k', & \operatorname{dn}(K+iK') = 0; \end{array}$$

where K, K' are the complete functions Fk, Fk'.

To prove these observe that writing

$$u = \int_0^\xi \frac{dx}{\sqrt{1-x^2 \, . \, 1-k^2x^2}},$$

we have $\operatorname{sn} u = \xi$, $\operatorname{cn} u = \sqrt{1-\xi^2}$, $\operatorname{dn} u = \sqrt{1-k^2\xi^2}$, whence writing $\xi = 0$ we have the first triad of formulæ, and writing $\xi = 1$ the second triad. For the third triad, writing $\xi = \frac{1}{k}$, we have

$$u = \int_0^{\frac{1}{k}} \frac{dx}{\sqrt{1-x^2 \,.\, 1-k^2x^2}}$$

$$= \left(\int_0^1 + \int_1^{\frac{1}{k}}\right) \frac{dx}{\sqrt{1-x^2 \,.\, 1-k^2x^2}}$$

$$= K + \int_1^{\frac{1}{k}} \frac{dx}{\sqrt{1-x^2 \,.\, 1-k^2x^2}},$$

and to transform the integral we write $\left(x=1, z=0;\ x=\frac{1}{k}, z=1\right)$,

$$x = \frac{1}{\sqrt{1-k'^2z^2}},$$

whence

$$dx = \frac{k'^2 z\,dz}{(1-k'^2z^2)^{\frac{3}{2}}},$$

$$\frac{1}{\sqrt{1-x^2}} = \frac{i\sqrt{1-k'^2z^2}}{k'z},$$

$$\frac{1}{\sqrt{1-k^2x^2}} = \frac{\sqrt{1-k'^2z^2}}{k'\sqrt{1-z^2}};$$

or multiplying

$$\frac{dx}{\sqrt{1-x^2 \,.\, 1-k^2x^2}} = \frac{i\,dz}{\sqrt{1-z^2 \,.\, 1-k'^2z^2}},$$

so that the integral is

$$i\int_0^1 \frac{dz}{\sqrt{1-z^2 \,.\, 1-k'^2z^2}}, = iK';$$

and the value of u is $= K + iK'$. Hence writing $u = K + iK'$, $\xi = \frac{1}{k}$, and observing that the value of $\sqrt{1-\xi^2}$ is

$$= \frac{k'z}{i\sqrt{1-k'^2z^2}}\ (z=1),$$

viz. that it is $= \frac{-ik'}{k}$, we have the required formulæ

$$\text{sn}(K+iK') = \frac{1}{k}, \quad \text{cn}(K+iK') = \frac{-ik'}{k}, \quad \text{dn}(K+iK') = 0.$$

Property arising from the transformation. Art. No. 95.

95. In the foregoing relation between x and s, write for a moment $x = \sin\phi$, $s = \sin\chi$, the differential equation is

$$\frac{d\phi}{\sqrt{1-k^2\sin^2\phi}} = \frac{id\chi}{\sqrt{1-k'^2\sin^2\chi}};$$

whence, assuming $\sin\phi = \text{sn}(v, k)$, $\sin\chi = \text{sn}(u, k')$, this is $dv = idu$, or we have $v = iu + \text{const.}$ But we have simultaneously $x = 1$, $s = 0$; and for $x = 1$, v is $= K$, and for $s = 0$, u is $= 0$: hence the constant is $= K$, or we have $v = iu + K$: consequently $x = \text{sn}(iu + K, k)$, $s = \text{sn}(u, k')$. Substituting in the integral equations between x, s, we have

$$\text{sn}(iu + K, k) = \frac{1}{\text{dn}(u, k')},$$

$$\text{cn}(iu + K, k) = \frac{-ik'\,\text{sn}(u, k')}{\text{dn}(u, k')},$$

$$\text{dn}(iu + K, k) = \frac{k'\,\text{cn}(u, k')}{\text{dn}(u, k')},$$

which are equivalent to the equations obtained in the next article.

Jacobi's imaginary transformation. Art. No. 96.

96. Write $\sin\phi = i\tan\psi$, whence also $\cos\phi = \frac{1}{\cos\psi}$, and $\sin\psi = -i\tan\phi$; consequently $d\phi = \frac{id\psi}{\cos\psi}$, and

$$\frac{d\phi}{\sqrt{1-k^2\sin^2\phi}} = \frac{id\psi}{\sqrt{1-k'^2\sin^2\psi}}.$$

Hence, putting $\sin\phi = \text{sn}(v, k)$, $\sin\psi = \text{sn}(u, k')$, we have $dv = i\,du$, or since v, u vanish together $v = iu$; that is

$$\sin\phi = \text{sn}(iu, k), \quad \sin\psi = \text{sn}(u, k').$$

The integral relations between ϕ, ψ give

$$\text{sn}(iu, k) = \frac{i\,\text{sn}(u, k')}{\text{cn}(u, k')},$$

$$\text{cn}(iu, k) = \frac{1}{\text{cn}(u, k')},$$

$$\text{dn}(iu, k) = \frac{\text{dn}(u, k')}{\text{cn}(u, k')}.$$

It may be observed that in this transformation writing $\phi = iu$ we have $\psi = \text{gd}\,u$. It is to be further observed that writing $\sin\psi = y$, and as before $\sin\phi = x$, we have

$$x = \frac{iy}{\sqrt{1-y^2}}, = \frac{1}{\sqrt{1-k'^2z^2}},$$

that is $\frac{1}{y} = k'z$, which exhibits the relation between this and the transformation in the preceding article.

Functions of $u + (0, 1, 2, 3)\,K + (0, 1, 2, 3)\,iK'$.
Art. Nos. 97 to 99.

97. It is easy from the foregoing values of the sn, cn, dn of K and $K + iK'$ to obtain the values given in the following table: for instance we have

$$\begin{aligned} \text{sn}(u+K) &= \text{sn}\,K\,\text{cn}\,u\,\text{dn}\,u \div 1 - k^2\,\text{sn}^2K\,\text{sn}^2u, \\ &= \text{cn}\,u\,\text{dn}\,u \div \text{dn}^2u, \\ &= \text{cn}\,u \div \text{dn}\,u; \end{aligned}$$

$$\text{sn}(u-K) = -\text{cn}\,u \div \text{dn}\,u, \text{ \&c.}$$

Similarly finding $\text{sn}(u + K + iK')$, and in the resulting formulæ substituting $u - K$ for u and reducing, we have $\text{sn}(u + iK')$: and so in the other cases.

Functions of $u+(0, 1, 2, 3)K+(0, 1, 2, 3)iK'$.

	$+0K$	$+K$	$+2K$	$+3K$
$+0K'$	$\text{sn}\,u$ $\text{cn}\,u$ $\text{dn}\,u$	$\text{cn}\,u\ (+)$ $-k'\,\text{sn}\,u\ (+)$ $k'\ (+)$ denom. $=\text{dn}\,u$	$-\text{sn}\,u$ $-\text{cn}\,u$ $\text{dn}\,u$	$-\ \text{cn}\,u\ (+)$ $k'\,\text{sn}\,u\ (+)$ $k'\ (+)$ denom. $=\text{dn}\,u$
$+iK'$	$1\ (+)$ $-i\ \text{dn}\,u\ (+)$ $-ik\,\text{cn}\,u\ (+)$ denom. $=k\,\text{sn}\,u$	$\text{dn}\,u\ (+)$ $-ik'\ (+)$ $ikk'\,\text{sn}\,u\ (+)$ denom. $=k\,\text{cn}\,u$	$-1\ (+)$ $i\ \text{dn}\,u\ (+)$ $-ik\,\text{cn}\,u\ (+)$ denom. $=k\,\text{sn}\,u$	$-\ \text{dn}\,u\ (+)$ $ik'\ (+)$ $ikk'\,\text{sn}\,u\ (+)$ denom. $=k\,\text{cn}\,u$
$+2iK'$	$\text{sn}\,u$ $-\text{cn}\,u$ $-\text{dn}\,u$	$\text{cn}\,u\ (+)$ $k'\,\text{sn}\,u\ (+)$ $-k'\ (+)$ denom. $=\text{dn}\,u$	$-\text{sn}\,u$ $\text{cn}\,u$ $-\text{dn}\,u$	$-\ \text{cn}\,u\ (+)$ $-k'\,\text{sn}\,u\ (+)$ $-k'\ (+)$ denom. $=\text{dn}\,u$
$+3iK'$	$1\ (+)$ $i\ \text{dn}\,u\ (+)$ $ik\,\text{cn}\,u\ (+)$ denom. $=k\,\text{sn}\,u$	$\text{dn}\,u\ (+)$ $ik'\ (+)$ $-ikk'\,\text{sn}\,u\ (+)$ denom. $=k\,\text{cn}\,u$	$-1\ (+)$ $-i\ \text{dn}\,u\ (+)$ $ik\,\text{cn}\,u\ (+)$ denom. $=k\,\text{sn}\,u$	$-\ \text{dn}\,u\ (+)$ $-ik'\ (+)$ $-ikk'\,\text{sn}\,u\ (+)$ denom. $=k\,\text{cn}\,u$

where the arrangement hardly requires explanation: the table shows for instance that

$$\text{sn}(u+iK') = 1, \qquad (+)$$

$$\text{cn}(u+iK') = -i\ \text{dn}\,u, \qquad (+)$$

$$\text{dn}(u+iK') = -ik\ \text{cn}\,u, \qquad (+)$$

where $\qquad$ denom. $= k\,\text{sn}\,u$;

it sometimes, as here for $\text{dn}(u+iK')$, happens that there is in the numerator and denominator a common factor k, this is of course to be omitted.

98. The table, writing therein $u=0$, gives the values of the functions of $mK+m'iK'$. In particular, where there is a denominator $k\,\text{sn}\,u$, the functions become infinite: it is necessary to attend to the ratios of these infinite values, and the convenient course is to write $\dfrac{1}{k\,\text{sn}\,u}=I$, where I is regarded as a definite infinite value. The table thus gives

$$
\begin{array}{llll}
\operatorname{sn}\ iK' = & I, & \operatorname{sn}(2K+\ iK') = - & I,\\
\operatorname{cn}\ iK' = - & iI, & \operatorname{cn}(2K+\ iK') = & iI,\\
\operatorname{dn}\ iK' = - & ikI, & \operatorname{dn}(2K+\ iK') = - & ikI,\\
\operatorname{sn} 3iK' = & I, & \operatorname{sn}(2K+3iK') = - & I,\\
\operatorname{cn} 3iK' = & iI, & \operatorname{cn}(2K+3iK') = - & iI,\\
\operatorname{dn} 3iK' = & ikI, & \operatorname{dn}(2K+3iK') = & ikI.
\end{array}
$$

We may from these reproduce the original formulæ which involve u; thus

$$\operatorname{sn}(u+iK') = \frac{\operatorname{sn} u\,(-kI^2) + I \operatorname{cn} u \operatorname{dn} u}{1 - k^2 I^2 \operatorname{sn}^2 u},$$

$$= \frac{-kI^2 \operatorname{sn} u}{-k^2 I^2 \operatorname{sn}^2 u}, \quad = \frac{1}{k \operatorname{sn} u},$$

and so in the other cases.

99. The table shows that the functions have $2K$, $2K'$ as half-periods: we in fact deduce

$$
\begin{array}{ll}
\operatorname{sn}(u + 2mK + 2m'iK') = (-)^{m} & \operatorname{sn} u,\\
\operatorname{cn}(\quad ,, \quad ,, \quad) = (-)^{m+m'} & \operatorname{cn} u,\\
\operatorname{dn}(\quad ,, \quad ,, \quad) = (-)^{m'} & \operatorname{dn} u;
\end{array}
$$

whence taking m, m' each even it appears that $4K$, $4K'$ are whole periods; viz. that increasing the argument by

$$4mK + 4m'iK',$$

the functions are severally unaltered.

Duplication. Art. No. 100.

100. Writing $v=u$, we deduce the functions of $2u$, or say the duplication-formulæ. We have

$$
\begin{array}{lll}
\operatorname{sn} 2u = 2 \operatorname{sn} u \operatorname{cn} u \operatorname{dn} u, & & (\div)\\
\operatorname{cn} 2u = \operatorname{cn}^2 u - \operatorname{sn}^2 u \operatorname{dn}^2 u, & = 1 - 2\operatorname{sn}^2 u + k^2 \operatorname{sn}^4 u, & (\div)\\
\operatorname{dn} 2u = \operatorname{dn}^2 u - k^2 \operatorname{sn}^2 u \operatorname{cn}^2 u, & = 1 - 2k^2 \operatorname{sn}^2 u + k^2 \operatorname{sn}^4 u, & (\div)
\end{array}
$$

where

denom. $= 1 - k^2 \operatorname{sn}^4 u$;

or if for convenience we write

$$\operatorname{sn} u = x,$$

$$\operatorname{cn} u = \sqrt{1 - x^2},$$

$$\operatorname{dn} u = \sqrt{1 - k^2x^2};$$

then the formulæ are

$$\operatorname{sn} 2u = 2x\sqrt{1 - x^2}\sqrt{1 - k^2x^2}, \qquad (\div)$$

$$\operatorname{cn} 2u = 1 - 2x^2 + k^2x^4, \qquad (\div)$$

$$\operatorname{dn} 2u = 1 - 2k^2x^2 + k^2x^4, \qquad (\div)$$

where $\quad$ denom. $= 1 - k^2x^4$.

It may be added that

$$1 - \operatorname{cn} 2u = 2x^2(1 - k^2x^2), = 2\operatorname{sn}^2 u \operatorname{dn}^2 u, \qquad (\div)$$

$$1 + \operatorname{cn} 2u = 2(1 - x^2), = 2\operatorname{cn}^2 u, \qquad (\div)$$

$$1 - \operatorname{dn} 2u = 2k^2x^2(1 - x^2), = 2k^2 \operatorname{sn}^2 u \operatorname{cn}^2 u, \qquad (\div)$$

$$1 + \operatorname{dn} 2u = 2(1 - k^2x^2), = 2\operatorname{dn}^2 u, \qquad (\div)$$

the denominator being as above $1 - k^2x^4$ or $1 - k^2\operatorname{sn}^4 u$. And we thence deduce

$$\operatorname{sn}^2 u = 1 - \operatorname{cn} 2u, \qquad (\div)$$

$$\operatorname{cn}^2 u = \operatorname{dn} 2u + \operatorname{cn} 2u, \qquad (\div)$$

$$\operatorname{dn}^2 u = k'^2 + \operatorname{dn} 2u + k^2 \operatorname{cn} 2u, \qquad (\div)$$

where $\quad$ denom. $= 1 + \operatorname{dn} 2u$.

Dimidiation. Art. Nos. 101 to 106.

101. In the expressions for the functions of $2u$, writing $\frac{1}{2}u$ instead of u, we have the functions of u expressed in terms of those of $\frac{1}{2}u$, and from these equations can obtain the ex-

pressions of the functions of $\tfrac{1}{2}u$ in terms of those of u. Thus, writing for a moment $x = \operatorname{sn} \tfrac{1}{2}u$, we have

$$\operatorname{sn} u = 2x\sqrt{1-x^2}\sqrt{1-k^2x^2}, \qquad (+)$$

$$\operatorname{cn} u = 1 - 2x^2 + k^2x^4, \qquad (+)$$

$$\operatorname{dn} u = 1 - 2k^2x^2 + k^2x^4, \qquad (+)$$

$$\text{denom.} = 1 - k^2x^4.$$

The last two equations may be written

$$(1 - \operatorname{cn} u) - 2x^2 + k^2(1 + \operatorname{cn} u)\,x^4 = 0,$$

$$(1 - \operatorname{dn} u) - 2k^2x^2 + k^2(1 + \operatorname{dn} u)\,x^4 = 0,$$

and from these eliminating x^4 we have x^2, that is $\operatorname{sn}^2 \tfrac{1}{2}u$, expressed rationally. Obtaining from it the expressions of $\operatorname{cn}^2 \tfrac{1}{2}u$ and $\operatorname{dn}^2 \tfrac{1}{2}u$, we have

$$\operatorname{sn}^2 \tfrac{1}{2}u = \operatorname{dn} u - \operatorname{cn} u, \qquad (\div)$$

$$\operatorname{cn}^2 \tfrac{1}{2}u = k'^2(1 + \operatorname{cn} u), \qquad (+)$$

$$\operatorname{dn}^2 \tfrac{1}{2}u = k'^2(1 + \operatorname{dn} u), \qquad (+)$$

where $$\text{denom.} = k'^2 + \operatorname{dn} u - k^2 \operatorname{cn} u.$$

102. But, *ante* No. 100, it appears that we have also the expressions

$$\operatorname{sn}^2 \tfrac{1}{2}u = 1 - \operatorname{cn} u,$$

$$\operatorname{cn}^2 \tfrac{1}{2}u = \operatorname{dn} u + \operatorname{cn} u,$$

$$\operatorname{dn}^2 \tfrac{1}{2}u = k'^2 + \operatorname{dn} u + k^2 \operatorname{cn} u,$$

where $$\text{denom.} = 1 + \operatorname{dn} u.$$

In passing to the expressions of $\operatorname{sn} \tfrac{1}{2}u$, $\operatorname{cn} \tfrac{1}{2}u$, $\operatorname{dn} \tfrac{1}{2}u$, the radicals must of course be taken with the proper sign.

We deduce the following special formulæ:

	sn =	cn =	dn =
$\tfrac{1}{2}K$	$\frac{1}{\sqrt{1+k'}}$	$\frac{\sqrt{k'}}{\sqrt{1+k'}}$	$\sqrt{k'}$
$\tfrac{3}{2}K$	$\frac{1}{\sqrt{1+k'}}$	$-\frac{\sqrt{k'}}{\sqrt{1+k'}}$	$\sqrt{k'}$
$\tfrac{1}{2}K+iK'$	$\frac{1}{\sqrt{1-k'}}$	$-\frac{i\sqrt{k'}}{\sqrt{1-k'}}$	$-i\sqrt{k'}$
$\tfrac{3}{2}K+iK'$	$\frac{1}{\sqrt{1-k'}}$	$-\frac{i\sqrt{k'}}{\sqrt{1-k'}}$	$i\sqrt{k'}$
$\tfrac{1}{2}iK'$	$\frac{i}{\sqrt{k}}$	$\frac{\sqrt{1+k}}{\sqrt{k}}$	$\sqrt{1+k}$
$K+\tfrac{1}{2}iK'$	$\frac{1}{\sqrt{k}}$	$-\frac{i\sqrt{1-k}}{\sqrt{k}}$	$\sqrt{1-k}$
$\tfrac{3}{2}iK'$	$-\frac{i}{\sqrt{k}}$	$-\frac{\sqrt{1+k}}{\sqrt{k}}$	$-\sqrt{1+k}$
$K+\tfrac{3}{2}iK'$	$\frac{1}{\sqrt{k}}$	$-\frac{i\sqrt{1-k}}{\sqrt{k}}$	$-\sqrt{1-k}$
$\tfrac{1}{2}K+\tfrac{1}{2}iK'$	$\frac{1}{\sqrt{2}\sqrt{k}}\{\sqrt{1+k}+i\sqrt{1-k}\}$	$-\frac{1-i}{\sqrt{2}}\frac{\sqrt{k'}}{\sqrt{k}}$	$\frac{\sqrt{k'}}{\sqrt{2}}\{\sqrt{1+k'}-i\sqrt{1-k'}\}$
$\tfrac{3}{2}K+\tfrac{1}{2}iK'$	$\frac{1}{\sqrt{2}\sqrt{k}}\{\sqrt{1+k}-i\sqrt{1-k}\}$	$-\frac{1+i}{\sqrt{2}}\frac{\sqrt{k'}}{\sqrt{k}}$	$\frac{\sqrt{k'}}{\sqrt{2}}\{\sqrt{1+k'}+i\sqrt{1-k'}\}$
$\tfrac{1}{2}K+\tfrac{3}{2}iK'$	$\frac{1}{\sqrt{2}\sqrt{k}}\{\sqrt{1+k}-i\sqrt{1-k}\}$	$-\frac{1+i}{\sqrt{2}}\frac{\sqrt{k'}}{\sqrt{k}}$	$-\frac{\sqrt{k'}}{\sqrt{2}}\{\sqrt{1+k'}+i\sqrt{1-k'}\}$
$\tfrac{3}{2}K+\tfrac{3}{2}iK'$	$\frac{1}{\sqrt{2}\sqrt{k}}\{\sqrt{1+k}+i\sqrt{1-k}\}$	$\frac{1-i}{\sqrt{2}}\frac{\sqrt{k'}}{\sqrt{k}}$	$-\frac{\sqrt{k'}}{\sqrt{k}}\{\sqrt{1+k'}-i\sqrt{1-k}\}$

where for the last set of formulæ we may substitute:

	$\mathrm{sn}^2 =$	$\mathrm{cn}^2 =$	$\mathrm{dn}^2 =$
$\tfrac{1}{2}K+\tfrac{1}{2}iK'$	$\frac{1}{k}(k+ik')$	$-\frac{ik'}{k}$	$k'(k'-ik)$
$\tfrac{3}{2}K+\tfrac{1}{2}iK'$	$\frac{1}{k}(k-ik')$	$\frac{ik'}{k}$	$k'(k'+ik)$
$\tfrac{1}{2}K+\tfrac{3}{2}iK'$	$\frac{1}{k}(k-ik')$	$\frac{ik'}{k}$	$k'(k'+ik)$
$\tfrac{3}{2}K+\tfrac{3}{2}iK'$	$\frac{1}{k}(k+ik')$	$-\frac{ik'}{k}$	$k'(k'-ik)$

103. We find

$$\operatorname{sn}(u+\tfrac{1}{2}K) = \frac{1}{\sqrt{1+k'}}\,\frac{k'\operatorname{sn}u+\operatorname{cn}u\operatorname{dn}u}{1-(1-k')\operatorname{sn}^2u},$$

$$= \frac{1}{\sqrt{1+k'}}\,\frac{\operatorname{dn}u+(1+k')\operatorname{sn}u\operatorname{cn}u}{\operatorname{cn}u+\operatorname{sn}u\operatorname{dn}u};$$

$$\operatorname{sn}(u+\tfrac{1}{2}iK') = \frac{1}{\sqrt{k}}\,\frac{(1+k)\operatorname{sn}u+i\operatorname{cn}u\operatorname{dn}u}{1+k\operatorname{sn}^2u},$$

$$= \frac{1}{\sqrt{k}}\sqrt{\frac{(1+k)\operatorname{sn}u+i\operatorname{cn}u\operatorname{dn}u}{(1+k)\operatorname{sn}u-i\operatorname{cn}u\operatorname{dn}u}};$$

$$\operatorname{sn}(u+\tfrac{1}{2}K+\tfrac{1}{2}iK') = \sqrt{\frac{k+ik'}{k}}\,\frac{-ik'\operatorname{sn}u+\operatorname{cn}u\operatorname{dn}u}{1-k(k+ik')\operatorname{sn}^2u},$$

$$= \sqrt{\frac{k+ik'}{k}}\,\frac{\operatorname{cn}u+(k-ik')\operatorname{sn}u\operatorname{dn}u}{\operatorname{dn}+k\operatorname{sn}u\operatorname{cn}u};$$

where the first expressions are those given at once by substitution in the general formula for $\operatorname{sn}(u+v)$.

104. To identify the two expressions of $\operatorname{sn}(u+\tfrac{1}{2}K)$, writing for convenience $\operatorname{sn}u=x$, observe that in the first expression the denominator is $1-(1-k')x^2$, and multiplying this by $1+(1-k')x^2$, the product is $1-2x^2+k^2x^4$. And in the second expression the denominator is $\sqrt{1-x^2}+x\sqrt{1-k^2x^2}$, which multiplied by $\sqrt{1-x^2}-x\sqrt{1-k^2x^2}$ gives $1-x^2-x^2(1-k^2x^2)$, $=$ same value, $1-2x^2+k^2x^4$: reducing in this manner the two expressions to a common denominator, the numerators would be found to be equal. Similarly as regards the two expressions of $\operatorname{sn}(u+\tfrac{1}{2}K+\tfrac{1}{2}iK')$, we have

$$\{1-k(k+ik')x^2\}\{1-k(k-ik')x^2\} = 1-2k^2x^2+k^2x^4,$$

and

$$\{\sqrt{1-k^2x^2}+kx\sqrt{1-x^2}\}\{\sqrt{1-k^2x^2}-kx\sqrt{1-x^2}\}$$

$$= 1-k^2x^2-k^2x^2(1-x^2), = \text{same value}.$$

As regards the two values of sn $(u + \tfrac{1}{2}iK')$, we have

$$[(1+k)x + i\sqrt{1-x^2}\sqrt{1-k^2x^2}][(1+k)x - i\sqrt{1-x^2}\sqrt{1-k^2x^2}]$$
$$= (1+k)^2x^2 + (1-x^2)(1-k^2x^2), \ = (1+kx^2)^2,$$

and the identity is at once established.

105. We deduce without difficulty from the second formulæ:

$$\text{sn}^2(u+\tfrac{1}{2}K) \quad = \frac{1}{1+k'}\cdot\frac{\text{dn}\, u + (1+k')\,\text{sn}\, u\,\text{cn}\, u}{\text{dn}\, u + (1-k')\,\text{sn}\, u\,\text{cn}\, u},$$

$$\text{sn}^2(u+\tfrac{1}{2}iK') \quad = \frac{1}{k}\quad\frac{(1+k)\,\text{sn}\, u + i\,\text{cn}\, u\,\text{dn}\, u}{(1+k)\,\text{sn}\, u - i\,\text{cn}\, u\,\text{dn}\, u},$$

$$\text{sn}^2(u+\tfrac{1}{2}K+\tfrac{1}{2}iK') = \frac{k+ik'}{k}\quad\frac{\text{cn}\, u + (k-ik')\,\text{sn}\, u\,\text{dn}\, u}{\text{cn}\, u + (k+ik')\,\text{sn}\, u\,\text{dn}\, u};$$

to these may be joined the formulæ obtained by considering $u + \tfrac{1}{2}K$, &c. as the halves of $2u + K$, &c., see No. 102, viz. we thus have

$$\text{sn}^2(u+\tfrac{1}{2}K) \quad = \quad \frac{\text{dn}\, 2u + k'\,\text{sn}\, 2u}{k' + \text{dn}\, 2u},$$

$$\text{sn}^2(u+\tfrac{1}{2}iK') \quad = \frac{1}{k}\ \frac{k\,\text{sn}\, 2u + i\,\text{dn}\, 2u}{\text{sn}\, 2u - i\,\text{cn}\, 2u},$$

$$\text{sn}^2(u+\tfrac{1}{2}K+\tfrac{1}{2}iK') = \frac{1}{k}\ \frac{k\,\text{cn}\, 2u + ik'}{\text{cn}\, 2u + ik'\,\text{sn}\, 2u}.$$

106. Observe that in the first expression the denominator multiplied by $1 - k^2x^4$ is

$$k'(1-k^2x^4) + 1 - 2k^2x^2 + k^2x^4,$$
$$= 1 + k' - 2k^2x^2 + (1-k')k^2x^4,$$
$$= (1+k')[1-(1-k')x^2]^2.$$

In the second expression, multiplying the numerator and denominator by $\text{sn}\, 2u + i\,\text{cn}\, 2u$, the expression becomes an integral function $(\text{sn}\, 2u, \text{cn}\, 2u, \text{dn}\, 2u)^2$; having therefore a denominator $(1-k^2x^4)^2, = (1+kx^2)^2(1-kx^2)^2$.

In the third expression the denominator multiplied by $1 - k^2x^4$ is

$$1 - 2x^2 + k^2x^4 + 2ik'x\sqrt{1-x^2}\sqrt{1-k^2x^2},$$
$$= [ik'x + \sqrt{1-x^2}\sqrt{1-k^2x^2}]^2,$$
$$= (ik' \operatorname{sn} u + \operatorname{cn} u \operatorname{dn} u)^2;$$

by aid of these remarks the identifications can be easily effected.

Triplication. Art. No. 107.

107. Writing $v = 2u$, and using the duplication-formulæ, we obtain the functions of $3u$. These are easily found to be

$$\operatorname{sn} 3u = 3x - (4 + 4k^2)x^3 + 6k^2x^5 - k^4x^9, \qquad (\div)$$
$$\operatorname{cn} 3u = (1 - 4x^2 + 6k^2x^4 - 4k^4x^6 + k^4x^8)\sqrt{1-x^2}, \qquad (\div)$$
$$\operatorname{dn} 3u = (1 - 4k^2x^2 + 6k^2x^4 - 4k^2x^6 + k^4x^8)\sqrt{1-k^2x^2}, \qquad (\div)$$

where

$$\text{denom.} = 1 - 6k^2x^4 + (4k^2 + 4k^4)x^6 - 3k^4x^8.$$

And we may add

$$1 - \operatorname{sn} 3u = (1 + x)\ [1 - 2x + 2k^2x^3 - k^2x^4]^2, \qquad (\div)$$
$$1 + \operatorname{sn} 3u = (1 - x)\ [1 + 2x - 2k^2x^3 - k^2x^4]^2, \qquad (\div)$$
$$1 - k \operatorname{sn} 3u = (1 + kx)\,[1 - 2kx + 2kx^3 - k^2x^4]^2, \qquad (\div)$$
$$1 + k \operatorname{sn} 3u = (1 - kx)\,[1 + 2kx - 2kx^3 - k^2x^4]^2, \qquad (\div)$$

the denominator as above.

The duplication and triplication formulæ possess various properties which are in fact particular cases of those for the multiplication by any even or odd integer n: and it will be convenient to defer the consideration of them until other instances of the formulæ are obtained.

Multiplication. Art. Nos. 108 to 116.

108. It has been seen how the functions of $2u$ and $3u$ are obtained: to consider the general question of determining the functions of nu, suppose $n=p+q$, and imagine that the functions of pu, qu are known. We may write

$$\begin{aligned} \operatorname{sn} pu &= A_p \quad (\div), & \operatorname{sn} qu &= A_q \quad (\div), \\ \operatorname{cn} pu &= B_p \quad (\div), & \operatorname{cn} qu &= B_q \quad (\div), \\ \operatorname{dn} pu &= C_p \quad (\div), & \operatorname{dn} qu &= C_q \quad (\div), \\ \text{denom.} &= D_p, & \text{denom.} &= D_q. \end{aligned}$$

The addition-formulæ give

$$\begin{aligned} \operatorname{sn}(p+q)u &= A_pD_pB_qC_q + \quad B_pC_pA_qD_q \quad (\div), \\ \operatorname{cn}(p+q)u &= B_pD_pB_qD_q - \quad A_pC_pA_qC_q \quad (\div), \\ \operatorname{dn}(p+q)u &= C_pD_pC_qD_q - k^2A_pB_pA_qB_q \quad (\div), \end{aligned}$$

where $\qquad \text{denom.} = D_p^2D_q^2 - k^2A_p^2A_q^2;$

and the functions on the right-hand side are consequently proportional to A_{p+q}, B_{p+q}, C_{p+q}, D_{p+q} respectively. We have $A_1=x$, $B_1=\sqrt{1-x^2}$, $C_1=\sqrt{1-k^2x^2}$, $D_1=1$; and hence writing $p=q=1$, we find four values which have no common divisor, and which may therefore be taken for the values of A_2, B_2, C_2, D_2 respectively: viz. we thus obtain

$$\begin{aligned} A_2 &= 2x\sqrt{1-x^2}\sqrt{1-k^2x^2}, \\ B_2 &= 1-2x^2 \quad +k^2x^4, \\ C_2 &= 1-2k^2x^2+k^2x^4, \\ D_2 &= 1-k^2x^4, \end{aligned}$$

the foregoing duplication-formulæ. And similarly, writing $p=2$, $q=1$, we obtain the triplication-formulæ. But at the next step, if we write $p=q=2$ we obtain four values, and if we write $p=3$, $q=1$ we obtain four other values of higher degrees; these are of course proportional to the former ones, and they contain a common factor, throwing which out they would coincide with them. And so in general, for a given value of $\overline{p+q}$ the degrees are lowest when p, q are as nearly as possible equal: that is $p+q$ even, when $p=q$, and $p+q$ odd, when $p-q=1$: or what is the same thing, the proper partitionments are $4=2+2$, $5=2+3$, $6=3+3$, &c. Taking the functions thus obtained for the values of A_{p+q}, B_{p+q}, C_{p+q}, D_{p+q}, we may write

$$p+q \text{ odd};\ p-q=1.$$

$$\begin{aligned}
A_{p+q} &= A_pD_pB_qC_q + B_pC_pA_qD_q,\\
B_{p+q} &= B_pD_pB_qD_q - A_pC_pA_qC_q,\\
C_{p+q} &= C_pD_pC_qD_q - k^2A_pB_pA_qB_q,\\
D_{p+q} &= D_p^2D_q^2 - k^2A_p^2A_q^2,
\end{aligned}$$

$$p+q \text{ even};\ p=q.$$

$$\begin{aligned}
A_{2p} &= 2A_pB_pC_pD_p,\\
B_{2p} &= B_p^2D_p^2 - A_p^2C_p^2,\\
C_{2p} &= C_p^2D_p^2 - k^2A_p^2B_p^2,\\
D_{2p} &= D_p^4 - k^2A_p^4.
\end{aligned}$$

109. The calculations for the cases 4 and 5 may be performed without difficulty: but for 6 and 7 they become very laborious: the results have however been calculated by Baehr, *Grunert's Archiv* XXXVI. (1861), pp. 125—176, and for convenience of reference I reproduce them here, partially verifying them as afterwards mentioned. The whole series of formulæ for the cases $n=2, 3, 4, 5, 6, 7$ are as follows:

sn 2u = $x\sqrt{1-x^2}\sqrt{1-k^2x^2}$ into	cn 2u =	dn 2u =	denom. =	1 − cn 2u = $1-k^2x^2$ into	1 + cn 2u = $1-x^2$ into	1 − dn 2u = $1-x^2$ into	1 + dn 2u = $1-k^2x^2$ into
2	1	1	1		2		2
	$-2x^2$	$-2k^2x^2$		$2x^2$		$2k^2x^2$	
	$+k^2x^4$	$+k^2x^4$	$-k^2x^4$				
(+)	(+)	(+)		(+)	(+)	(+)	(+)

sn 3u = x into	cn 3u = $\sqrt{1-x^2}$ into	dn 3u = $\sqrt{1-k^2x^2}$ into	denom. =	1 − sn 3u = $(1+x)$ into sq. of	1 − k sn 3u = $(1+kx)$ into sq. of
3	1	1	1	1	1
$-(4+4k^2)x^2$	$-4x^2$	$-4k^2x^2$	0	$-2x$	$-2kx$
$+6k^2x^4$	$+6k^2x^4$	$+6k^2x^4$	$-6k^2x^4$	0	0
0	$-4k^4x^6$	$-4k^2x^6$	$+(4k^2+4k^4)x^6$	$+2k^2x^3$	$+2kx^3$
$-k^4x^8$	$+k^4x^8$	$+k^4x^8$	$-3k^4x^8$	$-k^2x^4$	$-k^2x^4$
(+)	(+)	(+)		(+)	(+)

sn 4u = $x\sqrt{1-x^2}\sqrt{1-k^2x^2}$ into	cn 4u =	dn 4u =	denom. =
4	1	1	1
$-(8+8k^2)\ x^2$	$-8\ x^2$	$-8k^2\ x^2$	0
$+20k^2\ x^4$	$+(8+20k^2)\ x^4$	$+(20k^2+8k^4)\ x^4$	$-20k^2\ x^4$
0	$-(24k^2+32k^4)\ x^6$	$-(32k^2+24k^4)\ x^6$	$+(32k^2+32k^4)\ x^6$
$-20k^4\ x^8$	$+(56k^4+16k^6)\ x^8$	$+(16k^2+56k^4)\ x^8$	$-(16k^2+56k^4+16k^6)\ x^8$
$+(8k^4+8k^6)\ x^{10}$	$-(24k^4+32k^6)\ x^{10}$	$-(32k^4+24k^6)\ x^{10}$	$+(32k^4+32k^6)\ x^{10}$
$-4k^6\ x^{12}$	$+(8k^4+20k^6)\ x^{12}$	$+(20k^6+8k^8)\ x^{12}$	$-20k^6\ x^{12}$
	$-8k^6\ x^{14}$	$-8k^8\ x^{14}$	0
	$+k^8\ x^{16}$	$+k^8\ x^{16}$	$+k^8\ x^{16}$
(+)	(+)	(+)	

	sn 5u = x into	cn 5u = $\sqrt{1-x^2}$ into	dn 5u = $\sqrt{1-k^2x^2}$ into
x^0	5	1	1
x^2	$-(20+20k^2)$	-12	$-12k^2$
x^4	$+16+94k^2+16k^4$	$+16+50k^2$	$+50k^2+16k^4$
x^6	$-(80k^2+80k^4)$	$-80k^2-140k^4$	$-140k^2-80k^4$
x^8	$-105k^4$	$+335k^4+160k^6$	$+160k^2+335k^4$
x^{10}	$+360k^4+360k^6$	$-264k^4-464k^6-64k^8$	$-64k^2-464k^4-264k^6$
x^{12}	$-(240k^4+780k^6+240k^8)$	$+208k^4+508k^6+208k^8$	$+208k^4+508k^6+208k^8$
x^{14}	$+64k^4+560k^6+560k^8+64k^{10}$	$-64k^4-464k^6-264k^8$	$-264k^6-464k^8-64k^{10}$
x^{16}	$-(160k^6+445k^8+160k^{10})$	$+160k^6+335k^8$	$+335k^8+160k^{10}$
x^{18}	$+140k^8+140k^{10}$	$-140k^8-80k^{10}$	$-80k^8-140k^{10}$
x^{20}	$-50k^{10}$	$+50k^{10}+16k^{12}$	$+16k^8+50k^{10}$
x^{22}	0	$-12k^{12}$	$-12k^{10}$
x^{24}	$+k^{12}$	$+k^{12}$	$+k^{12}$
	(+)	(÷)	(+)

	1 − k sn 5u = (1 − x) into square of	1 − k sn 5u = (1 − kx) into square of	Denom. =
x^0	1	1	1
x^1	-2	$-2k$	0
x^2	-4	$-4k^2$	$-50k^2$
x^3	$+10k^2$	$+10k$	$+140k^2+140k^4$
x^4	$+5k^2$	$+5k^2$	$-(160k^2+445k^4+160k^6)$
x^5	$-12k^2-8k^4$	$-8k-12k^3$	$+64k^2+560k^4+560k^6+64k^8$
x^6	$+4k^2-4k^4$	$-4k^2+4k^4$	$-(240k^4+780k^6+240k^8)$
x^7	$+8k^2+12k^4$	$+12k^3+8k^5$	$+360k^6+360k^8$
x^8	$-5k^4$	$-5k^4$	$-105k^8$
x^9	$-10k^4$	$-10k^3$	$-(80k^8+80k^{10})$
x^{10}	$+4k^4$	$+4k^4$	$+16k^8+94k^{10}+16k^{12}$
x^{11}	$+2k^4$	$+2k^5$	$-(20k^{10}+20k^{12})$
x^{12}	$-k^4$	$-k^6$	$+5k^{12}$
	(+)	(+)	

In the Tables which follow, some obvious abbreviations are made use of. Thus we must read in the table for sn 6u

$$6+(-32-32k^2)\,x^2+(32+208k^2+32k^4)\,x^4-\&c.,$$

and in that for sn 7u,

$$7+(-56-56k^2)x^2+(112+532k^2+112k^4)\,x^4-\&c.,$$

the numerical coefficients in this last case being printed to the middle term only, −56: for (−56 −56), and +112 (+532) for (+112+532+112), the expressions being symmetrical as here shown. The numerical coefficients of denom. 7u are in a reverse order the same as for sn 7u, and those of dn 7u the same as for cn 7u, but in a reverse order, as is sufficiently indicated in the tables.

	$\text{sn}\,6u = x\sqrt{1-x^2}\sqrt{1-k^2x^2}$ into		Denom. =	
x^0	6	1	1	1
x^2	− 32 − 32	1, k^2	0	
x^4	+ 32 + 208 + 32	1, k^2, k^4	− 105	k^2
x^6	− 224 − 224	k^2, k^4	+ 448 + 448	k^2, k^4
x^8	− 728	k^4	− 864 − 2172 − 864	k^2, k^4, k^6
x^{10}	+ 2912 + 2912	k^4, k^6	+ 768 + 4608 + 4608 + 768	k^2, k^4, k^6, k^8
x^{12}	− 8360 − 9296 − 8360	k^4, k^6, k^8	− 256 − 4884 − 10740 − 4884 − 256	k^2, k^4, k^6, k^8, k^{10}
x^{14}	+ 2048 + 11680 + 11680 + 2048	k^4, k^6, k^8, k^{10}	+ 1536 + 10944 + 10944 + 1536	k^4, k^6, k^8, k^{10}
x^{16}	− 512 − 7680 − 16320 − 7680 − 512	k^4, k^6, k^8, k^{10}, k^{12}	− 4092 − 9964 − 4092	k^6, k^8, k^{10}
x^{18}	+ 2048 + 11680 + 11680 + 2048	k^6, k^8, k^{10}, k^{12}	0	
x^{20}	− 8360 − 9296 − 8360	k^8, k^{10}, k^{12}	+ 4092 + 9964 + 4092	k^8, k^{10}, k^{12}
x^{22}	+ 2912 + 2912	k^{10}, k^{12}	− 1536 − 10944 − 10944 − 1536	k^8, k^{10}, k^{12}, k^{14}
x^{24}	− 728	k^{12}	+ 256 + 4884 + 10740 + 4884 + 256	k^8, k^{10}, k^{12}, k^{14}, k^{16}
x^{26}	− 224 − 224	k^{12}, k^{14}	− 768 − 4608 − 4608 − 768	k^{10}, k^{12}, k^{14}, k^{16}
x^{28}	+ 32 + 208 + 32	k^{12}, k^{14}, k^{16}	+ 864 + 2172 + 864	k^{12}, k^{14}, k^{16}
x^{30}	− 32 − 32	k^{14}, k^{16}	− 448 − 448	k^{14}, k^{16}
x^{32}	+ 6	k^{16}	+ 105	k^{16}
x^{34}			0	
x^{36}			− 1	k^{18}

(+)

	cn 6u =		dn 6u =	
x^0	1	1	1	1
x^2	− 18	1	− 18	k^2
x^4	+ 48 + 105	1, k^2	+ 105 + 48	k^2, k^4
x^6	− 32 − 336 − 448	1, k^2, k^4	− 448 − 336 − 32	k^2, k^4, k^6
x^8	+ 240 + 1956 + 864	k^2, k^4, k^6	+ 864 + 1956 + 240	k^2, k^4, k^6
x^{10}	− 3384 − 4416 − 768	k^4, k^6, k^8	− 768 − 4416 − 3384	k^2, k^4, k^6
x^{12}	+ 4368 + 9020 + 4320 + 256	k^4, k^6, k^8, k^{10}	+ 256 + 4320 + 9690 + 4868	k^2, k^4, k^6, k^8
x^{14}	− 4320 − 15408 − 10560 − 1536	k^4, k^6, k^8, k^{10}	− 1536 − 10560 − 15408 − 4320	k^4, k^6, k^8, k^{10}
x^{16}	+ 2304 + 16848 + 20574 + 4032	k^4, k^6, k^8, k^{10}	+ 4032 + 20574 + 16848 + 2304	k^6, k^8, k^{10}, k^{12}
x^{18}	− 512 − 9728 − 26220 − 19160	k^4, k^6, k^8, k^{10}	− 19160 − 26320 − 9728 − 512	k^8, k^{10}, k^{12}, k^{14}
x^{20}	+ 2304 + 16848 + 20574 + 4032	k^6, k^8, k^{10}, k^{12}	+ 4032 + 20574 + 16848 + 2304	k^8, k^{10}, k^{12}, k^{14}
x^{22}	− 4320 − 15408 − 10560 − 1536	k^8, k^{10}, k^{12}, k^{14}	− 1536 − 10560 − 15408 − 4320	k^8, k^{10}, k^{12}, k^{14}
x^{24}	+ 4368 + 9630 + 4320 + 256	k^{10}, k^{12}, k^{14}, k^{16}	+ 256 + 4320 + 9620 + 4868	k^8, k^{10}, k^{12}, k^{14}
x^{26}	− 3384 − 4416 − 768	k^{12}, k^{14}, k^{16}	− 768 − 4416 − 3384	k^{10}, k^{12}, k^{14}
x^{28}	+ 240 + 1956 + 864	k^{12}, k^{14}, k^{16}	+ 864 + 1936 + 240	k^{12}, k^{14}, k^{16}
x^{30}	− 32 − 336 − 448	k^{12}, k^{14}, k^{16}	− 448 − 336 − 32	k^{14}, k^{16}, k^{18}
x^{32}	+ 48 + 105	k^{14}, k^{16}	+ 105 + 48	k^{16}, k^{18}
x^{34}	− 18	k^{16}	− 18	k^{18}
x^{36}	+ 1	k^{18}	+ 1	k^{18}

(+) denom. ut suprà. (+) denom. ut suprà.

	sn 7u = x into		Denom. =	
x^0	7	1	1	1
x^2	− 56 :	1, k^2	0	
x^4	+ 112 (+ 532)	1, k^2, k^4	− 196	k^2
x^6	− 64 − 1136 :	1, k^2, k^4, k^6	+ 1176 :	k^2, k^4
x^8	+ 672 (− 1610)	k^2, k^4, k^6	...	k^2, k^4, k^6
x^{10}	+ 19656 :	k^4, k^6	...	k^2, k^4, k^6, k^8
x^{12}	− 47040 (− 118272)	k^4, k^6, k^8	...	k^2, k^4, k^6, k^8, k^{10}
x^{14}	+ 60928 + 271040 :	k^4, k^6, k^8, k^{10}	...	k^2, k^4, k^6, k^8, k^{10}, k^{12}
x^{16}	− 48384 − 572736 (− 688985)	k^4, k^6, k^8, k^{10}, k^{12}	...	k^4, k^6, k^8, k^{10}, k^{12}
x^{18}	+ 21504 + 311808 + 1005200 :	k^4, k^6, k^8, k^{10}, k^{12}, k^{14}	...	k^6, k^8, k^{10}, k^{12}
x^{20}	− 4096 − 144896 − 890720 (− 1582104)	k^4, k^6, k^8, k^{10}, k^{12}, k^{14}, k^{16}	...	k^8, k^{10}, k^{12}
x^{22}	+ 28672 + 434560 + 1490272 :	k^6, k^8, k^{10}, k^{12}, k^{14}, k^{16}	...	k^8, k^{10}, k^{12}, k^{14}
x^{24}	− 89600 − 765632 (− 1405960)	k^8, k^{10}, k^{12}, k^{14}, k^{16}	...	k^8, k^{10}, k^{12}, k^{14}, k^{16}
x^{26}	+ 164864 + 789536 :	k^{10}, k^{12}, k^{14}, k^{16}	...	k^8, k^{10}, k^{12}, k^{14}, k^{16}, k^{18}
x^{28}	− 111552 (− 265272)	k^{12}, k^{14}, k^{16}	...	k^8, k^{10}, k^{12}, k^{14}, k^{16}, k^{18}, k^{20}
x^{30}	− 35840 − 108864 :	k^{12}, k^{14}, k^{16}, k^{18}	...	k^{10}, k^{12}, k^{14}, k^{16}, k^{18}, k^{20}
x^{32}	+ 6960 + 116480 (+ 230718)	k^{12}, k^{14}, k^{16}, k^{18}, k^{20}	...	k^{12}, k^{14}, k^{16}, k^{18}, k^{20}
x^{34}	− 1024 − 80208 − 145240 :	k^{12}, k^{14}, k^{16}, k^{18}, k^{20}, k^{22}	...	k^{14}, k^{16}, k^{18}, k^{20}
x^{36}	+ 8584 + 89536 (+ 86600)	k^{14}, k^{16}, k^{18}, k^{20}, k^{22}	...	k^{16}, k^{18}, k^{20}
x^{38}	− 4928 − 25200 :	k^{16}, k^{18}, k^{20}, k^{22}	...	k^{18}, k^{20}
x^{40}	+ 2360 (+ 8092)	k^{18}, k^{20}, k^{22}	...	k^{18}, k^{20}, k^{22}
x^{42}	− 1176 :	k^{20}, k^{22}	...	k^{18}, k^{20}, k^{22}, k^{24}
x^{44}	+ 196	k^{22}	− 112 (− 532)	k^{20}, k^{22}, k^{24}
x^{46}	0		+ 56 :	k^{22}, k^{24}
x^{48}	− 1	k^{24}	− 7	k^{24}

(+)

	cn $7u = \sqrt{1-x^2}$ into		dn $7u = \sqrt{1-k^2x^2}$ into	
x^0	1	1	1	1
x^2	$-\ 24$	1	$-\ 24$	k^2
x^4	$+\ 60 + 196$	$1, k^2$	$+\ 196 + 60$	k^2, k^4
x^6	$-\ 64 - 784 - 1176$	$1, k^2, k^4$	$-1176 - 784 - 64$	k^2, k^4, k^6
x^8	$+\ 672 + 6594 + 8560$	k^2, k^4, k^6	...	k^4, k^6, k^8
x^{10}	$-\ 15288 - 22288 - 4928$	k^4, k^6, k^8	...	k^4, k^6, k^8
x^{12}	$+\ 29120 + 65716 + 86176 + 8584$	k^4, k^6, k^8, k^{10}	...	k^4, k^6, k^8, k^{10}
x^{14}	$-\ 43520 - 153280 - 120120 - 28160 - 1024$	$k^4, k^6, k^8, k^{10}, k^{12}$	...	$k^4, k^6, k^8, k^{10}, k^{12}$
x^{16}	$+\ 89168 + 252416 + 834831 + 100608 + 8448$	$k^4, k^6, k^8, k^{10}, k^{12}$	...	$k^4, k^6, k^8, k^{10}, k^{12}$
x^{18}	$-\ 19456 - 245248 - 630448 - 390608 - 81744$	$k^4, k^6, k^8, k^{10}, k^{12}$	...	$k^6, k^8, k^{10}, k^{12}, k^{14}$
x^{20}	$+\ 4096 + 129536 + 674016 + 883336 + 270272$	$k^4, k^6, k^8, k^{10}, k^{12}$	...	$k^8, k^{10}, k^{12}, k^{14}, k^{16}$
x^{22}	$-\ 28672 - 882592 - 1069152 - 850864 - 164864$	$k^6, k^8, k^{10}, k^{12}, k^{14}$	...	$k^8, k^{10}, k^{12}, k^{14}, k^{16}$
x^{24}	$+\ 89600 + 660800 + 1303356 + 660800 + 89600$	$k^8, k^{10}, k^{12}, k^{14}, k^{16}$	...	$k^8, k^{10}, k^{12}, k^{14}, k^{16}$
x^{26}	$-164864 - 850864 - 1069152 - 882592 - 28672$	$k^{10}, k^{12}, k^{14}, k^{16}, k^{18}$	...	$k^8, k^{10}, k^{12}, k^{14}, k^{16}$
x^{28}	$+270272 + 883336 + 674016 + 129536 + 4096$	$k^{12}, k^{14}, k^{16}, k^{18}, k^{20}$	...	$k^8, k^{10}, k^{12}, k^{14}, k^{16}$
x^{30}	$-\ 81744 - 390608 - 630448 - 245248 - 19456$	$k^{12}, k^{14}, k^{16}, k^{18}, k^{20}$	...	$k^{10}, k^{12}, k^{14}, k^{16}, k^{18}$
x^{32}	$+\ 8448 + 100608 + 834831 + 252416 + 89168$	$k^{12}, k^{14}, k^{16}, k^{18}, k^{20}$	...	$k^{12}, k^{14}, k^{16}, k^{18}, k^{20}$
x^{34}	$-\ 1024 - 28160 - 120120 - 153280 - 43520$	$k^{12}, k^{14}, k^{16}, k^{18}, k^{20}$	...	$k^{14}, k^{16}, k^{18}, k^{20}, k^{22}$
x^{36}	$+\ 8584 + 86176 + 65716 + 29120$	$k^{14}, k^{16}, k^{18}, k^{20}$	...	$k^{16}, k^{18}, k^{20}, k^{22}$
x^{38}	$-\ 4928 - 22288 - 15288$	k^{16}, k^{18}, k^{20}	...	k^{16}, k^{20}, k^{22}
x^{40}	$+\ 8560 + 6594 + 672$	k^{18}, k^{20}, k^{22}	...	k^{16}, k^{20}, k^{22}
x^{42}	$-\ 1176 - 784 - 64$	k^{20}, k^{22}, k^{24}	$-64 - 784 - 1176$	k^{18}, k^{20}, k^{22}
x^{44}	$+\ 196 + 60$	k^{22}, k^{24}	$60 + 196$	k^{20}, k^{22}
x^{46}	$-\ 24$	k^{24}	-24	k^{22}
x^{48}	$+\ 1$	k^{24}	$+\ 1$	k^{24}

(+) denom. ut suprà. (÷) denom. ut suprà.

	$1 - \text{sn}\, 7u = (1+x)$ into square of		$1 - k\,\text{sn}\, 7u = (1+kx)$ into square of	
x^0	1	1	1	1
x^1	$-\ 4$	1	$-\ 4$	k
x^2	$-\ 4$	1	$-\ 4$	k^2
x^3	$+\ 8 + 28$	$1,\ k^2$	$+\ 28 + 8$	$k,\ k^3$
x^4	$-\ 14$	k^2	$-\ 14$	k^2
x^5	$-\ 84 - 56$	$k^2,\ k^4$	$-\ 56 - 84$	$k,\ k^3$
x^6	$+112 + 28$	$k^2,\ k^4$	$+\ 28 + 112$	$k^2,\ k^4$
x^7	$+\ 64 + 204 + 32$	$k^2,\ k^4,\ k^6$	$+\ 32 + 204 + 64$	$k,\ k^3,\ k^5$
x^8	$-144 - 305 - 16$	$k^2,\ k^4,\ k^6$	$-\ 16 - 305 - 144$	$k^2,\ k^4,\ k^6$
x^9	$-\ 32 - 200 - 128$	$k^2,\ k^4,\ k^6$	$-128 - 200 - 32$	$k^3,\ k^5,\ k^7$
x^{10}	$+\ 64 + 456 + 368$	$k^2,\ k^4,\ k^6$	$+368 + 456 + 64$	$k^4,\ k^6,\ k^8$
x^{11}	$+112 + 56$	$k^4,\ k^6$	$+\ 56 + 112$	$k^5,\ k^7$
x^{12}	$-224 - 644 - 224$	$k^4,\ k^6,\ k^8$	$-224 - 644 - 224$	$k^4,\ k^6,\ k^8$
x^{13}	$+\ 56 + 112$	$k^6,\ k^8$	$+112 + 56$	$k^5,\ k^7$
x^{14}	$+368 + 456 + 64$	$k^6,\ k^8,\ k^{10}$	$+\ 64 + 456 + 368$	$k^4,\ k^6,\ k^8$
x^{15}	$-128 - 200 - 32$	$k^6,\ k^8,\ k^{10}$	$-\ 32 - 200 - 128$	$k^5,\ k^7,\ k^9$
x^{16}	$-\ 16 - 305 - 144$	$k^6,\ k^8,\ k^{10}$	$-144 - 305 - 16$	$k^6,\ k^8,\ k^{10}$
x^{17}	$+\ 32 + 204 + 64$	$k^6,\ k^8,\ k^{10}$	$+\ 64 + 204 + 32$	$k^7,\ k^9,\ k^{11}$
x^{18}	$+\ 28 + 112$	$k^8,\ k^{10}$	$+112 + 28$	$k^8,\ k^{10}$
x^{19}	$-\ 56 - 84$	$k^8,\ k^{10}$	$-\ 84 - 56$	$k^9,\ k^{11}$
x^{20}	$-\ 14$	k^{10}	$-\ 14$	k^{10}
x^{21}	$+\ 28 + 8$	$k^{10},\ k^{12}$	$+\ 8 + 28$	$k^9,\ k^{11}$
x^{22}	$-\ 4$	k^{12}	$-\ 4$	k^{10}
x^{23}	$-\ 4$	k^{12}	$-\ 4$	k^{11}
x^{24}	$+\ 1$	k^{12}	$+\ 1$	k^{12}

(÷) denom. ut suprà. (÷) denom. ut suprà.

110. It will be observed that the forms are essentially different according as n is odd or even.

When n is odd, the numerators and denominators, say $A(x)$, $B(x)$, $C(x)$ and $D(x)$, are of the forms

$$x\,(1,\ x^2)^{\frac{1}{2}(n^2-1)},$$

$$(1,\ x^2)^{\frac{1}{2}(n^2-1)}\sqrt{1-x^2},$$

$$(1,\ x^2)^{\frac{1}{2}(n^2-1)}\sqrt{1-k^2x^2},$$

$$(1,\ x^2)^{\frac{1}{2}(n^2-1)},$$

viz., the degrees are n^2, n^2, n^2, n^2-1.

But, n even, the forms are

$$x\,(1,\ x^2)^{\frac{1}{2}(n^2-4)}\sqrt{1-x^2}\,\sqrt{1-k^2x^2},$$
$$(1,\ x^2)^{\frac{1}{2}n^2},$$
$$(1,\ x^2)^{\frac{1}{2}n^2},$$
$$(1,\ x^2)^{\frac{1}{2}n^2},$$

viz. the degrees are n^2-1, n^2, n^2, n^2.

The rational functions $(1,\ x^2)$ presenting themselves in the foregoing forms may be called $A'(x)$, $B'(x)$, $C'(x)$, $D'(x)$: the degrees in x^2 are $\frac{1}{2}(n^2-1)$, $\frac{1}{2}(n^2-1)$, $\frac{1}{2}(n^2-1)$, $\frac{1}{2}(n^2-1)$ or $\frac{1}{2}(n^2-4)$, $\frac{1}{2}n^2$, $\frac{1}{2}n^2$, $\frac{1}{2}n^2$ according as n is odd or even.

111. Whether n is odd or even, if we change k into $\frac{1}{k}$ and x into kx, the functions A', D' each remain unaltered, while the functions B', C' are interchanged: thus

$n=2$, A' becomes $=2$

$$B' \quad ,, \quad =1-2k^2x^2+\frac{1}{k^2}k^4x^4,$$

$$C' \quad ,, \quad =1-\frac{2}{k^2}k^2x^2+\frac{1}{k^2}k^4x^4,$$

$$D' \quad ,, \quad =1 \qquad -\frac{1}{k^2}k^4x^4.$$

$$n=3,\ A' \quad ,, \quad =3-\left(4+\frac{4}{k^2}\right)k^2x^2+\frac{6}{k^2}k^4x^4-\frac{1}{k^4}k^8x^8,$$

&c.

And the same is the case with the functions A, B, C, D, except that A is changed into kA.

112. But there is another change, x into $\frac{1}{kx}$, the effect of which is different according as n is odd or even.

If n be odd, then disregarding a monomial factor $k^{n}x^{n^2}$, the change x into $\frac{1}{kx}$ interchanges A', D' and also interchanges B', C': thus

$n=3$, A' becomes $3-(4+4k^2)\dfrac{1}{k^2x^2}+6k^2\dfrac{1}{k^4x^4}-k^4\dfrac{1}{k^8x^8}$,

$$=-\frac{1}{k^4x^8}\left(1-6k^2x^4+(4k^2+4k^4)x^6-3k^4x^8\right);$$

B' " $1-\dfrac{4}{k^2x^2}+6k^2\dfrac{1}{k^4x^4}-4k^4\dfrac{1}{k^6x^6}+k^4\dfrac{1}{k^8x^8}$,

$$=\frac{1}{k^4x^8}\left(1-4k^2x^2+6k^2x^4-4k^2x^6+k^4x^8\right);$$

C' " $1-4k^2\dfrac{1}{k^2x^2}+6k^2\dfrac{1}{k^4x^4}-4k^2\dfrac{1}{k^6x^6}+k^4\dfrac{1}{k^8x^8}$,

$$=\frac{1}{k^4x^8}\left(1-4x^2+6k^2x^4-4k^4x^6+k^4x^8\right);$$

D' " $1-6k^2\dfrac{1}{k^4x^4}+(4k^2+4k^4)\dfrac{1}{k^6x^6}-3k^4\dfrac{1}{k^8x^8}$,

$$=-\frac{1}{k^4x^8}\left(3-(4+4k^2)x^2+6k^2x^4-k^4x^8\right).$$

If passing to the functions A, D we write down the general formula, this is

$$D(x)=(-)^{\frac{1}{2}(n-1)}k^{\frac{1}{2}(n^2+1)}x^{n^2}A\left(\frac{1}{kx}\right),\text{ implying}$$

$$D\left(\frac{1}{kx}\right)=(-)^{\frac{1}{2}(n-1)}k^{-\frac{1}{2}(n^2-1)}x^{-n^2}A(x);$$

and we thence deduce

$$D(x)\,D\left(\frac{1}{kx}\right)=k\,A(x)\,A\left(\frac{1}{kx}\right),$$

that is $$A\left(\frac{1}{kx}\right)\div D\left(\frac{1}{kx}\right)=1\div[k\,A(x)\div D(x)],$$

viz. the change of x into $\dfrac{1}{kx}$ changes sn nu into $\dfrac{1}{k\,\text{sn}\,nu}$: and making this change in cn nu and dn nu considered as functions of sn nu ($=\sqrt{1-\text{sn}^2 nu}$ and $\sqrt{1-k^2\text{sn}^2 nu}$ respectively) it is obvious that the effect must be to interchange the numerators of these functions, that is, the functions B and C or B' and C' as above.

113. If n be even, the effect, to a factor près, is to leave the four functions unaltered; thus $n=2$,

A' becomes $=2$,

$$B' \quad " \quad 1-2\,\frac{1}{k^2x^2}+k^2\,\frac{1}{k^4x^4},\ =\frac{1}{k^2x^4}(1-2x^2+k^2x^4),$$

$$C' \quad " \quad 1-2k^2\,\frac{1}{k^2x^2}+k^2\,\frac{1}{k^4x^4},\ =\frac{1}{k^2x^4}(1-2k^2x^2+k^2x^4),$$

$$D' \quad " \quad 1-\ k^2\,\frac{1}{k^4x^4} \qquad ,\ =-\frac{1}{k^2x^4}(1-k^2x^4).$$

Hence, n being even, we cannot in either of the above ways effect an interchange of the functions A, D so as to derive one of them from the other, and it is in fact clear that they are essentially different functions. It is to be observed that A' is always a composite function, viz. writing $n=2p$ we have

$$A_n=2A_pB_pC_pD_p,$$

which is a product of rational functions into $\sqrt{1-x^2}\,\sqrt{1-k^2x^2}$: the numerator-function $A'(x)$ in the above values of sn $4u$ and sn $6u$ might therefore be expressed as a product of lower integral functions of x^2: in particular $n=4$, we have $A(x)$ $=4x\,(1-2x^2+k^2x^4)\,(1-2k^2x^2+k^2x^4)\,(1-k^2x^4)\,\sqrt{1-x^2}\,\sqrt{1-k^2x^2}$. As regards the denominator $D\,(=D')$ we have

$$D_n=D_p^4-k^2A_p^4,$$

which when p is odd, and therefore A_p and D_p each rational, breaks up into four rational factors (rational, that is, as regards x, but involving the radical $\sqrt{k}$). But if p be even, then $A_p=A'_p\,x\sqrt{1-x^2}\,\sqrt{1-k^2x^2}$, and the form is

$$D_n=D_p^4-k^2x^4\,(1-x^2)^2\,(1-k^2x^2)^2\,A'^4_p,$$

which breaks up into two rational factors only. That is, n being the double of an odd number the denominator is the product of four rational factors, but n being the double of an even number it is the product of two rational factors only: thus

$n=2$ $\qquad D(x)=1-k^2x^4,$

$n=4$ $\qquad D(x)=(1-k^2x^4)^4-16k^2x^4\,(1-x^2)^2\,(1-k^2x^2)^2.$

Although for many purposes the expressions thus obtained in the case in question (n even) would be in their original form more convenient than the completely developed expressions, yet for other purposes and in particular for the calculation of the functions of a higher uneven value of n those last are the more convenient.

114. When n is odd the numerator of $1 \pm \text{sn}\, nu$ is a rational and integral function of the order n^2, containing the factor $1+x$ or $1-x$, and the other factor being the square of a rational and integral function of the order $\frac{1}{2}(n^2-1)$: the two formulæ are derived one from the other by merely changing the sign of x: say they are

$$D(x) - A(x) = (1 \pm x)\,[P(x)]^2,$$

$$D(x) + A(x) = (1 \mp x)\,[P(-x)]^2,$$

giving when multiplied together

$$D^2(x) - A^2(x) = (1-x^2)\,[P(x)\,P(-x)]^2,$$

viz. the left-hand side is $B^2(x)$, $=(1-x^2)\,[B'(x)]^2$; and conversely the equation $(1-x^2)\,[A'(x)]^2 = D^2(x) - A^2(x)$ implies that the factors $D(x) - A(x)$, $D(x) + A(x)$ are of the forms in question. As regards the sign $\pm$ it is to be observed that $D(x) - A(x)$ contains the factor $1 - (-)^{\frac{1}{2}(n-1)}x$; viz. in the numerator of $1 - \text{sn}\, nu$, $n = 3, 7, \ldots$ or $4p+3$, the factor is $1+x$, but $n = 5, 9, \ldots$ or $4p+1$ the factor is $1-x$. The reason is obvious; $n = 4p+3$, $1 - \text{sn}\, nu$ vanishes for $u = -K$, that is $\text{sn}\, u = x = -1$ (but not for $u = +K$), while, $n = 4p+1$, $1 - \text{sn}\, nu$ vanishes for $u = K$, that is $\text{sn}\, u = x = 1$ (but not for $u = -K$): we have in fact $\text{sn}\,(4p+3)\,K = \text{sn}\, 3K = -1$, but $\text{sn}\,(4p+1)\,K = \text{sn}\, K = +1$.

The like considerations apply to the numerators of $1 \pm k\, \text{sn}\, nu$: the single factor is $1 \pm kx$, viz. for $1 - k\, \text{sn}\, nu$ this is $1 - (-)^{\frac{1}{2}(n-1)}kx$.

115. In the case n even, there are given $(n=2)$ formulæ for $1 \pm \text{cn}\, 2u$, $1 \pm \text{dn}\, 2u$, and from these may be deduced analogous formulæ for $1 \pm \text{cn}\, nu$, $1 \pm \text{dn}\, nu$, but I have not

thought it worth while to write these down for the cases 4 and 6. We in fact have

$$1-\text{cn}\,2pu = 2\text{sn}^2 pu\,\text{dn}^2 pu, \qquad (+)$$
$$1+\text{cn}\,2pu = 2\text{cn}^2 pu, \qquad (+)$$
$$1-\text{dn}\,2pu = 2k^2\,\text{sn}^2 pu\,\text{cn}^2 pu, \qquad (+)$$
$$1+\text{dn}\,2pu = 2\text{dn}^2 pu, \qquad (+)$$
$$\text{denom.} = 1-k^2\text{sn}^4 pu,$$

and substituting for the functions of pu their values we have

$$1-\text{cn}\,2pu = \quad 2A_p^2C_p^2, \qquad (+)$$
$$1+\text{cn}\,2pu = \quad 2D_p^2D_p^2, \qquad (+)$$
$$1-\text{dn}\,2pu = 2k^2A_p^2B_p^2, \qquad (+)$$
$$1+\text{dn}\,2pu = \quad 2C_p^2D_p^2, \qquad (+)$$

where $\quad$ denom. $= D_{2p}$, as for the other $2pu$-functions.

116. We may in the multiplication-formulæ write $k=0$, viz. we then have $x=\sin u$, and sn nu, cn nu, dn $nu = \sin nu$, $\cos nu$, 1 respectively: this however affords a verification only of the terms not multiplied by any power of k. A more complete verification is obtained by writing $k=1$, we then have $x = \text{sg}\,u$, $\sqrt{1-x^2}$ and $\sqrt{1-k^2x^2}$ each $= \text{cg}\,u$; and sn nu, cn nu, dn nu = sg nu, cg nu, cg nu respectively. Recalling the formulæ

$$\text{sg}\,nu = \tfrac{1}{2}[(1+x)^n-(1-x)^n], \qquad (+)$$
$$\text{cg}\,nu = \quad (1-x^2)^{\frac{1}{2}n}, \qquad (+)$$
$$\text{denom.} = \tfrac{1}{2}[(1+x)^n+(1-x)^n], \qquad (+)$$

the terms of the fractions require to be each multiplied by $(1-x^2)^{\frac{1}{2}(n^2-n)}$, viz. the formulæ then are

$$\text{sg}\,nu = \tfrac{1}{2}[(1+x)^n-(1-x)^n]\,(1-x^2)^{\frac{1}{2}(n^2-n)}, \qquad (+)$$
$$\text{cg}\,nu = \quad (1-x^2)^{\frac{1}{2}n^2}, \qquad (+)$$
$$\text{denom.} = \tfrac{1}{2}[(1+x)^n+(1-x)^n]\,(1-x^2)^{\frac{1}{2}(n^2-n)}.$$

Thus $n=3$, the formulæ are

$$\text{sg}\,3u = x(3+x^2)(1-x^2)^3, = x(3-8x^2+6x^4+0x^6-x^8), \qquad (+)$$
$$\text{cg}\,3u = (1-x^2)^4\sqrt{1-x^2}, = (1-4x^2+6x^4-4x^6+x^8)\sqrt{1-x^2}, (+)$$
$$\text{denom.} = (1+3x^2)(1-x^2)^3, = (1+0x^2-6x^4+8x^6-3x^8),$$

agreeing with the foregoing values of sn $3u$, cn $3u$, dn $3u$ on putting therein $k=1$.

Factorial-formulæ. Art. Nos. 117 to 122.

117. In the expressions for the numerators and denominator of the functions of nu, the rational functions of x^2 may be decomposed into their simple factors. This may be effected *à priori* by considering what are the values of x (that is sn u) which make these functions respectively vanish. But in the particular case $n=2$, it may be done *à posteriori*, by means of the duplication-formulæ, and the formulæ obtained for the dimidiation of the periods.

Write

$$(m, m') = \quad 2mK \quad + 2m'iK',$$

$$(\bar{m}, m') = (2m+1)K + 2m'iK',$$

$$(m, \bar{m}') = \quad 2mK \quad + (2m'+1)iK',$$

$$(\bar{m}, \bar{m}') = (2m+1)K + (2m'+1)iK'.$$

Then, using [ ] to denote a product, as explained by the appended values of m, m', we have

$$\text{sn } 2u = 2x\sqrt{1-x^2}\,\sqrt{1-k^2x^2}, \qquad (+)$$

$$\text{cn } 2u = \left\{1 + \frac{x}{\text{sn } \tfrac{1}{2}(\bar{m}, m')}\right\}, \qquad (+) \qquad \begin{matrix} m = 0, 2 \\ m' = 0, 1 \end{matrix}$$

$$\text{dn } 2u = \left\{1 + \frac{x}{\text{sn } \tfrac{1}{2}(m, \bar{m}')}\right\}, \qquad (+) \qquad \begin{matrix} m = 0, 2 \\ m' = 0, 1 \end{matrix}$$

$$\text{denom.} = \left\{1 + \frac{x}{\text{sn } \tfrac{1}{2}(\bar{m}, \bar{m}')}\right\}. \qquad \begin{matrix} m = 0, 1 \\ m' = 0, 1. \end{matrix}$$

118. Thus in cn $2u$ the product is

$$\left(1 + \frac{x}{\text{sn } \tfrac{1}{2}K}\right)\left(1 + \frac{x}{\text{sn } \tfrac{5}{2}K}\right)$$

$$\left(1 + \frac{x}{\text{sn } (\tfrac{1}{2}K + iK')}\right)\left(1 + \frac{x}{\text{sn } (\tfrac{5}{2}K + iK')}\right),$$

which is

$$=\left(1+\frac{x}{\operatorname{sn}\frac{1}{2}K}\right)\left(1-\frac{x}{\operatorname{sn}\frac{1}{2}K}\right)$$

$$\left(1+\frac{x}{\operatorname{sn}(\frac{1}{2}K+iK')}\right)\left(1-\frac{x}{\operatorname{sn}(\frac{1}{2}K+iK')}\right),$$

$$=[1-(1+k')x^2]\,[1-(1-k')x^2],$$

$$=1-2x^2+k^2x^4.$$

So in dn $2u$ tho product is

$$\left(1+\frac{x}{\operatorname{sn}(\frac{1}{4}K+\frac{1}{4}iK')}\right)\left(1+\frac{x}{\operatorname{sn}(\frac{3}{4}K+\frac{1}{4}iK')}\right)$$

$$\left(1+\frac{x}{\operatorname{sn}(\frac{1}{4}K+\frac{3}{4}iK')}\right)\left(1+\frac{x}{\operatorname{sn}(\frac{3}{4}K+\frac{3}{4}iK')}\right),$$

which is

$$=\left(1+\frac{x}{\operatorname{sn}(\frac{1}{4}K+\frac{1}{4}iK')}\right)\left(1-\frac{x}{\operatorname{sn}(\frac{1}{4}K+\frac{1}{4}iK')}\right)$$

$$\left(1+\frac{x}{\operatorname{sn}(\frac{1}{4}K+\frac{3}{4}iK')}\right)\left(1-\frac{x}{\operatorname{sn}(\frac{1}{4}K+\frac{3}{4}iK')}\right),$$

$$=\left(1-kx^2(k-2k')\right)\left(1-kx^2(k+2k')\right),$$

$$=1-2k^2x^2+k^2x^4.$$

And in the denominator the product is

$$\left(1+\frac{x}{\operatorname{sn}\frac{1}{4}iK'}\right)\left(1+\frac{x}{\operatorname{sn}\frac{3}{4}iK'}\right)$$

$$\left(1+\frac{x}{\operatorname{sn}(K+\frac{1}{4}iK')}\right)\left(1+\frac{x}{\operatorname{sn}(K+\frac{3}{4}iK')}\right),$$

which is

$$=(1-i\sqrt{k}x)(1+i\sqrt{k}x)(1+\sqrt{k}x)(1-\sqrt{k}x),$$

$$=(1+kx^2)(1-kx^2),$$

$$=1-k^2x^4.$$

119. Considering next the case where n is an odd number, $=2p+1$ suppose,

write as before

$$\begin{aligned}
(m, m') &= \quad 2mK \quad + 2m'iK',\\
(\bar{m}, m') &= (2m+1)\,K + 2m'iK',\\
(m, \bar{m}') &= \quad 2mK \quad + (2m'+1)\,iK',\\
(\bar{m}, \bar{m}') &= (2m+1)\,K + (2m'+1)\,iK';
\end{aligned}$$

then the distinct values of $\operatorname{sn}\frac{1}{n}(m, m')$ are those obtained by giving to m the values $-p, -(p-1), \ldots -1, 0, 1 \ldots p$; and to m' the same values; viz. there are in all $(2p+1)^2, = n^2$ values of $\operatorname{sn}\frac{1}{n}(m, m')$.

For take μ, μ' values of the form in question, any other values are $\mu + n\theta, \mu' + n\theta'$ (θ and θ' integers).

$$\frac{1}{n}(m, m') = \frac{1}{n}(\mu, \mu') + (\theta, \theta'),$$

where $(\theta, \theta') = 2\theta K + 2\theta' iK'$. Hence

$$\operatorname{sn}\frac{1}{n}(m, m') = (-)^{\theta+\theta'} \operatorname{sn}\frac{1}{n}(\mu, \mu');$$

which, when $\theta + \theta'$ is even, is

$$= \qquad \sin\frac{1}{n}(\mu, \mu'),$$

and, when $\theta + \theta'$ is odd, it is

$$= \qquad -\sin\frac{1}{n}(\mu, \mu'),$$

which is $\qquad = \qquad \operatorname{sn}\frac{1}{n}(-\mu, -\mu'),$

where $-\mu, -\mu'$ are of the form in question.

One of the foregoing values is $\operatorname{sn}\frac{1}{n}(0, 0), = 0$; and if we exclude this there remains a system of $n^2 - 1$ values.

120. Consider next the distinct values of $\text{sn}\,\frac{1}{n}(\overline{m}, \overline{m}')$. Suppose in the first instance that m, m' each extend from $-(p+1), -p, \ldots -1, 0, 1 \ldots p$ (viz. that each has $n+1$ values). I call the values $-(p+1), p$ extreme values and the others intermediate; so that m, m' have respectively 2 extreme and $n-1$ intermediate values. We have in all a system of $(n+1)^2$ terms, viz. these are

m, m' both extreme	4
m extreme, m' mean	$2n-2$
m' extreme, m mean	$2n-2$
m, m' both mean	n^2-2n+1
	n^2+2n+1

Now m, m' both extreme the values are $\text{sn}(\pm K \pm iK')$, and these are excluded from consideration.

If m is extreme, m' mean, the values are

$$\text{sn}\left(\pm K + \frac{2m'+1}{n}\, iK'\right),$$

say for shortness $\text{sn}(\pm K + a)$, that is $\text{sn}(K+a)$ and $\text{sn}(-K+a)$, where a has $\frac{1}{2}(n-1)$ pairs of equal and opposite values.

But $\text{sn}(K+a) = -\text{sn}(-K+a)$, $= \text{sin}(K-a)$; hence $\text{sn}(K+a)$ has $\frac{1}{2}(n-1)$ values; similarly $\text{sn}(-K+a)$ has $\frac{1}{2}(n-1)$ values; or $\text{sn}(\pm K + a)$ has $(n-1)$ values.

And in like manner, m' being extreme, the value is

$$= \text{sn}\left(\pm iK' + \frac{2m+1}{n} K\right), \quad = \text{sn}(\pm iK' + \beta),$$

which has $(n-1)$ values. We have thus in all

$$(n-1) + (n-1) + (n-1)^2, \quad = n^2 - 1 \text{ values.}$$

And as for $\text{sn}\,\frac{1}{n}(m, m')$, it may be shown that these are all the values.

121. Consider in like manner $\text{sn}\,\frac{1}{n}(\bar{m}, m')$: here m has the values $-(p+1), -p, \ldots -1, 0, 1, \ldots p$, say $-(p+1), p$ are extreme values and the others mean; and m' has the values $-p, \ldots -1, 0, 1, \ldots p$, say 0 is the extreme value and the others mean. The cases are

m, m' both extreme	2	exclude	
m extreme, m' mean	$2n-2$	reduce to	$n-1$
m mean, m' extreme	$n-1$	is	$n-1$
m, m' both mean	n^2-2n+1		n^2-2n+1
	n^2+n		n^2-1

or number in resulting system is $= n^2-1$.

And so $\text{sn}\,\frac{1}{n}(m, \bar{m}')$ has same number $= n^2-1$ of values.

122. We now obtain, n odd,

$$\text{sn}\,nu = nx \quad \left\{1 + \frac{x}{\text{sn}\,\frac{1}{n}(m, m')}\right\}, \qquad (+)$$

$$\text{cn}\,nu = \sqrt{1-x^2} \left\{1 + \frac{x}{\text{sn}\,\frac{1}{n}(\bar{m}, m')}\right\}, \qquad (+)$$

$$\text{dn}\,nu = \sqrt{1-k^2x^2} \left\{1 + \frac{x}{\text{sn}\,\frac{1}{n}(\bar{m}, m')}\right\}, \qquad (+)$$

where denom. $=$

$$\left\{1 + \frac{x}{\text{sn}\,\frac{1}{n}(m, \bar{m}')}\right\},$$

the number of factors being in each case n^2-1, viz. the values of m, m' are those belonging to the several systems of (n^2-1) values as above explained.

New Form of the Factorial-Formulæ.

Art. Nos. 123 to 126.

123. The formulæ may be presented in a different form: observing that to each term $1+\frac{x}{a}$ in the numerator or denominator there corresponds a term $1-\frac{x}{a}$, and combining together the pair of factors, also making an easy change of form, we have

$$\text{sn}\, nu = \quad nx \quad \left\{1-\frac{x^2}{\text{sn}^2\frac{1}{n}(m, m')}\right\}, \qquad (+)$$

$$\text{cn}\, nu = \sqrt{1-x^2}\left\{1-\frac{x^2}{\text{sn}^2\left(K-\frac{1}{n}(m, m')\right)}\right\}, \qquad (+)$$

$$\text{dn}\, nu = \sqrt{1-k^2x^2}\left\{1-k^2\,\text{sn}^2\left(K-\frac{1}{n}(m, m')\right)x^2\right\}, \qquad (+)$$

$$\text{denom.} = \qquad \left\{1-k^2\,\text{sn}^2\frac{1}{n}(m, m')\,x^2\right\},$$

where as regards the values of m, m' observe that these are

$$m=0, \quad m'= \quad 1, \quad 2 \ldots \quad \tfrac{1}{2}(n-1);$$

$$m=1, 2, \ldots \text{ or } \tfrac{1}{2}(n-1), \quad m'=0, \pm 1, \pm 2 \ldots \pm \tfrac{1}{2}(n-1);$$

viz. there are in all $\frac{1}{2}(n-1)+\frac{1}{2}(n-1)\,n$, $=\frac{1}{2}(n^2-1)$ combinations.

124. Restoring for x its value sn u, and observing that

$$\frac{1-\dfrac{\text{sn}^2 u}{\text{sn}^2 a}}{1-k^2\,\text{sn}^2 u\,\text{sn}^2 a}=\frac{\text{sn}\,(u+a)\,\text{sn}\,(u-a)}{\text{sn}\, a\,\text{sn}\,(-a)},$$

and combining all the constant factors (that is factors independent of u) into a single factor A, we find

$$\text{sn}\, nu = A\,\text{sn}\, u\left\{\text{sn}\left[u+\frac{1}{n}(m, m')\right]\text{sn}\left[u-\frac{1}{n}(m, m')\right]\right\},$$

or as this may be written

$$\operatorname{sn} nu = A \left\{ \operatorname{sn} \left[u + \frac{1}{n} (m, m') \right] \right\},$$

where m, m' have now each of them all the values

$$0, \pm 1, \pm 2 \ldots \pm \tfrac{1}{2}(n-1).$$

Proceeding in the same manner with the other equations, we have, with the same limits for (m, m'), the system

$$\operatorname{sn} nu = A \left\{ \operatorname{sn} \left[u + \frac{1}{n} (m, m') \right] \right\},$$

$$\operatorname{cn} nu = B \left\{ \operatorname{cn} \left[u + \frac{1}{n} (m, m') \right] \right\},$$

$$\operatorname{dn} nu = C \left\{ \operatorname{dn} \left[u + \frac{1}{n} (m, m') \right] \right\},$$

where the coefficients A, B, C have to be determined. The values are

$$A = (-)^{\frac{1}{2}(n-1)} k^{\frac{1}{2}(n^2-1)}, \quad B = \left(\frac{k}{k'}\right)^{\frac{1}{2}(n^2-1)}, \quad C = \left(\frac{1}{k'}\right)^{\frac{1}{2}(n^2-1)}.$$

125. To show this, write in the formulæ $u+K$ in place of u. Observing that we have

$$\operatorname{sn}(nu+nK) = (-)^{\frac{1}{2}(n-1)} \operatorname{sn}(nu+K),$$

$$\operatorname{cn}(nu+nK) = (-)^{\frac{1}{2}(n-1)} \operatorname{cn}(nu+K),$$

$$\operatorname{dn}(nu+nK) = \qquad \operatorname{dn}(nu+K),$$

and that the products on the right-hand sides contain n^2 terms, n^2-1 being evenly even, or $(-)^{n^2-1} = +$, we obtain

$$(-)^{\frac{1}{2}(n-1)} \frac{\operatorname{cn} nu}{\operatorname{dn} nu} = A \left\{ \frac{\operatorname{cn} \left[u + \frac{1}{n} (m, m') \right]}{\operatorname{dn} \left[u + \frac{1}{n} (m, m') \right]} \right\},$$

$$(-)^{\frac{1}{2}(n-1)}\frac{\text{sn}\,nu}{\text{dn}\,nu} = (k')^{n^2-1}B\left\{\frac{\text{sn}\left[u+\frac{1}{n}(m,m')\right]}{\text{dn}\left[u+\frac{1}{n}(m,m')\right]}\right\},$$

$$\frac{1}{\text{dn}\,nu} = (k')^{n^2-1}C\left\{\frac{1}{\text{dn}\left[u+\frac{1}{n}(m,m')\right]}\right\},$$

agreeing with the original equations if only

$$(-)^{\frac{1}{2}(n-1)}\frac{B}{C} = A,$$

$$(-)^{\frac{1}{2}(n-1)}\frac{A}{C} = k'^{n^2-1}B,$$

$$\frac{1}{C} = k'^{n^2-1}C,$$

but these reduce themselves to the two independent equations

$$C^2 = \left(\frac{1}{k'}\right)^{n^2-1},\quad B = (-)^{\frac{1}{2}(n-1)}AC.$$

The change of u into $u+iK'$ gives in like manner two independent equations, one of which is

$$A^2 = k^{n^2-1},$$

and we thus have A, B, C subject to an indetermination of the signs of A and C.

126. But it may be shown that the signs of A, B, C are $(-)^{\frac{1}{2}(n-1)}$, $+$, $+$. For this purpose recurring to the original equations and writing therein $u=0$, we find, observing that then $\frac{\text{sn}\,nu}{\text{sn}\,u} = n$, $\text{cn}\,u = 1$, $\text{dn}\,u = 1$,

$$n = A\left\{\text{sn}\,\frac{1}{n}(m,m')\right\},\quad 1 = B\left\{\text{cn}\,\frac{1}{n}(m,m')\right\},\quad 1 = C\left\{\text{dn}\,\frac{1}{n}(m,m')\right\},$$

where in each case the combination $m=0$, $m'=0$ is to be omitted, viz. the products each contain n^2-1 terms. Grouping together the opposite terms sn a, sn $(-a)$, &c., and recollecting that $\frac{1}{2}(n^2-1)$ is even, we may write

$$n = A\left\{\mathrm{sn}^2\frac{1}{n}(m, m')\right\},\ 1 = B\left\{\mathrm{cn}^2\frac{1}{n}(m, m')\right\},\ 1 = C\left\{\mathrm{dn}^2\frac{1}{n}(m, m')\right\},$$

and we may in each product consider separately the $\frac{1}{2}(n-1)$ terms in which m' is $=0$, the $\frac{1}{2}(n-1)$ terms in which m is $=0$, and the $\frac{1}{2}(n-1)^2$ terms in which neither m nor m' is $=0$. As regards these last we may consider that m has any value whatever from 1 to $\frac{1}{2}(n-1)$, and m' any value whatever from ± 1, to $\pm\frac{1}{2}(n-1)$; uniting together the terms which belong to the same value of m but to opposite values of m' these are conjugate imaginaries and their product is positive: hence the whole third product is positive. Taking next the terms for which m' is $=0$, each term is real and positive; hence the whole first product is positive. There remains only the second product; viz. as regards A this is $\left\{\mathrm{sn}^2\frac{1}{n}(0, m')\right\}$, where m' has the values 1, 2, ... $\frac{1}{2}(n-1)$. Each term is the square of a pure imaginary, viz. it is real and negative; and the sign is thus $(-)^{\frac{1}{2}(n-1)}$. But as regards B the product is $\left\{\mathrm{cn}^2\frac{1}{n}(0, m')\right\}$, where each term is positive (since cn $\frac{1}{n}(0, m')$ is real): hence the product is positive. And so as regards C the product is $\left\{\mathrm{dn}^2\frac{1}{n}(0, m')\right\}$, which is in like manner positive. Hence in the three cases respectively the sign of the first product is $(-)^{\frac{1}{2}(n-1)}$, +, +. And the required quantities A, B, C have these signs accordingly; wherefore we have

$$A = (-)^{\frac{1}{2}(n-1)}k^{-\frac{1}{2}(n^2-1)},\ B = \left(\frac{k}{k'}\right)^{\frac{1}{2}(n^2-1)},\ C = \left(\frac{1}{k'}\right)^{\frac{1}{2}(n^2-1)},$$

as mentioned above.

Anticipation of the doubly-infinite-product Forms of the Elliptic Functions. Art. No. 127.

127. In the formulæ No. 122 for u write $\frac{u}{n}$, then $x = \text{sn}\,\frac{u}{n}$, $= \frac{u}{n}$ when n is very large. Moreover *when* m, m' *are finite*, then in like manner $\text{sn}\,\frac{1}{n}(m, m')$ is $= \frac{1}{n}(m, m')$: and substituting these values and writing $n = \infty$ we obtain the following formulæ:

$$\text{sn}\,u = u\left\{1 + \frac{u}{(m, m')}\right\}, \qquad (\div)$$

$$\text{cn}\,u = \left\{1 + \frac{u}{(m, \overline{m'})}\right\}, \qquad (\div)$$

$$\text{dn}\,u = \left\{1 + \frac{u}{(m, m')}\right\}, \qquad (\div)$$

$$\text{denom.} = \left\{1 + \frac{u}{(m, \overline{m'})}\right\},$$

where m, m' have each of them every integer value from $-\infty$ to $+\infty$, the simultaneous values $m = 0$, $m' = 0$ being excluded from the numerator of sn u. I defer the further consideration of these formulæ, only remarking that not only they are not as yet proved, but that, *in the absence of further definition as to the limits*, they are wholly meaningless.

Derivatives of sn u, cn u, dn u *in regard to* k. Art. No. 128.

128. We have seen, Chap. III. No. 79, that

$$\frac{dF}{dk} = \frac{1}{kk'^2}(E - k'^2 F) - \frac{k \sin\phi\cos\phi}{k'^2\Delta},$$

where F, E, Δ stand for $F(k, \phi)$, $E(k, \phi)$, $\Delta(k, \phi)$ respectively. But we have $u = F$, giving $\text{sn}\,u = \sin\phi$, $\text{cn}\,u = \cos\phi$, $\text{dn}\,u = \Delta$; also

$$E = \int_0 \Delta\, d\phi = \int_0 \text{dn}^2 u\, du, = \int_0 du(1 - k^2 + k^2\,\text{cn}^2 u) = k'^2 u + k^2 \int_0 \text{cn}^2 u\, du,$$

and therefore $\qquad E - k'^2 F = k^2 \int_0 \mathrm{cn}^2 u \, du$;

hence

$$\frac{dF}{dk} = \frac{k}{k'^2}\left\{\int_0 \mathrm{cn}^2 u \, du - \frac{\mathrm{sn}\, u \,\mathrm{cn}\, u}{\mathrm{dn}\, u}\right\}.$$

But $\mathrm{sn}\, u = \sin\phi$, viz. considering sn u as a function of u, k, where $u, = F(k, \phi)$ is a function of k and ϕ, we have sn u, a function of ϕ only without k; and we hence obtain

$$\frac{d\,\mathrm{sn}\, u}{du}\,\frac{dF}{dk} + \frac{d\,\mathrm{sn}\, u}{dk} = 0,$$

that is

$$\frac{d\,\mathrm{sn}\, u}{dk} = -\,\mathrm{cn}\, u \,\mathrm{dn}\, u\,\frac{dF}{dk},$$

or finally

$$\frac{d\,\mathrm{sn}\, u}{dk} = -\frac{k}{k'^2}\,\mathrm{cn}\, u\,\mathrm{dn}\, u \int_0 \mathrm{cn}^2 u \, du + \frac{k}{k'^2}\,\mathrm{sn}\, u\,\mathrm{cn}^2 u,$$

and thence

$$\frac{d\,\mathrm{cn}\, u}{dk} = \frac{k}{k'^2}\,\mathrm{sn}\, u\,\mathrm{dn}\, u \int_0 \mathrm{cn}^2 u \, du - \frac{k}{k'^2}\,\mathrm{sn}^2 u\,\mathrm{cn}\, u,$$

and

$$\frac{d\,\mathrm{dn}\, u}{dk} = \frac{k^3}{k'^2}\,\mathrm{sn}\, u\,\mathrm{cn}\, u \int_0 \mathrm{cn}^2 u \, du - \frac{k}{k'^2}\,\mathrm{sn}^2 u\,\mathrm{dn}\, u.$$

And it will be convenient to repeat here from Nos. 73 and 74 the following formulæ, in which we now write K, K', E, E' for the complete functions $F_1 k$, $F_1 k'$, $E_1 k$, $E_1 k'$;

$$\frac{dE}{dk} = \frac{1}{k}(E - K), \qquad \frac{dE'}{dk} = -\frac{k}{k'^2}(E' - K'),$$

$$\frac{dK}{dk} = \frac{1}{kk'^2}(E - k'^2 K), \qquad \frac{dK'}{dk} = -\frac{1}{kk'^2}(E' - k^2 K'),$$

giving $\qquad EK' + E'K - KK' = \tfrac{1}{2}\pi.$

CHAPTER V.

THE THREE KINDS OF ELLIPTIC INTEGRALS.

129. In the present and following Chapters we revert to the notation of the elliptic integrals $F\phi$, $E\phi$, $\Pi\phi$, bringing up the theory to the point at which it is expedient to introduce the elliptic functions sn u, cn u, dn u: and explaining the resulting new notations.

The Addition-Theory. Art. Nos. 130 to 134.

130. We have throughout ϕ, ψ, μ connected by the addition-equation: regarding herein μ as a constant, this gives $\frac{d\phi}{\Delta\phi}+\frac{d\psi}{\Delta\psi}=0$: hence if U be any function of ϕ, ψ, μ, such that in virtue of the addition-equation we have

$$\frac{dU}{d\phi}\Delta\phi-\frac{dU}{d\psi}\Delta\psi=0,$$

or (what is the same thing) if this last equation be a form of the addition-equation, we hence derive $\frac{dU}{d\phi}d\phi+\frac{dU}{d\psi}d\psi=0$, that is $dU=0$, or by integration $U=$ funct. μ: and if moreover the function U is such that it vanishes for $\phi=0$, $\psi=\mu$, then the constant of integration, funct. μ, is $=0$; and we have $U=0$ as a consequence of the addition-equation. For instance the function $U=F\phi+F\psi-F\mu$, satisfies the conditions in question, and we thus have

$$F\phi+F\psi-F\mu=0,$$

as the addition-theorem for the first kind of elliptic integrals.

131. Again we have

$$E\phi + E\psi - E\mu - k^2 \sin\phi \sin\psi \sin\mu = 0$$

as the addition-theorem for the second kind of integrals. In fact the equation to be verified is

$$(\Delta\phi - k^2 \sin\psi \cos\phi \sin\mu)\,\Delta\phi - (\Delta\psi - k^2 \sin\phi \cos\psi \sin\mu)\,\Delta\psi = 0,$$

that is

$$\Delta^2\phi - \Delta^2\psi + k^2 \sin\mu\,(\sin\phi \cos\psi\,\Delta\psi - \sin\psi \cos\phi\,\Delta\phi) = 0;$$

which in virtue of

$$\sin\mu = \frac{\sin\phi \cos\psi \Delta\psi + \sin\psi \cos\phi \Delta\phi}{1 - k^2 \sin^2\phi \sin^2\psi},$$

and the identity

$$\sin^2\phi \cos^2\psi\,\Delta^2\psi - \sin^2\psi \cos^2\phi\,\Delta^2\phi$$
$$= (\sin^2\phi - \sin^2\psi)\,(1 - k^2 \sin^2\phi \sin^2\psi),$$

reduces itself to

$$\Delta^2\phi - \Delta^2\psi + k^2 (\sin^2\phi - \sin^2\psi) = 0,$$

which is an identity.

132. Again for the third kind of integrals, writing

$$a = (1+n)\left(1 + \frac{k^2}{n}\right), \quad R = \frac{n \sin\phi \sin\psi \sin\mu}{1 + n - n \cos\phi \cos\psi \cos\mu},$$

and for convenience retaining

$$\int \frac{dR}{1 + aR^2} \left(= \frac{1}{\sqrt{a}} \tan^{-1} R\sqrt{a}, \text{ or } = \frac{1}{2\sqrt{-a}} \log \frac{1 + R\sqrt{-a}}{1 - R\sqrt{-a}}\right.$$

according as a is positive or negative$\Big)$, to denote the one or other of these expressions as the case may be, then the addition-theorem is

$$\Pi\phi + \Pi\psi - \Pi\mu - \int \frac{dR}{1 + aR^2} = 0.$$

In fact the equation to be verified is here

$$\left\{\frac{1}{(1+n\sin^2\phi)\,\Delta\phi}-\frac{1}{1+aR^2}\frac{dR}{d\phi}\right\}\Delta\phi$$
$$-\left\{\frac{1}{(1+n\sin^2\psi)\,\Delta\psi}-\frac{1}{1+aR^2}\frac{dR}{d\psi}\right\}\Delta\psi=0,$$

where, writing $R=\frac{P}{Q}$, we have

$$\frac{1}{1+aR^2}\frac{dR}{d\phi}=\frac{1}{Q^2+aP^2}\left(Q\frac{dP}{d\phi}-P\frac{dQ}{d\phi}\right),$$
$$\frac{1}{1+aR^2}\frac{dR}{d\psi}=\frac{1}{Q^2+aP^2}\left(Q\frac{dP}{d\psi}-P\frac{dQ}{d\psi}\right),$$

and the equation thus becomes

$$\left\{\frac{1}{1+n\sin^2\phi}-\frac{1}{1+n\sin^2\psi}\right\}(Q^2+aP^2)$$
$$=\left(Q\frac{dP}{d\phi}-P\frac{dQ}{d\phi}\right)\Delta\phi-\left(Q\frac{dP}{d\psi}-P\frac{dQ}{d\psi}\right)\Delta\psi.$$

But in virtue of the addition-equation, as shown in the next No.,

$$Q^2+aP^2=(1+n\sin^2\mu)\,(1+n\sin^2\phi)\,(1+n\sin^2\psi),$$

and the equation to be verified thus becomes

$$(1+n\sin^2\mu)\,n\,(\sin^2\psi-\sin^2\phi)$$
$$=\left(Q\frac{dP}{d\phi}-P\frac{dQ}{d\phi}\right)\Delta\phi-\left(Q\frac{dP}{d\psi}-P\frac{dQ}{d\psi}\right)\Delta\psi.$$

133. In regard to the expression for Q^2+aP^2, observe that this is

$$=(1+n-n\cos\mu\cos\phi\cos\psi)^2+n^2a\sin^2\mu\sin^2\phi\sin^2\psi,$$

or putting herein $\cos\phi\cos\psi=\cos\mu+\sin\phi\sin\psi\Delta\mu$, this is

$$=(1+n\sin^2\mu-n\cos\mu\Delta\mu\sin\phi\sin\psi)^2+n^2a\sin^2\mu\sin^2\phi\sin^2\psi,$$
$$=(1+n\sin^2\mu)^2-2\,(1+n\sin^2\mu)\,n\cos\mu\Delta\mu\sin\phi\sin\psi$$
$$+n^2\left[(1-\sin^2\mu)(1-k^2\sin^2\mu)+\left(1+n+\frac{k^2}{n}+k^2\right)\sin^2\mu\right]\sin^2\phi\sin^2\psi,$$
$$=(1+n\sin^2\mu)\,\{1+n\sin^2\mu-2n\cos\mu\Delta\mu\sin\phi\sin\psi$$
$$+(n^2+nk^2\sin^2\mu)\sin^2\phi\sin^2\psi\}.$$

But we have

$$(1-\sin^2\phi)(1-\sin^2\psi)-\cos^2\mu-\sin^2\phi\sin^2\psi\,\Delta^2\mu$$
$$=2\cos\mu\Delta\mu\sin\phi\sin\psi,$$

or what is the same thing,

$$2\cos\mu\Delta\mu\sin\phi\sin\psi=\sin^2\mu-\sin^2\phi-\sin^2\psi+k^2\sin^2\mu\sin^2\phi\sin^2\psi,$$

and substituting this value within the { } we obtain the above expression for Q^2+aP^2.

134. We have

$$Q\frac{dP}{d\phi}-P\frac{dQ}{d\phi}=(1+n-n\cos\mu\cos\phi\cos\psi)\,n\sin\mu\cos\phi\sin\psi$$
$$-n\cos\mu\sin\phi\cos\psi\,.\,n\sin\mu\sin\phi\sin\psi,$$
$$=n\sin\mu\sin\psi\,[(1+n)\cos\phi-n\cos\mu\cos\psi\,(\cos^2\phi+\sin^2\phi)]$$
$$=n\sin\mu\sin\psi\,[\cos\phi+n\,(\cos\phi-\cos\mu\cos\psi)],$$

that is

$$Q\frac{dP}{d\phi}-P\frac{dQ}{d\phi}=n\sin\mu\sin\psi\,(\cos\phi+n\sin\mu\sin\psi\Delta\phi),$$

and similarly

$$Q\frac{dP}{d\psi}-P\frac{dQ}{d\psi}=n\sin\mu\sin\phi\,(\cos\psi+n\sin\mu\sin\phi\Delta\psi).$$

The equation to be verified is thus

$$(1+n\sin^2\mu)(\sin^2\psi-\sin^2\phi)=\sin\mu\sin\psi(\cos\phi+n\sin\mu\sin\psi\Delta\phi)\Delta\phi$$
$$-\sin\mu\sin\phi(\cos\psi+n\sin\mu\sin\phi\Delta\psi)\Delta\psi,$$

breaking up into the two equations

$$\sin^2\psi-\sin^2\phi=\sin\mu\,(\sin\psi\cos\phi\Delta\phi-\sin\phi\cos\psi\Delta\psi),$$

and $$\sin^2\psi-\sin^2\phi=\sin^2\psi\,\Delta^2\phi-\sin^2\phi\,\Delta^2\psi,$$

the former of which is equivalent to the equation which gives the addition-theorem for the second kind of integrals, and the latter is obviously true.

New Notations for the Integrals of the Second and Third Kinds. Art. Nos. 135 to 137.

135. If in the equation $E\phi = \int_0 \Delta\phi\, d\phi$ we write $\sin\phi = \operatorname{sn} u$, and consequently $d\phi = \operatorname{dn} u\, du$, $\Delta\phi = \operatorname{dn} u$, the value of $E\phi$ becomes $= \int_0 \operatorname{dn}^2 u\, du$, or what is the same thing $\int_0 (1 - k^2 \operatorname{sn}^2 u)\, du$. Jacobi, *changing the original signification* of the functional symbol E, calls this Eu, viz. he writes

$$Eu = \int_0 \operatorname{dn}^2 u\, du,$$

the effect being to throw the addition-theorem into the form

$$Eu + Ev - E(u+v) = k^2 \operatorname{sn} u \operatorname{sn} v \operatorname{sn}(u+v).$$

He further considers in place of Eu a new function Zu, differing from it only by a multiple of u, viz. we have

$$Zu = Eu - \frac{E}{K} u, \quad = u\left(1 - \frac{E}{K}\right) - k^2 \int_0 \operatorname{sn}^2 u\, du,$$

where E is the complete integral of the second kind. Substituting for E its expression in terms of Z, we thus have

$$Zu + Zv - Z(u+v) = k^2 \operatorname{sn} u \operatorname{sn} v \operatorname{sn}(u+v).$$

136. If similarly in the equation $\Pi\phi = \int_0 \frac{d\phi}{(1 + n \sin^2\phi)\,\Delta\phi}$, we write $\sin\phi = \operatorname{sn} u$, and therefore $\frac{d\phi}{\Delta\phi} = du$, the value of $\Pi\phi$ becomes $= \int_0 \frac{du}{1 + n \operatorname{sn}^2 u}$; and if, *changing the notation*, this were called Πu, we should have

$$\Pi u = \frac{du}{1 + n \operatorname{sn}^2 u},$$

or what is the same thing,

$$\Pi u - u = \int_0 \frac{-n \operatorname{sn}^2 u\, du}{1 + n \operatorname{sn}^2 u}.$$

The effect would be to change the addition-theorem into

$$\Pi u + \Pi v - \Pi (u+v) = \int \frac{dR}{1 + aR^2},$$

where
$$R = \frac{n \operatorname{sn} u \operatorname{sn} v \operatorname{sn}(u+v)}{1 + n - n \operatorname{cn} u \operatorname{cn} v \operatorname{cn}(u+v)}.$$

137. Jacobi makes however a *further change of notation*, viz. expressing the parameter n in the form $-k^2 \operatorname{sn}^2 a$*, he omits from Πu the term u and multiplies the remaining term by a constant factor; he writes in fact

$$\Pi (u, a) = \int_0 \frac{k^2 \operatorname{sn} a \operatorname{cn} a \operatorname{dn} a \operatorname{sn}^2 u \, du}{1 - k^2 \operatorname{sn}^2 a \operatorname{sn}^2 u}.$$

The full advantages of the change will appear in the sequel, but it is convenient to mention here that the addition-theorem takes the form

$$\Pi (u, a) + \Pi (v, a) - \Pi (u+v, a) = \tfrac{1}{2} \log \frac{1 - k^2 \operatorname{sn} u \operatorname{sn} v \operatorname{sn}(u+v-a) \operatorname{sn} a}{1 + k^2 \operatorname{sn} u \operatorname{sn} v \operatorname{sn}(u+v+a) \operatorname{sn} a}.$$

The Third Kind of Elliptic Integral. Outline of the further Theory. Art. Nos. 138 to 147.

138. We have a theory for the addition of the parameters, including in it a theory of the reduction of the parameter to the forms $-1 + k'^2 \sin^2\theta$, and $-k^2 \sin^2\theta$ respectively. This is derived from the consideration of the function

$$\varpi = \frac{\sin\phi \cos\phi}{(1 + \zeta \sin^2\phi)\,\Delta},$$

where ζ is an arbitrary constant. Taking also ρ an arbitrary constant, we obtain

$$\frac{d\varpi}{1 + \rho\varpi^2} = \frac{d\phi}{\Delta} \frac{1 - (2+\zeta)\sin^2\phi + (1+2\zeta) k^2 \sin^4\phi - \zeta k^2 \sin^6\phi}{(1 + \zeta\sin^2\phi)^2 (1 - k^2 \sin^2\phi) + \rho \sin^2\phi (1 - \sin^2\phi)},$$

* Regarding a as real, this would imply that the parameter is negative and in absolute magnitude $< k^2$; but a is regarded as susceptible of imaginary values, and the other forms of parameter are thus included.

and putting the denominator

$$= (1 + n \sin^2\phi)(1 + n' \sin^2\phi)(1 + m \sin^2\phi),$$

and decomposing into an integer part and partial fractions, we have

$$\frac{d\varpi}{1+\rho\varpi^2} = \frac{d\phi}{\Delta}\left\{\frac{1}{\zeta} + \frac{A}{1+n\sin^2\phi} + \frac{A'}{1+n'\sin^2\phi} + \frac{B}{1+m\sin^2\phi}\right\},$$

whence integrating from $\phi = 0$,

$$A\Pi n + A'\Pi n' + B\Pi m + \frac{1}{\zeta} F = \int \frac{d\varpi}{1+\rho\varpi^2},$$

where the integral on the right-hand side is

$$= \frac{1}{\sqrt{\rho}} \tan^{-1}\varpi\sqrt{\rho}, \text{ or } \frac{1}{2\sqrt{-\rho}} \log \frac{1+\varpi\sqrt{-\rho}}{1-\varpi\sqrt{-\rho}},$$

according as ρ is positive or negative.

139. There are two particular cases, $\zeta = -1$, that is $\varpi = \frac{\tan\phi}{\Delta}$, and $\zeta = 0$, that is $\varpi = \frac{\sin\phi\cos\phi}{\Delta}$, in each of which one of the terms, say $B\Pi m$, disappears, and the formula takes the more simple form

$$A\Pi n + A'\Pi n' + CF = \int \frac{d\varpi}{1+\rho\varpi^2},$$

establishing a relation between only the two functions Πn, $\Pi n'$. In the first of these the parameters n, n' are connected by the relation $nn' = k^2$: in the second by the relation

$$(1+n)(1+n') = k'^2.$$

Hence by the first relation the function Πn, where n is positive or negative, but in absolute value greater than 1, is expressed by means of the function $\Pi n'$, where n' (having the same sign as n) is in absolute magnitude less than k^2: in particular n being negative and greater than 1, that is, between $(-1, -\infty)$, n' will be between 0 and $-k^2$. If n be positive and greater than 1, then in the second relation $n', = -1 + \frac{k'^2}{n+1}$, will be negative and between $-1, -k^2$: and it thus appears that the only values of n which need be considered are the negative values between $(0, -1)$: viz. we may have n between $(0, -k^2)$ say $n = -k^2 \sin^2\theta$, or else n between $(-k^2, -1)$ say $n = -1 + k'^2 \sin^2\theta$. Observe that in the former case $a, = (1+n)\left(1+\frac{k^2}{n}\right)$, is negative, and in the addition-theorem we have a logarithm; in the second case a is positive and we have a circular function (the inverse function $\tan^{-1}$). This leads to the consideration of two kinds of functions Π, viz. $n = -k^2 \sin^2\theta$, logarithmic functions, and $n = -1 + k'^2 \sin^2\theta$, circular functions: there is a convenience in taking $n = -k^2 \sin^2\theta$, as a universal form, allowing θ to assume imaginary values.

140. Recurring to the general form

$$A\Pi n + A'\Pi n' + B\Pi m + \frac{1}{\zeta} F = \int \frac{d\varpi}{1+p\varpi^2},$$

we may consider herein n, n' as arbitrary quantities, m as a determinate function of n, n'; it is in fact obtained by means of a quadratic equation. Taking n, n' to be real the value of m will in certain cases be imaginary: and conversely taking m to be a given imaginary quantity it is possible to determine real values for n, n': and thus the function Πm of a given imaginary parameter is made to depend upon two functions Πn, $\Pi n'$ of real parameters. This may be effected in a different manner: viz. taking n, n' to be conjugate imaginaries we obtain two real values of m, and thence two formulæ each involving the conjugate imaginary functions Πn, $\Pi n'$: the combination of these would lead to the expressions of Πn, $\Pi n'$ in terms of the two real functions $\Pi m_{,}$, $\Pi m_{,,}$

In either of the ways just referred to we in effect obtain an expression for the function of the third kind Πn, of imaginary parameter, in terms of two functions with real parameters: but it will presently appear that the solution admits of a very considerable simplification.

141. Introducing in the formula for n, n', m the values $-k^2 \sin^2 p$, $-k^2 \sin^2 q$, $-k^2 \sin^2\theta$, the relation between n, n', m gives a relation between p, q, θ, viz. this is found to be

$$k'(1 - k^2 \sin p \sin q \sin\theta) = \Delta p\, \Delta q\, \Delta\theta,$$

being in fact equivalent to the relation

$$Fp + Fq - F\theta - F_1 = 0,$$

or (what is the same thing) p, q, θ being connected by this equation, and writing Πp, &c. in place of $\Pi(-k^2\sin^2 p)$, &c., we have between the three functions the foregoing relation

$$A\Pi p + A'\Pi q + B\Pi\theta + \frac{1}{\zeta}F = \int \frac{d\varpi}{1+\rho\varpi^2}.$$

142. It is natural in place of θ to introduce a new angle θ' such that $F\theta + F_1 = F\theta'$, the relation between p, q, θ' being consequently the algebraical relation answering to $Fp + Fq - F\theta' = 0$. The function $\Pi\theta$ can be expressed in terms of $\Pi\theta'$, and the resulting equation is found to be

$$\frac{\cos p\,\Delta p}{\sin p}(\Pi p - F) + \frac{\cos q\,\Delta q}{\sin q}(\Pi q - F) - \frac{\cos\theta'\Delta\theta'}{\sin\theta'}(\Pi\theta' - F)$$

$$= k^2 \sin p \sin q \sin\theta' . F + \tfrac{1}{2}\log\frac{[A + k^2\sin p\sin q\sin\phi\,.\,B]\,A'}{[A' + k^2\sin p\sin q\sin\phi\,.\,B']\,A},$$

where A, B, A', B' are certain functions of θ' and ϕ.

143. This equation assumes a very simple form on writing therein $\sin\phi = \text{sn}\,u$, $\sin p = \text{sn}\,a$, $\sin q = \text{sn}\,b$, and therefore (by reason of $Fp + Fq - F\theta' = 0$) $\sin\theta' = \text{sn}(a+b)$: and by introducing Jacobi's notation for the function Π; viz. making the changes in question the three terms on the left-hand side are to a common factor près $\Pi(u, a)$, $\Pi(u, b)$, $\Pi(u, a+b)$: the logarithmic term is considerably simplified and the final equation is

$$\Pi(u, a) + \Pi(u, b) - \Pi(u, a+b)$$

$$= k^2\text{sn}\,a\,\text{sn}\,b\,\text{sn}(a+b)\,.\,u + \tfrac{1}{2}\log\frac{1 - k^2\text{sn}\,u\,\text{sn}\,a\,\text{sn}\,b\,\text{sn}(a+b-u)}{1 + k^2\text{sn}\,u\,\text{sn}\,a\,\text{sn}\,b\,\text{sn}(a+b+u)},$$

viz. we thus see the theorem in its true point of view as a theorem for the addition of the parameters.

144. We also gain a further insight into the problem of the determination of the function Πu with an imaginary value

of n: viz. any such value is expressible in the form $-k^2\,\mathrm{sn}^2(a+bi)$, where a and b are real; the function Πn, or say $\Pi\,(u,\,a+bi)$, is then made to depend on the two functions $\Pi\,(u,\,a)$, $\Pi\,(u,\,bi)$, which have each of them a real parameter, viz. in the second function the parameter is $n=-k^2\,\mathrm{sn}^2\,bi$, which is a real positive quantity.

145. There is another theory, the interchange of amplitude and parameter. Starting with the equation

$$\Pi=\int_0\frac{d\phi}{(1+n\sin^2\phi)\,\Delta}$$

we have $\dfrac{d\Pi}{dn}$ depending on the integral $\displaystyle\int\frac{d\phi}{(1+n\sin^2\phi)^2\,\Delta}$, which is expressible in terms of F, E and Π. The terms involving $\dfrac{d\Pi}{dn}$ and Π combine together into a term $\dfrac{d}{dn}\,\Pi\sqrt{a}$, where as before $a=(1+n)\left(1+\dfrac{k^2}{n}\right)$, and we have this term equal to a function of n, ϕ, where ϕ enters through the functions $\sin\phi$, $\cos\phi$, Δ, E, F, but which is algebraical in regard to n, viz. the actual equation is

$$\frac{d}{dn}\,\Pi\sqrt{a}=-\tfrac{1}{2}k^2F\,\frac{1}{n^2\sqrt{a}}-\tfrac{1}{2}(F-E)\,\frac{1}{n\sqrt{a}}$$

$$+\tfrac{1}{2}\Delta\sin\phi\cos\phi\,\frac{1}{(1+n\sin^2\phi)\sqrt{a}},$$

$a=(1+n)\left(1+\dfrac{k^2}{n}\right)$ as just mentioned: so that integrating in regard to n, we have $\Pi\sqrt{a}$ depending on the integrals $\displaystyle\int\frac{dn}{n^2\sqrt{a}}$, $\displaystyle\int\frac{dn}{n\sqrt{a}}$, $\displaystyle\int\frac{dn}{(1+n\sin^2\phi)\sqrt{a}}$; these are really elliptic integrals as at once appears by writing therein $n=-k^2\sin^2\theta$ (viz. we thus adopt for the parameter n the before mentioned form $-k^2\sin^2\theta$), reducing the integrals to the forms

$$\int\frac{d\theta}{\sin^2\theta\,\Delta\theta},\quad\int\frac{d\theta}{\Delta\theta},\quad\int\frac{\sin^2\theta\,d\theta}{(1-k^2\sin^2\theta\sin^2\phi)\,\Delta\theta},$$

or ultimately to the forms

$$\int \frac{d\theta}{\Delta\theta},\ \int \Delta\theta\, d\theta, \text{ and } \int \frac{d\theta}{(1-k^2\sin^2\phi\sin^2\theta)\Delta\theta};$$

viz. the first two of these are $F\theta$ and $E\theta$, and the last is the integral of the third kind $\Pi(n', \theta)$ with parameter $n', = -k^2\sin^2\phi$: the final result is

$$\cot\theta\,\Delta\theta\,[\Pi(n, \phi) - F\phi] - \cot\phi\,\Delta\phi\,[\Pi(n', \theta) - F\theta]$$
$$= E\theta\, F\phi - E\phi\, F\theta,$$

which is the equation for the interchange of amplitude and parameter.

146. If as before $\phi = \text{sn}\,u$, $\theta = \text{sn}\,a$, then using Jacobi's notation, the functions on the left-hand side are $\Pi(u, a)$, $\Pi(a, u)$; and the functions $E\phi$, $E\theta$ are in the same notation Eu, Ea: the equation therefore is

$$\Pi(u, a) - \Pi(a, u) = uEa - aEu,$$

or what is the same thing

$$\Pi(u, a) - \Pi(a, u) = uZa - aZu.$$

Fund. Nova. p. 146.

147. The foregoing outline of the theory of the elliptic integral of the third kind brings up the theory to the point immediately preceding the introduction of Jacobi's function Θ: viz. his functions $\Pi(u, a)$, Zu are in fact each of them expressed in terms of the new transcendent Θ, by the equations

$$Zu = \frac{\Theta' u}{\Theta u},\quad \Pi(u, a) = uZa + \tfrac{1}{2}\log\frac{\Theta(u-a)}{\Theta(u+a)},$$

the second of these leads at once to the just-mentioned equation $\Pi(u, a) - \Pi(a, u) = uZa - aZu$ (interchange of amplitude and parameter): and by means of this theorem we can from either of the addition-theorems (for the amplitudes, and the parameters respectively) at once derive the other theorem.

Reduction of a given imaginary quantity to the form $\text{sn}(\alpha+\beta i)$. Art. Nos. 148—152.

148. By what precedes it appears that there is an advantage in bringing the parameter to the form $-k^2\,\text{sn}^2\alpha$: this however cannot always be done so long as α is restricted to be real: but we have to show that it can be done, admitting imaginary values of α: or what is the same thing, that a given real or imaginary quantity n can always be expressed in the form $-k^2\,\text{sn}^2(\alpha+\beta i)$: that is $\sqrt{\frac{-n}{k^2}}$ can always be expressed in the form $\text{sn}(\alpha+\beta i)$: and the theorem thus is, that taking as usual the modulus k to be a given positive real quantity less than unity, then any given real or imaginary quantity whatever can be expressed in the form $\text{sn}(\alpha+\beta i)$.

149. It is to be observed that x and y being given real quantities, the former of them equal to or less than ± 1, we can find the real values α, β such that $x=\text{sn}\,\alpha$, $iy=\text{sn}\,i\beta$: in fact these equations give

$$x=\text{sn}\,\alpha,\quad y=\frac{\text{sn}(\beta, k')}{\text{cn}(\beta, k')},$$

and as α passes continuously from 0 to $\pm K$, $\text{sn}\,\alpha$ passes from 0 to ± 1, that is through every real value x whatever between these limits; and similarly as β passes continuously from 0 to $\pm iK'$, $\frac{\text{sn}(\beta, k')}{\text{cn}(\beta, k')}$ passes from 0 to $\pm\infty$, that is, through every real value y whatever.

Hence writing

$$\lambda+\mu i=\text{sn}(\alpha+\beta i)$$

$$=\frac{x\sqrt{1+y^2\,.\,1+k'^2y^2}+iy\sqrt{1-x^2\,.\,1-k^2x^2}}{1+k'^2x^2y^2},$$

we have $\lambda=\frac{x\sqrt{1+y^2\,.\,1+k'^2y^2}}{1+k'^2x^2y^2},\quad \mu=\frac{y\sqrt{1-x^2\,.\,1-k^2x^2}}{1+k'^2x^2y^2},$

or what is the same thing

$$\lambda^2 (1 + k^2x^2y^2)^2 = x^2 (1 + y^2)(1 + k'^2y^2),$$

$$\mu^2 (1 + k^2x^2y^2)^2 = y^2 (1 - x^2)(1 - k^2x^2),$$

and it only remains to be shown that λ, μ being any given real values whatever, those equations are satisfied by real values of x, y, that of x not greater than ± 1; or what is the same thing, that the two curves have real intersections within the limits $x = \pm 1$.

150. This is at once seen by tracing the curves; the first curve has one or other of the two forms shown by the dotted lines; and the second curve has the form shown by the con-

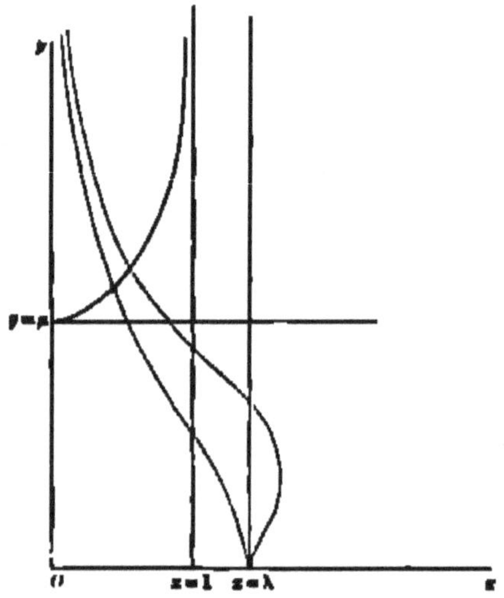

tinuous line (the curves are obviously symmetrical as to the four quadrants, and only a single quadrant is drawn): and there is thus in each quadrant a real intersection for which $x < \pm 1$. The original irrational equations show that x, y have the same signs as λ, μ respectively: there is thus for any given values of λ, μ a single intersection satisfying the condition in question.

151. But we may further develop the analytical theory: for this purpose instead of the original equations, consider for greater simplicity and symmetry the equations

$$\lambda (ab - xy)^2 = abx (a - y) (b - y),$$

$$\mu (ab - xy)^2 = aby (a - x) (b - x).$$

Each of these represents a quartic curve passing through the points (a, b), (b, a), and the two curves have besides at infinity 10 common points, 5 on the line $x = 0$ and 5 on the line $y = 0$: there remain therefore 4 intersections. To find these assume $xy = ab\omega$, the equations become

$$\lambda (1 - \omega)^2 = x + \omega y - (a + b) \omega,$$

$$\mu (1 - \omega)^2 = \omega x + y - (a + b) \omega,$$

giving x, y linearly in terms of ω. Solving the equations there is a factor $1 - \omega$ which divides out ($\omega = 1$ is in fact a solution answering to the two points (a, b), (b, a)), and the equations then become

$$(\lambda - \mu\omega) (1 - \omega) + (a + b) \omega = (1 + \omega) x,$$

$$(\mu - \lambda\omega) (1 - \omega) + (a + b) \omega = (1 + \omega) y,$$

and multiplying together these values and for xy writing $ab\omega$, we have a quartic equation in ω; this is a reciprocal equation, solvable by a quadric; or if for greater convenience we write $\theta = \left(\frac{\omega - 1}{\omega + 1}\right)^2$, then θ is also determined by a quadric equation; and putting

$$A = a^2 - 2a (\lambda + \mu) + (\lambda - \mu)^2,$$

$$B = b^2 - 2b (\lambda + \mu) + (\lambda - \mu)^2,$$

we find $\theta = \frac{(a - b)^2}{(\sqrt{A} - \sqrt{B})^2}$, that is $\theta_1 = \frac{\omega - 1}{\omega + 1} = \frac{a - b}{\sqrt{A} - \sqrt{B}}$.

and then after some reductions

$$x = \frac{(a-b)(\mu-\lambda)+b\sqrt{A}-a\sqrt{B}}{\sqrt{A}-\sqrt{B}-a+b},$$

$$= \frac{1}{4\lambda}(a+\lambda-\mu+\sqrt{A})(b+\lambda-\mu+\sqrt{B}),$$

$$y = \frac{(a-b)(\lambda-\mu)+b\sqrt{A}-a\sqrt{B}}{\sqrt{A}-\sqrt{B}-a+b},$$

$$= \frac{1}{4\mu}(a-\lambda+\mu+\sqrt{A})(b-\lambda+\mu+\sqrt{B}),$$

and changing herein the signs of $\sqrt{A}$, $\sqrt{B}$, we have of course the coordinates of the four points of intersection: it will be observed that A and B being real, these are all real or all imaginary.

152. To return to the original problem we must

for $\quad a,\ b,\ \lambda,\ \mu,\ x,\ y,$

write $\quad 1,\ \frac{1}{k^2},\ \lambda^2,\ -\mu^2,\ x^2,\ -y^2,$

whence if now

$$A = 1 - 2(\lambda^2-\mu^2) + (\lambda^2+\mu^2)^2,$$

$$B = \frac{1}{k^4} - \frac{2}{k^2}(\lambda^2-\mu^2) + (\lambda^2+\mu^2)^2,$$

(A and B being therefore, as is easily seen, each real), and choosing the root for which the radicals have the sign $-$, we have

$$x^2 = \frac{1}{4\lambda^2}(1+\lambda^2+\mu^2-\sqrt{A})\left(\frac{1}{k^2}+\lambda^2+\mu^2-\sqrt{B}\right),$$

$$y^2 = \frac{1}{4\mu^2}(1-\lambda^2-\mu^2-\sqrt{A})\left(\frac{1}{k^2}-\lambda^2-\mu^2-\sqrt{B}\right),$$

where x^2 is positive and less than 1, y^2 is positive. By the original investigation it was in fact shown that there was one such set of values of (x^2, y^2), and admitting this it is easy to see that the set just written down (wherein x^2 has its least value) must be the set in question: but it must admit of being shown independently that these values of x^2, y^2 do satisfy the conditions in question.

We have thus, by means of the quantities $\text{sn}\,\alpha, = x$ and $\text{sn}\,i\beta, = iy$, determined analytically the functions α, β, which are such that $\text{sn}\,(\alpha + \beta i) = \lambda + \mu i$, a given imaginary quantity; it may be remarked that the solution, although under a somewhat different form, is substantially identical with that given by Richelot, *Crelle*, t. 45 (1853), p. 225.

In the remainder of the present chapter we work out the foregoing theories.

Addition of Parameters, and Reduction to Standard Forms.
Art. Nos. 153 to 172.

153. We have

$$\Pi\,(n, k, \phi) = \int_0 \frac{d\phi}{(1 + n\sin^2\phi)\sqrt{1 - k^2\sin^2\phi}},$$

or expressing only the parameter, and writing for shortness $\sqrt{1 - k^2\sin^2\phi} = \Delta$,

$$\Pi n = \int_0 \frac{d\phi}{(1 + n\sin^2\phi)\,\Delta}.$$

Consider the function

$$\varpi = \frac{\sin\phi\cos\phi}{(1 + \zeta\sin^2\phi)\,\Delta},$$

where ζ is a constant. Taking also ρ a constant, we form the equation

$$\frac{d\varpi}{1 + \rho\varpi^2} = \frac{d\phi}{\Delta}\,\frac{1 - (2 + \zeta)\sin^2\phi + (1 + 2\zeta)\,k^2\sin^4\phi - \zeta k^2\sin^6\phi}{(1 + \zeta\sin^2\phi)^2(1 - k^2\sin^2\phi) + \rho\sin^2\phi\,(1 - \sin^2\phi)},$$

where the denominator, being a cubic function of $\sin^2\phi$, may be put $= (1 + n\sin^2\phi)\,(1 + n'\sin^2\phi)\,(1 + m\sin^2\phi)$. The expression which multiplies $\frac{d\phi}{\Delta}$ is then a fraction with this denominator, and breaking it up into partial fractions there is an integral part $\frac{1}{\zeta}$, and we have

$$\frac{d\varpi}{1+\rho\varpi^2} = \frac{d\phi}{\Delta}\left\{\frac{1}{\zeta} + \frac{A}{1+n\sin^2\phi} + \frac{A'}{1+n'\sin^2\phi} + \frac{B}{1+m\sin^2\phi}\right\},$$

whence integrating from $\phi = 0$, we have

$$A\Pi n + A'\Pi n' + B\Pi m + \frac{1}{\zeta}F = \int\frac{d\varpi}{1+\rho\varpi^2},$$

where the integral expression on the right-hand side is retained to stand for

$$\frac{1}{\sqrt{\rho}}\tan^{-1}\varpi\sqrt{\rho}, \quad (\rho \text{ positive})$$

or

$$\tfrac{1}{2}\log\frac{1+\varpi\sqrt{-\rho}}{1-\varpi\sqrt{-\rho}}, \quad (\rho \text{ negative});$$

and we have thus an identical relation between three functions Π each with the same modulus k and amplitude ϕ, but with the parameters n, n', m respectively.

154. The relations between these quantities and the values of the coefficients A, A', B are given by the equations

$$nn'm = -k^2\zeta^2,$$

$$(1+n)(1+n')(1+m) = k'^2(1+\zeta)^2,$$

$$(k^2+n)(k^2+n')(k^2+m) = -k^2k'^2\rho,$$

or, what is the same thing,

$$n+n'+m = 2\zeta - k^2 + \rho,$$

$$nn' + m(n+n') = \zeta^2 - 2k^2\zeta - \rho,$$

$$nn'm = -k^2\zeta^2.$$

Hence considering n, n' as given, we have

$$1 = \frac{k^2\zeta^2}{nn'} + \frac{k'^2(1+\zeta)^2}{(1+n)(1+n')},$$

or, what is the same thing,

$$\zeta = \frac{-k'^2 \pm \sqrt{\dfrac{1+n}{n}\,\dfrac{1+n'}{n'}\,(k^2+n)\,(k^2+n')}}{k'^2 + k^2\,\dfrac{1+n}{n}\,\dfrac{1+n'}{n'}},$$

and then $\qquad m = \dfrac{-k^2\zeta^2}{nn'},$

$$\rho = -\,(k^2+n)\,(k^2+n')\,\frac{k^2+m}{k^2k'^2},$$

and then further

$$A = [n^3 + (2+\zeta)\,n^2 + (1+2\zeta)\,k^2n + \zeta k^2] \div n\;(n-n')\,(n-m),$$

$$A' = [n'^3 + (2+\zeta)\,n'^2 + (1+2\zeta)\,k^2n' + \zeta k^2] \div n'\,(n'-n)\,(n'-m),$$

$$B = [m^3 + (2+\zeta)\,m^2 + (1+2\zeta)\,k^2m + \zeta k^2] \div m\,(m-n)\,(m-n').$$

The reduction of the general formula is somewhat laborious; there are two important particular cases which it is as well to discuss separately: these are

$$\zeta = -1\left(\varpi = \frac{\tan\phi}{\Delta}\right),\ \text{and}\ \zeta = 0\left(\varpi = \frac{\sin\phi\cos\phi}{\Delta}\right).$$

The Case $\zeta = -1,\ \varpi = \dfrac{\tan\phi}{\Delta}$.

135. We have here

$$m = -1,$$

$$nn' = k^2,$$

$$(k^2+n)\,(k^2+n') = k^2\rho,$$

which last equation may be written

$$(k^2+n)\left(k^2 + \frac{k^2}{n}\right) = k^2\rho,$$

or what is the same thing,

$$\rho = (1+n)\left(1+\frac{k^2}{n}\right),\ = \alpha.$$

We then have

$$A = [n^2 + n^2 - k^2n - k^2] + n(n-n')(n+1), = \frac{n^2 - k^2}{n(n-n')}, = 1,$$

and similarly $A' = 1$; also $B = 0$; so that the parameters n, n' being connected by the relation $nn' = k^2$, we have between the two functions Π the relation

$$\Pi n + \Pi n' = F + \int \frac{d\varpi}{1 + a\varpi^2}, \qquad (f')^*$$

viz. a being positive, the integral, substituting therein for ϖ its value, is

$$= \frac{1}{\sqrt{a}} \tan^{-1} \frac{\sqrt{a} \tan\phi}{\Delta},$$

and a being negative it is

$$= \frac{1}{2\sqrt{-a}} \log \frac{\Delta + \sqrt{-a} \tan\phi}{\Delta - \sqrt{-a} \tan\phi}.$$

The Case $\zeta = 0$, $\varpi = \dfrac{\sin\phi\cos\phi}{\Delta}$.

156. The general expressions are not immediately applicable: they give $m = 0$ and then $B = \dfrac{0}{0}$, but the two terms $\dfrac{1}{\zeta}$ and $\dfrac{B}{1 + m\sin^2\phi}$ are together equal to a determinate constant, the value of which, $= -\dfrac{k^2}{\rho}$, can be found by writing in the first instance $\zeta = 0$: the formula becomes

$$\frac{d\varpi}{1 + \rho\varpi^2} = \left(-\frac{k^2}{\rho} + \frac{A}{1 + n\sin^2\phi} + \frac{A'}{1 + n'\sin^2\phi}\right) \frac{d\phi}{\Delta},$$

or what is the same thing

$$A\Pi n + A'\Pi n' - \frac{k^2}{\rho} F = \int \frac{d\varpi}{1 + \rho\varpi^2},$$

* (f') and, next page, (g') are the formulæ thus designated, Legendre, *Traité des Fonctions Elliptiques*, Chap. XV.

where

$$n + n' = -k^2 + \rho,$$
$$nn' = \quad -\rho,$$
$$A = \frac{n^2 + 2n + k^2}{n(n - n')}, \quad A' = \frac{n'^2 + 2n' + k^2}{n'(n' - n)}.$$

We have therefore $n + n' + nn' = -k^2$, or what is the same thing

$$(1 + n)(1 + n') = k'^2,$$

which is the relation between n, n'. And then writing

$$A = \frac{(n+1)^2 - k'^2}{n(n - n')},$$

and substituting for k'^2 its value we find

$$A = \frac{n+1}{n}, \text{ and similarly } A' = \frac{n'+1}{n'}.$$

Moreover writing for ρ its value the formula becomes

$$\frac{n+1}{n}\Pi n + \frac{n'+1}{n'}\Pi n' + \frac{k^2}{nn'}F = \int \frac{d\varpi}{1 - nn'\varpi^2}, \qquad (g)$$

which is the relation between two functions Π.

157. The two formulæ (f) and (g) enable us to perform the reduction of functions of real parameter. We may consider the four cases

I. n positive, $= \cot^2\theta$;

$$\sqrt{a} = \frac{\Delta(k', \theta)}{\sin\theta\cos\theta}.$$

II. n negative and between $0, -k^2$, $= -k^2\sin^2\theta$;

$$\sqrt{-a} = \cot\theta\,\Delta(k, \theta).$$

III. n negative and between $-k^2$ and -1, $= -1 + k'^2\sin^2\theta$;

$$\sqrt{a} = \frac{k'^2\sin\theta\cos\theta}{\Delta(k', \theta)}.$$

IV. n negative and between $-1, -\infty$, $= -\frac{1}{\sin^2\theta}$;

$$\sqrt{-a} = \cot\theta\,\Delta(k, \theta);$$

where in each case the value is annexed of $\sqrt{a}$ or $\sqrt{-a}$ as the case may be. Observe that in the cases I. and III. a is positive, or the function is circular: in II. and IV. a is negative or the function is logarithmic.

158. It is very noticeable how the formulæ (f') and (g') give each of them a relation between two circular functions or two logarithmic functions, but not in any case a relation between a circular function and a logarithmic function. Treating n, n' as coordinates, we shade by vertical lines the spaces

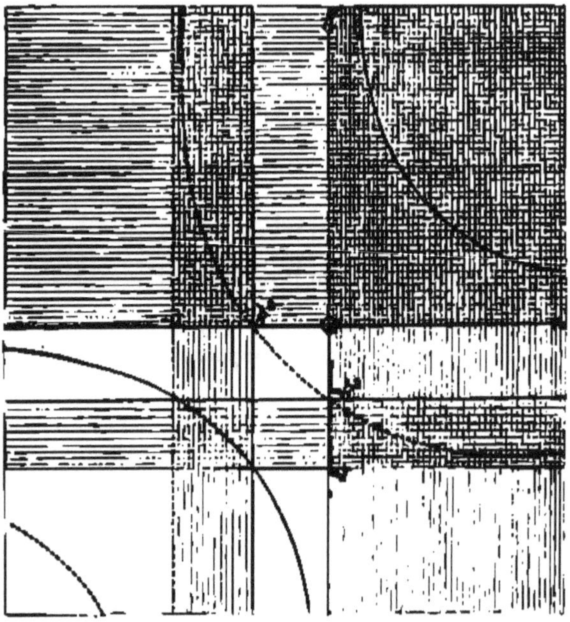

for which n is circular and by horizontal lines those for which n' is circular: the two curves $nn' = k^2$ and $(1+n)(1+n') = k'^2$, are then hyperbolas lying wholly in the spaces which are either cross-shaded or else white, viz. the corresponding values n, n' are both circular or both logarithmic.

99. In the formula (f), taking $n = -\dfrac{1}{\sin^2\theta}$ we have $n' = -k^2\sin^2\theta$, and thence, substituting for a its value,

$$\Pi\left(-\frac{1}{\sin^2\theta}\right) + \Pi\left(-k^2\sin^2\theta\right)$$

$$= F + \frac{1}{2\cot\theta\,\Delta(k,\theta)}\log\frac{\Delta + \cot\theta\,\Delta(k,\theta)\tan\phi}{\Delta - \cot\theta\,\Delta(k,\theta)\tan\phi},$$

or as this is better written

$$= F + \frac{1}{2\cot\theta\,\Delta(k,\theta)}\log\frac{\cot\phi\,\Delta(k,\phi) + \cot\theta\,\Delta(k,\theta)}{\cot\phi\,\Delta(k,\phi) - \cot\theta\,\Delta(k,\theta)}.$$

This equation shows that a logarithmic function of parameter which is negative and in absolute magnitude greater than 1, may be reduced to depend on a like function where the parameter is negative and in absolute magnitude less than k^2. The first-mentioned kind of logarithmic functions presents the difficulty that the function under the integral sign becomes infinite in the course of the integration $\left(\text{viz. for the real value } \sin^2\phi = -\dfrac{1}{n}\right)$: we therefore always consider the reduction as made, and attend only to the case where the parameter is of the form $-k^2\sin^2\theta$.

100. The formula (f) gives also a relation between two circular functions of positive parameter, viz. writing therein $n = \cot^2\theta$ we have $n' = k^2\tan^2\theta$. And the relation is

$$\Pi(\cot^2\theta) + \Pi(k^2\tan^2\theta) = F + \frac{\sin\theta\cos\theta}{\Delta(k',\theta)}\tan^{-1}\frac{\Delta(k',\theta)\tan\phi}{\Delta(k,\phi)\sin\theta\cos\theta},$$

which in fact serves to reduce a circular function of positive parameter greater than k to a like function of parameter less than k: but the original form

$$\Pi n + \Pi\left(\frac{k^2}{n}\right) = F + \frac{1}{\sqrt{a}}\tan^{-1}\frac{\sqrt{a}\tan\phi}{\Delta},$$

is for this purpose equally if not more convenient.

161. The formula (g') gives in like manner a relation between two logarithmic functions, or two circular functions: as regards the first case observe that if n, n' are both negative they are both in absolute magnitude greater than 1, (viz. $1+n$, $1+n'$ are each negative); and we have thus a relation between two logarithmic functions with parameters of this form; but such functions being excluded from consideration, the formula is not written down. There remains the case where the parameters (being by supposition logarithmic) are each negative and in absolute magnitude less than k^2: viz. writing $n = -k^2\sin^2\theta$, $n' = -k^2\sin^2\lambda$, the relation between the parameters is $(1-k^2\sin^2\theta)(1-k^2\sin^2\lambda) = k'^2$, or what is the same thing $(\cos^2\theta + k'^2\sin^2\theta)(\cos^2\lambda + k'^2\sin^2\lambda) = k'^2$, or as this may be written $(1+k'^2\tan^2\theta)(1+k'^2\tan^2\lambda) = k'^2(1+\tan^2\theta)(1+\tan^2\lambda)$, whence finally the relation is $1 = k'\tan\lambda\tan\theta$, (answering it will be observed to the transcendental relation $F\theta + F\lambda = F_1$).

We then have

$$\frac{1+n}{n} = \frac{1+k'^2\tan^2\theta}{-k^2\tan^2\theta}, = \frac{k'^2}{k^2}(-1-\tan^2\lambda), = -\frac{k'^2}{k^2\cos^2\lambda},$$

and completing the substitution, the formula becomes

$$\cos^2\theta\,\Pi(-k^2\sin^2\theta) + \cos^2\lambda\,\Pi(-k^2\sin^2\lambda)$$

$$= F + \tfrac{1}{2}\sin\theta\sin\lambda\log\left(\frac{\Delta\phi - k^2\sin\theta\sin\lambda\sin\phi\cos\phi}{\Delta\phi + k^2\sin\theta\sin\lambda\sin\phi\cos\phi}\right),$$

where as above $1 = k'\tan\theta\tan\lambda$. The formula enables the reduction of a logarithmic function of parameter $-k^2\sin^2\theta$ in absolute magnitude greater than $(1-k')$ $\left(\text{or for which } \tan\theta > \frac{1}{\sqrt{k'}}\right)$ to a like function of parameter in absolute magnitude less than $(1-k')$ $\left(\text{or for which } \tan\theta < \frac{1}{\sqrt{k'}}\right)$. But it is convenient, *not* using the formula, and therefore without thus restricting the value of θ, to retain $-k^2\sin^2\theta$ as the expression for the parameter.

162. In the same formula (g) if the parameters are both circular they may be taken to be $n = \cot^2\theta$ and $n' = -1 + k'^2 \sin^2\theta$; and the formula becomes

$$\Pi(\cot^2\theta) - \frac{k'^2 \sin^2\theta \cos^2\theta}{1 - k'^2 \sin^2\theta}\, \Pi(-1 + k'^2 \sin^2\theta)$$
$$= \frac{k^2 \sin^2\theta}{1 - k'^2 \sin^2\theta} F + \frac{\sin\theta\cos\theta}{\Delta(k', \theta)} \tan^{-1} \frac{\sin\phi\cos\phi\,\Delta(k', \theta)}{\tan\theta\,\Delta(k, \phi)},$$

which is a formula for the reduction of a circular function of positive parameter $\cot^2\theta$ to a circular function of negative parameter $-1 + k'^2 \sin^2\theta$.

163. The above formula

$$\cos^2\theta\,\Pi(-k^2 \sin^2\theta) + \cos^2\lambda\,\Pi(-k^2 \sin^2\lambda)$$
$$= F + \tfrac{1}{2} \sin\theta \sin\lambda \log \left(\frac{\Delta\phi - k^2 \sin\theta \sin\lambda \sin\phi \cos\phi}{\Delta\phi + k^2 \sin\theta \sin\lambda \sin\phi \cos\phi}\right)$$

may be written under a slightly different form: viz. expressing it first in the form

$$\cos^2\theta\,[\Pi(-k^2 \sin^2\theta) - F] + \cos^2\lambda\,[\Pi(-k^2 \sin^2\lambda) - F]$$
$$= (1 - \cos^2\theta - \cos^2\lambda)\, F + \tfrac{1}{2} \sin\theta \sin\lambda \log \Omega,$$

and dividing the whole by $\sin\theta \sin\lambda$; then reducing the coefficients of the several terms by means of the relation $1 = k' \tan\lambda \tan\theta$, and finally restoring the value of Ω under a slightly altered form the equation becomes

$$\frac{\cos\theta\Delta\theta}{\sin\theta}\,[\Pi(-k^2 \sin^2\theta) - F] + \frac{\cos\lambda\Delta\lambda}{\sin\lambda}\,[\Pi(-k^2 \sin^2\lambda) - F]$$
$$= \frac{k^2}{k'} \cos\lambda \cos\theta \,.\, F + \tfrac{1}{2} \log \left(\frac{k'\Delta\phi - k^2 \cos\lambda \cos\theta \sin\phi \cos\phi}{k'\Delta\phi + k^2 \cos\lambda \cos\theta \sin\phi \cos\phi}\right),$$

where as before $1 = k' \tan\lambda \tan\theta$.

164. If to fix the ideas we consider herein θ, λ as positive and less than $\frac{1}{2}\pi$, then writing $\theta' = \pi - \lambda$, the relation between θ, θ' will be $k' \tan\theta \tan\theta' = -1$, ($\theta$ and θ' each positive but

$\theta < \frac{1}{2}\pi$, $\theta' > \frac{1}{2}\pi$). Substituting for λ its value $\pi - \theta'$ the formula becomes

$$\frac{\cos\theta\Delta\theta}{\sin\theta}\left[\Pi\left(-k^2\sin^2\theta\right) - F\right] - \frac{\cos\theta'\Delta\theta'}{\sin\theta'}\left[\Pi\left(-k^2\sin^2\theta'\right) - F\right]$$

$$= -\frac{k^2}{k'}\cos\theta\cos\theta' \,.\, F + \tfrac{1}{2}\log\frac{k'\Delta\phi + k^2\cos\theta\cos\theta'\sin\phi\cos\phi}{k'\Delta\phi - k^2\cos\theta\cos\theta'\sin\phi\cos\phi},$$

which is a form used in the sequel.

The general Case resumed.

165. Returning now to the general equation

$$A\Pi n + A'\Pi n' + B\Pi m + \frac{1}{\zeta}F = \int\frac{d\varpi}{1+\rho\varpi^2},$$

write
$$n = -k^2\sin^2 p,$$
$$n' = -k^2\sin^2 q,$$
$$m = -k^2\sin^2\theta;$$

then introducing these values we have

$$\zeta = -k^2\sin p\sin q\sin\theta,$$
$$\rho = -\frac{k^4}{k'^2}\cos^2 p\cos^2 q\cos^2\theta,$$
$$k'^2(1+\zeta)^2 = (1-k^2\sin^2 p)(1-k^2\sin^2 q)(1-k^2\sin^2\theta);$$

or writing this last under the form

$$k'(1+\zeta) = \Delta p\,\Delta q\,\Delta\theta,$$

we have $\quad k'(1 - k^2\sin p\sin q\sin\theta) = \Delta p\,\Delta q\,\Delta\theta$

as the relation between the parametric angles p, q, θ. This is in fact equivalent to the transcendental equation

$$Fp + Fq - F\theta - F_1 = 0,$$

and it suggests the introduction into the formulæ in place of θ, of a new angle θ', such that $F\theta + F_1 = F\theta'$ and consequently

$$Fp + Fq - F\theta' = 0.$$

166. But let us first express B in terms of the original angles p, q, θ. We have

$$B=\frac{m^3+(2+\zeta)\,m^2+(1+2\zeta)\,k^2m+\zeta k^2}{m\,(m-n)\,(m-n')},$$

the numerator is

$$\begin{aligned}
&-k^6\sin^6\theta\\
&+(2-k^2\sin\theta\sin p\sin q)\,k^4\sin^4\theta\\
&+(1-2k^2\sin\theta\sin p\sin q)\,.-k^4\sin^2\theta\\
&-k^4\sin\theta\sin p\sin q,\\
=&-k^4\sin\theta\,[\sin\theta\,(1-2\sin^2\theta+k^2\sin^4\theta)\\
&\qquad\qquad+\sin p\sin q\,(1-2k^2\sin^2\theta+k^2\sin^4\theta)],\\
=&-k^4\sin\theta\,[(\sin\theta+\sin p\sin q)\cos^2\theta\,\Delta^2\theta\\
&\qquad\qquad-k'^2\sin^2\theta\,(\sin\theta-\sin p\sin q)],
\end{aligned}$$

and the denominator is

$$-k^6\sin^2\theta\,(\sin^2\theta-\sin^2p)\,(\sin^2\theta-\sin^2q),$$

whence

$$B=\frac{\cos^2\theta\,\Delta^2\theta\,(\sin\theta+\sin p\sin q)-k'^2\sin^2\theta\,(\sin\theta-\sin p\sin q)}{k^2\sin\theta\,(\sin^2\theta-\sin^2p)\,(\sin^2\theta-\sin^2q)}.$$

167. The relation between θ, θ' may be written under the forms

$$\begin{array}{r|l}
\sin\theta=-\dfrac{\cos\theta'}{\Delta\theta'}, & \sin\theta'=\dfrac{\cos\theta}{\Delta\theta},\\[2ex]
\cos\theta=\dfrac{k'\sin\theta'}{\Delta\theta'}, & \cos\theta'=-\dfrac{k'\sin\theta}{\Delta\theta},\\[2ex]
\Delta\theta=\dfrac{k'}{\Delta\theta'}, & \Delta\theta'=\dfrac{k'}{\Delta\theta}.
\end{array}$$

Hence in the last-mentioned expression of B, the numerator is

$$\frac{k'^4\sin^2\theta'}{\Delta^4\theta'}\left(-\frac{\cos\theta'}{\Delta\theta'}+\sin p\sin q\right)+\frac{k'^2\cos^2\theta'}{\Delta^2\theta'}\left(\frac{\cos\theta'}{\Delta\theta'}+\sin p\sin q\right),$$

which is

$$\begin{aligned}
=\frac{k'^2}{\Delta^5\theta'}\,[&k'^2\sin^2\theta'\,(-\cos\theta'+\sin p\sin q\,\Delta\theta')\\
&+\cos^2\theta'\,\Delta^2\theta'\,(\cos\theta'+\sin p\sin q\,\Delta\theta')].
\end{aligned}$$

But in virtue of the relation between p, q, θ' we have

$$\cos\theta' = \cos p\cos q - \sin p\sin q\,\Delta\theta',$$

or the numerator is

$$= \frac{k'^2}{\Delta^2\theta'}\left[k'^2\sin^2\theta'\,(\cos p\cos q - 2\cos\theta') + \cos^2\theta'\,\Delta^2\theta' \,.\, \cos p\cos q\right],$$

say this is

$$= \frac{k'^2}{\Delta^2\theta'}\,\Omega.$$

Then we have

$$\Omega\cos p\cos q = (\cos^2\theta'\,\Delta^2\theta' + k'^2\sin^2\theta')\cos^2 p\cos^2 q$$
$$- k'^2\sin^2\theta' \,.\, 2\cos p\cos q\cos\theta'.$$

But

$$1 - \cos^2 p - \cos^2 q - \cos^2\theta' - k^2\sin^2 p\sin^2 q\sin^2\theta'$$
$$= -2\cos p\cos q\cos\theta',$$

say $\qquad R\sin^2\theta' - \cos^2 p - \cos^2 q = -2\cos p\cos q\cos\theta',$

where $\qquad R = 1 - k^2\sin^2 p\sin^2 q.$

Hence

$$\Omega\cos p\cos q = (\cos^2\theta'\,\Delta^2\theta' + k'^2\sin^2\theta')\cos^2 p\cos^2 q$$
$$+ k'^2\sin^2\theta'\,(R\sin^2\theta' - \cos^2 p - \cos^2 q),$$
$$= (1 - 2k^2\sin^2\theta' + k^2\sin^4\theta')\cos^2 p\cos^2 q$$
$$+ k'^2\sin^2\theta'\,(R\sin^2\theta' - \cos^2 p - \cos^2 q),$$
$$= \cos^2 p\cos^2 q$$
$$+ \sin^2\theta'\left[-2k^2\cos^2 p\cos^2 q - k'^2(\cos^2 p + \cos^2 q)\right]$$
$$+ \sin^4\theta'\left[k^2\cos^2 p\cos^2 q + k'^2(1 - k^2\sin^2 p\sin^2 q)\right],$$

which is

$$= \cos^2 p\cos^2 q$$
$$- \sin^2\theta'\,(\cos^2 p\,\Delta^2 q + \cos^2 q\,\Delta^2 p)$$
$$+ \sin^4\theta'\,\Delta^2 p\,\Delta^2 q,$$
$$= (\cos^2 p - \sin^2\theta'\,\Delta^2 p)(\cos^2 q - \sin^2\theta'\,\Delta^2 q),$$

so that numerator is

$$= \frac{k'^2}{\Delta^2\theta'}\,\frac{1}{\cos p\cos q}\,(\cos^2 p - \sin^2\theta'\,\Delta^2 p)(\cos^2 q - \sin^2\theta'\,\Delta^2 q).$$

108. Denominator is

$$-\frac{k^2\cos\theta'}{\Delta\theta'}\left(\sin^2 p-\frac{\cos^2\theta'}{\Delta^2\theta'}\right)\left(\sin^2 q-\frac{\cos^2\theta'}{\Delta^2\theta'}\right),$$

which is

$$=-\frac{k^2\cos\theta'}{\Delta^5\theta'}(\cos^2\theta'-\sin^2 p\,\Delta^2\theta')(\cos^2\theta'-\sin^2 q\,\Delta^2\theta'),$$

$$=-\frac{k^2\cos\theta'}{\Delta^5\theta'}(\cos^2 p-\sin^2\theta'\Delta^2 p)(\cos^2 q-\sin^2\theta'\Delta^2 q);$$

whence

$$B=\frac{-k'^2}{k^2\cos p\cos q\cos\theta'}.$$

Write

$$M=\frac{-\Delta\theta'}{k^2\sin\theta'\cos p\cos q}\left(=\frac{-k'}{k^2\cos p\cos q\cos\theta}\right),$$

then

$$B=M\frac{k'^2\sin\theta'}{\cos\theta'\Delta\theta'}\left(=-M\frac{\cos\theta\Delta\theta}{\sin\theta}\right);$$

and similarly

$$A=M\frac{\cos p\Delta p}{\sin p},$$

$$A'=M\frac{\cos q\Delta q}{\sin q}.$$

109. The equation is

$$A\Pi n+A'\Pi n'+B\Pi m+\frac{1}{\zeta}F=\int\frac{d\varpi}{1+\rho\varpi^2},$$

or since

$$A+A'+B=1-\frac{1}{\zeta},^*$$

* We have

$$\zeta=-k^2\sin p\sin q\sin\theta,$$

$$=k^2\sin p\sin q\frac{\cos\theta'}{\Delta\theta'},$$

and hence this equation is

$$\frac{-\Delta\theta'}{k^2\sin\theta'\cos p\cos q}\left\{\frac{\cos p\Delta p}{\sin p}+\frac{\cos q\Delta q}{\sin q}+\frac{k'^2\sin\theta'}{\cos\theta'\Delta\theta'}\right\}$$

$$=1-\frac{\Delta\theta'}{k^2\sin p\sin q\cos\theta'},$$

an identity which may be verified.

this is

$$A(\Pi n - F) + A'(\Pi n' - F) + B(\Pi m - F) + F = \int \frac{d\varpi}{1+\rho\varpi^2},$$

that is

$$\frac{\cos p \Delta p}{\sin p}(\Pi p - F) + \frac{\cos q \Delta q}{\sin q}(\Pi q - F) - \frac{\cos\theta\Delta\theta}{\sin\theta}(\Pi\theta - F)$$

$$+ \frac{1}{M}F = \frac{1}{M}\int\frac{d\varpi}{1+\rho\varpi^2},$$

where Πp, &c. are written in place of $\Pi(-k^2\sin^2 p)$, &c. We have

$$M = \frac{-k'}{k^2\cos p\cos q\cos\theta}, \quad \rho = -\frac{k^4}{k'^2}\cos^2 p\cos^2 q\cos^2\theta,$$

hence

$$\frac{1}{M}\int\frac{d\varpi}{1+\rho\varpi^2} = \frac{1}{2}\,\frac{k'}{Mk^2\cos p\cos q\cos\theta}\left(=-\tfrac{1}{2}\right)\times$$

$$\log\frac{1+\dfrac{k^2\cos p\cos q\cos\theta}{k'}\,\dfrac{\sin\phi\cos\phi}{(1+\zeta\sin^2\phi)\,\Delta}}{1-\dfrac{k^2\cos p\cos q\cos\theta}{k'}\,\dfrac{\sin\phi\cos\phi}{(1+\zeta\sin^2\phi)\,\Delta}},$$

and the formula thus becomes

$$\frac{\cos p\Delta p}{\sin p}(\Pi p - F) + \frac{\cos q\Delta q}{\sin q}(\Pi q - F) - \frac{\cos\theta\Delta\theta}{\sin\theta}(\Pi\theta - F)$$

$$= \frac{k^2}{k'}\cos p\cos q\cos\theta\,.\,F$$

$$+ \tfrac{1}{2}\log\frac{k'(1+\zeta\sin^2\phi)\,\Delta\phi - k^2\cos p\cos q\cos\theta\sin\phi\cos\phi}{k'(1+\zeta\sin^2\phi)\,\Delta\phi + k^2\cos p\cos q\cos\theta\sin\phi\cos\phi}.$$

170. Representing, for convenience, the logarithmic term by $\frac{1}{2}\log\Omega$, so that

$$\Omega = \frac{k'(1-k^2\sin p\sin q\sin\theta\sin^2\phi)\,\Delta\phi - k^2\cos p\cos q\cos\theta\sin\phi\cos\phi}{k'(1-k^2\sin p\sin q\sin\theta\sin^2\phi)\,\Delta\phi + k^2\cos p\cos q\cos\theta\sin\phi\cos\phi},$$

$$= \frac{P-Q}{P+Q} \text{ suppose,}$$

we have, *ante* No. 163, writing θ' for λ, and $-\theta$ for θ (thereby passing from the relation $F_1 = F\lambda + F\theta$ to the actual relation $F_1 = F\theta' - F\theta$),

$$\frac{\cos\theta\Delta\theta}{\sin\theta}(\Pi\theta - F) - \frac{\cos\theta'\Delta\theta'}{\sin\theta'}(\Pi\theta' - F) = -\frac{k^2}{k'}\cos\theta\cos\theta' . F$$

$$+\tfrac{1}{2}\log\frac{k'\Delta\phi + k^2\cos\theta'\cos\theta\sin\phi\cos\phi}{k'\Delta\phi - k^2\cos\theta'\cos\theta\sin\phi\cos\phi}\left(= \tfrac{1}{2}\log\frac{R+S}{R-S}\text{ suppose}\right);$$

and using this to introduce into the formula

$$\frac{\cos\theta'\Delta\theta'}{\sin\theta'}(\Pi\theta' - F)$$

in place of

$$\frac{\cos\theta\Delta\theta}{\sin\theta}(\Pi\theta - F),$$

the formula thus becomes

$$\frac{\cos p\Delta p}{\sin p}(\Pi p - F) + \frac{\cos q\Delta q}{\sin q}(\Pi q - F) - \frac{\cos\theta'\Delta\theta'}{\sin\theta'}(\Pi\theta' - F)$$

$$= \frac{k^2}{k'}(\cos p\cos q - \cos\theta')\cos\theta . F + \tfrac{1}{2}\log\frac{(P-Q)(R+S)}{(P+Q)(R-S)},$$

where the coefficient of F on the right-hand side is

$$\frac{k^2}{k'}\sin p\sin q\,\Delta\theta'\cos\theta, = k^2\sin p\sin q\sin\theta' .$$

171. Introducing into the logarithmic terms θ' instead of θ, we have

$$P - Q = k'\left(1 + k^2\frac{\sin p\sin q\cos\theta'}{\Delta\theta'}\sin^2\phi\right)\Delta\phi$$

$$-\frac{k^2k'}{\Delta\theta'}\sin\theta'\cos p\cos q\sin\phi\cos\phi,$$

or, multiplying by $\Delta\theta'$ and omitting the factor k', say

$$P - Q = (\Delta\theta' + k^2\sin p\sin q\cos\theta'\sin^2\phi)\Delta\phi - k^2\sin\theta'\cos p\cos q\sin\phi\cos\phi,$$

$$= [\Delta\theta'\Delta\phi - k^2\sin\theta'\cos\theta'\sin\phi\cos\phi]$$

$$+ k^2\sin\theta'\sin\phi\cos\phi\,[(\cos\theta' - \cos p\cos\theta), = -\sin p\sin q\,\Delta\theta']$$

$$+ k^2\sin p\sin q\cos\theta'\sin^2\phi\Delta\phi,$$

$$= \Delta\theta'\Delta\phi - k^2\sin\theta'\cos\theta'\sin\phi\cos\phi$$

$$+ k^2\sin p\sin q\sin\phi\,[\cos\theta'\sin\phi\Delta\phi - \cos\phi\sin\theta'\Delta\theta'];$$

and similarly

$$P+Q= \Delta\theta'\Delta\phi + k^2 \sin\theta' \cos\theta' \sin\phi\cos\phi$$
$$+ k^2 \sin p \sin q \sin\phi\, (\cos\theta' \sin\phi\Delta\phi + \cos\phi \sin\theta'\Delta\theta').$$

Also

$$R+S=k'\Delta\phi + \frac{k^2k' \sin\theta' \cos\theta' \sin\phi\cos\phi}{\Delta\theta'},$$

or, multiplying by $\Delta\theta'$ and omitting the factor k', say

$$R+S=\Delta\theta'\Delta\phi + k^2 \sin\theta' \cos\theta' \sin\phi \cos\phi;$$

and similarly,

$$R-S=\Delta\theta'\Delta\phi - k^2 \sin\theta' \cos\theta' \sin\phi\cos\phi.$$

172. Write for shortness

$$\Delta\theta'\Delta\phi - k^2 \sin\theta' \cos\theta' \sin\phi\cos\phi = A,$$
$$\Delta\theta'\Delta\phi + k^2 \sin\theta' \cos\theta' \sin\phi\cos\phi = A',$$
$$\cos\theta' \sin\phi\Delta\phi - \cos\phi \sin\theta'\Delta\theta' = B,$$
$$\cos\theta' \sin\phi\Delta\phi + \cos\phi \sin\theta'\Delta\theta' = B',$$

then the logarithmic term is at once expressible in terms of these quantities, and substituting in the formula, we have

$$\frac{\cos p\Delta p}{\sin p}(\Pi p - F) + \frac{\cos q\Delta q}{\sin q}(\Pi q - F) - \frac{\cos\theta'\Delta\theta'}{\sin\theta'}(\Pi\theta' - F)$$
$$= k^2 \sin p \sin q \sin\theta' \,.\, F + \tfrac{1}{2}\log\frac{[A + k^2 \sin p \sin q \sin\phi \,.\, B]\, A'}{[A' + k^2 \sin p \sin q \sin\phi \,.\, B']\, A},$$

which is in fact the formula connecting the three functions Πp, Πq, $\Pi\theta'$, or in the original notation

$$\Pi(-k^2 \sin^2 p),\ \Pi(-k^2 \sin^2 q),\ \Pi(-k^2 \sin^2\theta');$$

the angles p, q, θ' being, it will be remembered, connected by the algebraical equivalent of the equation

$$Fp + Fq - F\theta' = 0.$$

173. This apparently complicated formula is wonderfully simplified by introducing into it Jacobi's notation; viz. writing $\sin p = \operatorname{sn} a$, $\sin q = \operatorname{sn} b$; and therefore $\sin\theta' = \operatorname{sn}(a+b)$; also $\sin\phi = \operatorname{sn} u$; then omitting a common factor $1 - k^2 \sin^2\theta' \sin^2\phi$, we have

$$A = \operatorname{dn}(a+b+u),$$
$$A' = \operatorname{dn}(a+b-u),$$
$$-B = \operatorname{sn}(a+b-u)\operatorname{dn}(a+b+u),$$
$$B' = \operatorname{sn}(a+b+u)\operatorname{dn}(a+b-u),$$

and the formula becomes

$$\Pi(u, a) + \Pi(u, b) - \Pi(u, a+b) = k^2 \operatorname{sn} a \operatorname{sn} b \operatorname{sn}(a+b) \, . \, u$$
$$+ \tfrac{1}{2} \log \frac{1 - k^2 \operatorname{sn} a \operatorname{sn} b \operatorname{sn}(a+b-u) \operatorname{sn} u}{1 + k^2 \operatorname{sn} a \operatorname{sn} b \operatorname{sn}(a+b+u) \operatorname{sn} u},$$

viz. it is in fact a formula for the addition of the parameters.

Interchange of Amplitude and Parameter. Art. Nos. 174 to 180.

174. Starting with

$$\Pi = \int_0 \frac{d\phi}{(1 + n \sin^2\phi)\,\Delta},$$

and taking throughout the integrals in regard to ϕ from this inferior limit 0, we have

$$n \frac{d\Pi}{dn} = \int \frac{-n \sin^2\phi \, d\phi}{(1 + n \sin^2\phi)^2\,\Delta},$$
$$= \int \frac{d\phi}{(1 + n \sin^2\phi)^2\,\Delta} - \Pi.$$

But writing $\alpha = (1+n)\left(1 + \frac{k^2}{n}\right)$ we have

$$\frac{2\alpha}{n} \int \frac{d\phi}{(1 + n \sin^2\phi)^2\,\Delta} = \frac{\Delta \sin\phi \cos\phi}{1 + n \sin^2\phi} - \frac{k^2}{n^2} \int (1 + n \sin^2\phi) \frac{d\phi}{\Delta}$$
$$+ \left(1 + \frac{2 + 2k^2}{n} + \frac{3k^2}{n^2}\right) \int \frac{d\phi}{(1 + n \sin^2\phi)\,\Delta},$$

as may be verified by differentiation: or since

$$k^2 \int (1 + n \sin^2\phi) \frac{d\phi}{\Delta} = (k^2 + n) F - nE,$$
$$1 + \frac{2 + 2k^2}{n} + \frac{3k^2}{n^2} = \frac{1}{n}\left(2\alpha - n + \frac{k^2}{n}\right) = \frac{1}{n}\left(2\alpha - n \frac{d\alpha}{dn}\right),$$

this is

$$\frac{2\alpha}{n} \int \frac{d\phi}{(1 + n \sin^2\phi)^2\,\Delta} = \frac{\Delta \sin\phi \cos\phi}{1 + n \sin^2\phi} - \frac{k^2}{n^2} F - \frac{1}{n}(F - E)$$
$$+ \frac{1}{n}\left(2\alpha - n \frac{d\alpha}{dn}\right) \Pi,$$

viz. we have

$$\frac{2\alpha}{n} \left\{ \int \frac{d\phi}{(1 + n \sin^2\phi)^2\,\Delta} - \Pi \right\} = \frac{\Delta \sin\phi \cos\phi}{1 + n \sin^2\phi} - \frac{k^2}{n^2} F$$
$$- \frac{1}{n}(F - E) - \frac{d\alpha}{dn} \Pi.$$

175. This equation may be written

$$2a\frac{d\Pi}{dn} = \frac{\Delta\sin\phi\cos\phi}{1+n\sin^2\phi} - \frac{k^2}{n^2}F - \frac{1}{n}(F-E) - \Pi\frac{da}{dn},$$

or, what is the same thing,

$$2a\,d\Pi + \Pi\,da = \Delta\sin\phi\cos\phi\,\frac{dn}{1+n\sin^2\phi} - k^2F\frac{dn}{n^2} - (F-E)\frac{dn}{n},$$

viz. multiplying each side by $\frac{1}{2}a^{-\frac{1}{2}}$, this is

$$d\,.\,\Pi\sqrt{a} = \tfrac{1}{2}\Delta\sin\phi\cos\phi\,\frac{dn}{(1+n\sin^2\phi)\sqrt{a}} - \tfrac{1}{2}k^2F\frac{dn}{n^2\sqrt{a}} - \tfrac{1}{2}(F-E)\frac{dn}{n\sqrt{a}},$$

where of course $a, = (1+n)\left(1+\frac{k^2}{n}\right)$, is regarded as a function of n.

Integrating each side we have

$$\Pi\sqrt{a} = C + \tfrac{1}{2}\Delta\sin\phi\cos\phi\int\frac{dn}{(1+n\sin^2\phi)\sqrt{a}} - \tfrac{1}{2}k^2F\int\frac{dn}{n^2\sqrt{a}} - \tfrac{1}{2}(F-E)\int\frac{dn}{n\sqrt{a}},$$

where the constant of integration may of course be a function of k, ϕ, but it is independent of n.

The formula is simplified by representing the parameter n under any one of the foregoing forms $\cot^2\theta$, $-1+k'^2\sin^2\theta$, $-k^2\sin^2\theta$. The last is the most interesting case, but it is proper to consider them all three.

First case, $n = \cot^2\theta$.

176. Here

$$dn = -\frac{2\cos\theta}{\sin^3\theta}d\theta,\quad \sqrt{a} = \frac{\Delta(k',\theta)}{\sin\theta\cos\theta},$$

and the equation becomes

$$\frac{\Delta(k',\theta)}{\sin\theta\cos\theta}\Pi = C + k^2F\int\frac{\sin^2\theta\,d\theta}{\cos^2\theta\,\Delta(k',\theta)} + (F-E)\int\frac{d\theta}{\Delta(k',\theta)} - \Delta\sin\phi\cos\phi\int\frac{\cos^2\theta\,d\theta}{(\sin^2\theta+\sin^2\phi\cos^2\theta)\,\Delta(k',\theta)}.$$

where the integrals in regard to θ may be taken from the inferior limit 0.

The integral

$$\int \frac{d\theta}{\Delta(k', \theta)} \text{ is } = F(k', \theta)$$

and we have

$$k^2 \int \frac{d\theta \sin^2\theta}{\cos^2\theta\, \Delta(k', \theta)} = \frac{\sin\theta}{\cos\theta} \Delta(k', \theta) - E(k', \theta).$$

Moreover, writing $\cot^2\phi = n'$ we have

$$\Delta \sin\phi \cos\phi \int \frac{\cos^2\theta\, d\theta}{(\sin^2\theta + \sin^2\phi \cos^2\theta)\, \Delta(k', \theta)}$$

$$= -\frac{\Delta \sin\phi}{\cos\phi} \int \frac{d\theta}{\Delta(k', \theta)} + \frac{\Delta}{\sin\phi \cos\phi} \int \frac{d\theta}{(1 + n' \sin^2\theta)\, \Delta(k', \theta)},$$

$$= -\frac{\Delta \sin\phi}{\cos\phi} F(k', \theta) + \frac{\Delta}{\sin\phi \cos\phi} \Pi(n', k', \theta).$$

Substituting these values and for greater clearness writing $\Pi(n, k, \phi)$, $\Delta(k, \phi)$, $F(k, \phi)$, $E(k, \phi)$ instead of Π, Δ, F, E, putting also for C its value $= \frac{1}{2}\pi$, determined as presently mentioned, the formula is, ($n = \cot^2\theta$, $n' = \cot^2\phi$)

$$\frac{\Delta(k', \theta)}{\sin\theta \cos\theta} \Pi(n, k, \phi) + \frac{\Delta(k, \phi)}{\sin\phi \cos\phi} \Pi(n', k', \theta)$$

(1) *Leg.* p. 133.

$$= \tfrac{1}{2}\pi + \frac{\sin\theta}{\cos\theta} \Delta(k', \theta) F(k, \phi) + \frac{\sin\phi}{\cos\phi} \Delta(k, \phi) F(k', \theta)$$

$$+ F(k, \phi) F(k', \theta) - F(k, \phi) E(k', \theta) - E(k, \phi) F(k', \theta).$$

177. If in the formula, instead of $\frac{1}{2}\pi$, the term had been C, then C is independent of θ, and by the symmetry of the formula it must be independent also of ϕ: it is thus an absolute constant: to determine its value take θ, ϕ each indefinitely small: then

$$F(k, \phi) = E(k, \phi) = \phi, \quad F(k', \theta) = E(k', \theta) = \theta,$$

$$\Pi(n, k, \phi) = \int \frac{d\phi}{1 + n \sin^2\phi} = \frac{1}{\sqrt{n}} \tan^{-1}\phi\sqrt{n} = \theta \tan^{-1}\frac{\phi}{\theta},$$

and similarly

$$\Pi(n', k', \theta) = \phi \tan^{-1}\frac{\theta}{\phi}:$$

we have therefore

$$C = \tan^{-1}\frac{\phi}{\theta} + \tan^{-1}\frac{\theta}{\phi}, = \tfrac{1}{2}\pi,$$

which is thus the value of C.

178. Write $\phi = \frac{1}{2}\pi - \omega$, ω being indefinitely small: then $n' = \cot^2\phi = \tan^2\omega = \omega^2$ is indefinitely small, and therefore

$$\Pi(n', k', \theta) = \int\frac{d\theta}{(1 + n'\sin^2\theta)\,\Delta(k', \theta)} = \int\frac{d\theta}{\Delta(k', \theta)} - n'\int\frac{\sin^2\theta\,d\theta}{\Delta(k', \theta)},$$

$$= F(k', \theta) - n'\int\frac{\sin^2\theta\,d\theta}{\Delta(k', \theta)},$$

and thence

$$\frac{\Delta(k, \phi)}{\sin\phi\cos\phi}\Pi(n', k', \theta) - \frac{\sin\phi\Delta(k, \phi)}{\cos\phi}F(k', \theta)$$

$$= \frac{\Delta(k, \phi)}{\cos\phi}\left(\frac{1}{\sin\phi} - \sin\phi\right)F(k', \theta) - \frac{n'}{\cos\phi}\,\frac{\Delta(k, \phi)}{\sin\phi}\int\frac{\sin^2\theta\,d\theta}{\Delta(k', \theta)}.$$

The coefficients on the right-hand side are

$$\frac{1}{\cos\phi}\left(\frac{1}{\sin\phi} - \sin\phi\right), = \frac{\cos\phi}{\sin\phi} \text{ and } \frac{n'}{\cos\phi}, = \frac{\omega^2}{\cos\phi}:$$

viz. writing $\cos\phi = \omega$, these each contain the factor ω, and the function on the left-hand side is thus $= 0$; hence the equation becomes

$$\frac{\Delta(k', \theta)}{\sin\theta\cos\theta}\Pi_1(n, k) = \tfrac{1}{2}\pi + \frac{\sin\theta}{\cos\theta}F_1k\,\Delta(k', \theta) \quad (k')\,\textit{Leg.}\ \text{p. 134.}$$

$$+ F_1kF(k', \theta) - F_1kE(k', \theta) - E_1kF(k', \theta),$$

viz. we have thus an expression for the complete function $\Pi_1(n, k)$ in terms of the functions of the first and second kinds.

179. If in this equation we write $\frac{k^2}{n}$ in place of n, and

assume also $\frac{k^2}{n} = \cot^2\lambda$, so as to write λ in place of θ, we have

$$\frac{\Delta(k', \lambda)}{\sin\lambda\cos\lambda}\Pi_1\left(\frac{k^2}{n}, k\right) = \tfrac{1}{2}\pi + \frac{\sin\lambda}{\cos\lambda}F_1k\,\Delta(k', \lambda)$$
$$+ F_1kF(k', \lambda) - F_1kE(k', \lambda) - E_1kF(k', \lambda).$$

180. Adding the two equations together, we have

$$\frac{\Delta(k', \theta)}{\sin\theta\cos\theta}\Pi_1(n, k) + \frac{\Delta(k', \lambda)}{\sin\lambda\cos\lambda}\Pi_1\left(\frac{k^2}{n}, k\right)$$
$$= \pi + F_1k\left\{\frac{\sin\theta}{\cos\theta}\Delta(k', \theta) + \frac{\sin\lambda}{\cos\lambda}\Delta(k', \lambda)\right\}$$
$$+ F_1k\{F(k', \theta) + F(k', \lambda) - E(k', \theta) - E(k', \lambda)\}$$
$$- E_1k\{F(k', \theta) + F(k', \lambda)\}.$$

But in virtue of $\cot^2\theta\cot^2\lambda = k^2$, or $k\tan\theta\tan\lambda = 1$, we have

$$F(k', \theta) + F(k', \lambda) = F_1k',$$
$$E(k', \theta) + E(k', \lambda) = E_1k' + k'^2\sin\lambda\sin\theta,$$

and further

$$\sin\lambda = \frac{\cos\theta}{\Delta(k', \theta)},\quad \cos\lambda = \frac{k\sin\theta}{\Delta(k', \theta)},\quad \Delta(k', \lambda) = \frac{k}{\Delta(k', \theta)},$$
$$\frac{\Delta(k', \lambda)}{\sin\lambda\cos\lambda} = \frac{\Delta(k', \theta)}{\sin\theta\cos\theta}, = \sqrt{a},$$
$$\frac{\sin\lambda\,\Delta(k', \lambda)}{\cos\lambda} = \sqrt{a}\sin^2\lambda,\quad \frac{\sin\theta\,\Delta(k', \theta)}{\cos\theta} = \sqrt{a}\sin^2\theta;$$

hence the equation becomes

$$\sqrt{a}\left\{\Pi_1(n, k) + \Pi_1\left(\frac{k^2}{n}, k\right)\right\}$$
$$= \pi + F_1k\{\sqrt{a}(\sin^2\theta + \sin^2\lambda) - k'^2\sin\theta\sin\lambda\}$$
$$+ F_1k(F_1k' - E_1k') - E_1kF_1k'.$$

181. But we have

$$1 - \sin^2\theta - \sin^2\lambda = \cos^2\theta - \frac{\cos^2\theta}{\Delta^2(k', \theta)} = -\frac{k'^2\sin^2\theta\cos^2\theta}{\Delta^2(k', \theta)}, = -\frac{k'^2}{a},$$

that is

$$\sin^2\theta + \sin^2\lambda = 1 + \frac{k'^2}{a} \text{ or } \sqrt{a}\,(\sin^2\theta + \sin^2\lambda) = \sqrt{a} + \frac{k'^2}{\sqrt{a}},$$

$$\sin\lambda\sin\theta = \frac{\sin\theta\cos\theta}{\Delta(k',\theta)}, = \frac{1}{\sqrt{a}},$$

whence on the right-hand side the second term is $F_1k\sqrt{a}$, or the equation may be written

$$\sqrt{a}\left\{\Pi_1(n, k) + \Pi_1\left(\frac{k^2}{n}, k\right) - F_1k\right\} = \pi + F_1kF_1k' - F_1kE_1k' - E_1kF_1k',$$

which is in fact a consequence of two former results

$$\Pi_1(n, k) + \Pi_1\left(\frac{k^2}{n}, k\right) - F_1k = \frac{\frac{1}{2}\pi}{\sqrt{a}},$$

and

$$F_1kF_1k' - F_1kE_1k' - E_1kF_1k' = -\tfrac{1}{2}\pi,$$

viz. the equation is hereby reduced to the identity $\frac{1}{2}\pi = \pi - \frac{1}{2}\pi$.

Second case, $n = -1 + k'^2\sin^2\theta$.

182. Here $\sqrt{a} = \frac{k'^2\sin\theta\cos\theta}{\Delta(k',\theta)}$, $\frac{dn}{\sqrt{a}} = 2d\theta\,\Delta(k',\theta)$,

and taking as before the integrals in regard to θ from the inferior limit 0, the general equation, *ante* No. 175, becomes

$$\frac{k'^2\sin\theta\cos\theta}{\Delta(k',\theta)}\Pi = C + (F - E)\int\frac{d\theta}{\Delta(k',\theta)} - k^2F\int\frac{d\theta}{(1 - k'^2\sin^2\theta)^{\frac{3}{2}}}$$
$$+ \Delta\sin\phi\cos\phi\int\frac{d\theta\,.\,\Delta(k',\theta)}{\cos^2\phi + k'^2\sin^2\theta\sin^2\phi}.$$

We have

$$\int\frac{d\theta}{\Delta(k',\theta)} = F(k',\theta),$$

and

$$\int\frac{d\theta}{(1 - k'^2\sin^2\theta)^{\frac{3}{2}}} = \frac{1}{k^2}E(k',\theta) - \frac{k'^2}{k^2}\cdot\frac{\sin\theta\cos\theta}{\Delta(k',\theta)};$$

moreover, writing

$$k'^2 \tan^2\phi = n',$$

we have

$$\Delta \sin\phi\cos\phi \int \frac{\Delta(k', \theta)\,d\theta}{\cos^2\phi + k'^2\sin^2\theta\sin^2\phi}, = \frac{\Delta\sin\phi}{\cos\phi}\int\frac{\Delta(k', \theta)\,d\theta}{1 + n'\sin^2\phi}$$

$$= -\frac{\Delta\cos\phi}{\sin\phi}F(k', \theta) + \frac{\Delta}{\sin\phi\cos\phi}\Pi(n', k', \theta).$$

Making these substitutions, and writing also $\Delta(k, \phi)$ instead of Δ, &c., the formula becomes

$$(n = -1 + k'^2\sin^2\theta,\ \ n' = k'^2\tan^2\phi,)$$

$$\frac{k'^2\sin\theta\cos\theta}{\Delta(k', \theta)}[\Pi(n, k, \phi) - F(k, \phi)] \qquad (l')\ \textit{Leg.}\ \text{p. 138.}$$

$$= \frac{\Delta(k, \phi)}{\sin\phi\cos\phi}[\Pi(n', k', \theta) - \cos^2\phi\, F(k', \theta)]$$

$$+ F(k, \phi)\,F(k', \theta) - E(k, \phi)\,F(k', \theta) - F(k, \phi)\,E(k', \theta),$$

the value of the constant C being here $= 0$, since the two sides each vanish for $\theta = 0$.

183. Write $\phi = \frac{1}{2}\pi - \omega$, ω being indefinitely small; it is to be shown that

$$\frac{\Delta(k, \phi)}{\sin\phi\cos\phi}[\Pi(n', k', \theta) - \cos^2\phi\, F(k', \theta)] = \tfrac{1}{2}\pi,$$

which being so, the equation becomes

$$\frac{k'^2\sin\theta\cos\theta}{\Delta(k', \theta)}[\Pi_1(n, k) - F_1 k] \qquad (m')\ \textit{Leg.}\ \text{p. 138.}$$

$$= \tfrac{1}{2}\pi + F_1 k F(k', \theta) - E_1 k F(k', \theta) - F_1 k E(k', \theta),$$

giving the value of the complete function $\Pi_1(n, k)$ for the form in hand $n = -1 + k'^2\sin^2\theta.$

184. As regards the subsidiary proposition, observe that $\phi = \frac{1}{2}\pi - \omega$ gives $n' = k'^2\tan^2\phi\ (= k'^2\cot^2\omega)$, that is $\frac{k'^2}{n'} = \cot^2\phi$,

$=\tan^2\omega$; so that from the first formula in No. 160, writing therein k' for k, and interchanging θ and ϕ, we have

$$\Pi(n', k', \theta) + \Pi(\tan^2\omega, k', \theta) = F(k', \theta)$$
$$+ \frac{\sin\phi\cos\phi}{\Delta(k, \phi)}\tan^{-1}\frac{\Delta(k, \phi)\tan\theta}{\sin\phi\cos\phi\,\Delta(k', \theta)};$$

but $\tan\omega$ being indefinitely small,

$$\Pi(\tan^2\omega, k', \theta) = F(k', \theta) + \omega^2 M,$$

where M is finite: the equation thus is

$$\Pi(n', k', \theta) + \omega^2 M = \frac{\sin\phi\cos\phi}{\Delta(k, \phi)}\tan^{-1}\frac{\Delta(k, \phi)\tan\theta}{\sin\phi\cos\phi\,\Delta(k', \theta)},$$

that is

$$\frac{\Delta(k, \phi)}{\sin\phi\cos\phi}\Pi(n', k', \theta) + \frac{\omega^2\Delta(k, \phi)M}{\sin\phi\cos\phi} = \tan^{-1}\frac{\Delta(k, \phi)\tan\theta}{\sin\phi\cos\phi\,\Delta(k', \theta)},$$

or putting herein $\phi = \frac{1}{2}\pi - \omega$, then since $\dfrac{\omega^2}{\cos\phi}$ vanishes, and the function under the $\tan^{-1}$ becomes infinite, we have

$$\frac{\Delta(k, \phi)}{\sin\phi\cos\phi}\Pi(n', k', \theta) = \tfrac{1}{2}\pi,$$

whence the required relation.

Third case, $n = -k^2\sin^2\theta$.

165. Here $\sqrt{a} = i\Delta(k, \theta)\cot\theta$, $\dfrac{dn}{\sqrt{a}} = -\dfrac{2k^2\sin^2\theta}{\Delta(k, \theta)}$, and the general formula becomes

$$\frac{\cos\theta}{\sin\theta}\Delta(k, \theta)\Pi = C + (F - E)\int\frac{d\theta}{\Delta(k, \theta)} - F\int\frac{d\theta}{\sin^2\theta\,\Delta(k, \theta)}$$
$$+ \Delta\sin\phi\cos\phi\int\frac{k^2\sin^2\theta\,d\theta}{(1 - k^2\sin^2\theta\sin^2\phi)\Delta(k, \theta)},$$

where as before the integrals in regard to θ may be taken from the inferior limit 0. Since in the present case the modulus as regards both the ϕ and the θ functions is k, we may instead of $\Delta(k, \theta)$ write simply $\Delta\theta$ and so $F\theta$, $E\theta$. We have

$$\int \frac{d\theta}{\Delta\theta} = F\theta, \quad \int \frac{d\theta}{\sin^2\theta\,\Delta\theta} = F\theta - E\theta - \cot\theta\,\Delta\theta,$$

and moreover writing $n' = -k^2\sin^2\phi$, then

$$\int \frac{k^2\sin^2\theta\,d\theta}{(1 - k^2\sin^2\theta\sin^2\phi)\,\Delta\theta} = \int \frac{k^2\sin^2\theta\,d\theta}{(1 + n'\sin^2\theta)\,\Delta\theta}$$

$$= \frac{1}{\sin^2\phi}\left[\Pi(n', \theta) - F\theta\right];$$

whence the formula becomes ($n = -k^2\sin^2\theta$, $n' = -k^2\sin^2\phi$),

$$\cot\theta\,\Delta\theta\left[\Pi(n, \phi) - F\phi\right] = \qquad (n') \text{ Leg. p. 141.}$$

$$\cot\phi\,\Delta\phi\left[\Pi(n', \theta) - F\theta\right] + E\theta\,F\phi - F\theta\,E\phi,$$

the constant in this case being evidently $= 0$.

186. Writing $\phi = \frac{1}{2}\pi$ we have

$$\Pi_1 n - F_1 = \frac{\tan\theta}{\Delta\theta}(F_1 E\theta - E_1 F\theta), \qquad (p') \text{ Leg. p. 141,}$$

which is the value of the complete function $\Pi_1 n$, or say of $\Pi_1(n, k)$, for the form $n = -k^2\sin^2\theta$.

The formulæ marked (i'), (k'), (l'), (m'), (n'), (p') are those given by Legendre, *Traité des Fonctions Elliptiques*, t. I. ch. XXIII., in the pages severally referred to.

CHAPTER VI.

THE FUNCTIONS $\Pi(u, a)$, Zu, Θu, Hu.

THE functions referred to all depend on the modulus k, which may be expressed when necessary: as regards $\Pi(u, a)$ this is seldom required, but the other functions will be frequently written $Z(u, k)$, $\Theta(u, k)$, $H(u, k)$, so as to put the modulus in evidence.

Introductory. Art. No. 187.

187. The function $\Pi(u, a)$ has already been defined as

$$= \int_0 \frac{k^2 \operatorname{sn} a \operatorname{cn} a \operatorname{dn} a \operatorname{sn}^2 u \, du}{1 - k^2 \operatorname{sn}^2 a \operatorname{sn}^2 u},$$

and the several properties already obtained for the function $\Pi(n, k, \phi)$ admit of being translated into this new notation. But in the present chapter the theory is established in a different manner, by expressing this function $\Pi(u, a)$ in terms of the new function Θu. This function Θu may be considered as originating from the function Zu, which has already been mentioned as introduced in place of the function $E(k, \phi)$, viz. writing E to denote the complete function $E_1 k$, we have

$$Zu = u\left(1 - \frac{E}{K}\right) - k^2 \int_0 \operatorname{sn}^2 u \, du,$$

or, what is the same thing,

$$= -\frac{E}{K} u + \int_0 \operatorname{dn}^2 u \, du.$$

The new function Θu is in fact

$$=\sqrt{\frac{2k'K}{\pi}}\,e^{\int_0 Zu\,du},\ =\sqrt{\frac{2k'K}{\pi}}\,e^{-\frac{1}{2}\frac{E}{K}u^2+\int_0 du\int_0 du\,\mathrm{dn}^2u},$$

where the exterior factor $\sqrt{\frac{2k'K}{\pi}}$ is fixed upon for reasons which will appear: the original function Zu is thus expressible in terms of the new function Θu and its derived function, viz. we have $Zu=\frac{\Theta' u}{\Theta u}$, but the employment of the symbol Z as a separate notation is nevertheless convenient.

The function Θu is one of a series of four functions, Θu, $\Theta(u+iK')$, $\Theta(u+K)$, $\Theta(u+K+iK')$; but it is found convenient instead of $\Theta(u+iK')$ to introduce a new function Hu, and write the four as Θu, Hu, $\Theta(u+K)$, $H(u+K)$.

The following article is in the nature of a lemma.

Values of $\Pi(u+a, a)$ *in the three cases* $a=\frac{1}{2}iK'$, $a=\frac{1}{2}K$, $a=\frac{1}{2}K+\frac{1}{2}iK'$ *respectively.* Art. Nos. 188 to 190.

188. We have

$$\frac{d}{du}\Pi(u+a, a)=\frac{k^2\,\mathrm{sn}\,a\,\mathrm{cn}\,a\,\mathrm{dn}\,a\,\mathrm{sn}^2(u+a)}{1-k^2\,\mathrm{sn}^2a\,\mathrm{sn}^2(u+a)}:$$

first if $a=\frac{1}{2}iK'$, we have

$$\mathrm{sn}\,\tfrac{1}{2}iK',\ \mathrm{cn}\,\tfrac{1}{2}iK',\ \mathrm{dn}\,\tfrac{1}{2}iK'=\frac{i}{\sqrt{k}},\ \frac{\sqrt{1+k}}{\sqrt{k}},\ \sqrt{1+k},$$

$$\mathrm{sn}^2(u+\tfrac{1}{2}iK')=\frac{1}{k}\frac{(1+k)\,\mathrm{sn}\,u+i\,\mathrm{cn}\,u\,\mathrm{dn}\,u}{(1+k)\,\mathrm{sn}\,u-i\,\mathrm{cn}\,u\,\mathrm{dn}\,u},\ (\textit{ante}\ \text{No. } 103).$$

The right-hand side of the foregoing equation is therefore a fraction, the numerator of which is

$$ik(1+k)\,[(1+k)\,\mathrm{sn}\,u+i\,\mathrm{cn}\,u\,\mathrm{dn}\,u],$$

its denominator being

$$k\,[(1+k)\,\mathrm{sn}\,u-i\,\mathrm{cn}\,u\,\mathrm{dn}\,u]+k\,[(1+k)\,\mathrm{sn}\,u+i\,\mathrm{cn}\,u\,\mathrm{dn}\,u],$$

viz. this is $= 2k(1+k)\,\mathrm{sn}\,u$, a mere multiple of $\mathrm{sn}\,u$: and we thus have

$$\frac{d}{du}\Pi(u+\tfrac{1}{2}iK', \tfrac{1}{2}iK') = \tfrac{1}{2}i(1+k)\left\{1+\frac{i}{1+k}\,\frac{\mathrm{cn}\,u\,\mathrm{dn}\,u}{\mathrm{sn}\,u}\right\},$$

or observing that $\dfrac{\mathrm{cn}\,u\,\mathrm{dn}\,u}{\mathrm{sn}\,u} = \dfrac{d}{du}\log\mathrm{sn}\,u$, and integrating so that the value may vanish for $u = -\tfrac{1}{2}iK'$, we have

$$\Pi(u+\tfrac{1}{2}iK', \tfrac{1}{2}iK') = \tfrac{1}{2}i(1+k)(u+\tfrac{1}{2}iK') - \tfrac{1}{2}\log\mathrm{sn}\,u + \tfrac{1}{2}\log\left(\frac{-i}{\sqrt{k}}\right).$$

189. *Secondly* $a = \tfrac{1}{2}K$,

$$\mathrm{sn}\,\tfrac{1}{2}K,\ \mathrm{cn}\,\tfrac{1}{2}K,\ \mathrm{dn}\,\tfrac{1}{2}K = \frac{1}{\sqrt{1+k'}},\ \frac{\sqrt{k'}}{\sqrt{1+k'}},\ \sqrt{k'},$$

$$\mathrm{sn}^2(u+\tfrac{1}{2}K) = \frac{1}{1+k'}\,\frac{\mathrm{dn}\,u + (1+k')\,\mathrm{sn}\,u\,\mathrm{cn}\,u}{\mathrm{dn}\,u + (1-k')\,\mathrm{sn}\,u\,\mathrm{cn}\,u},$$

and then

$$\frac{d}{du}\Pi(u+\tfrac{1}{2}K, \tfrac{1}{2}K) = \tfrac{1}{2}(1-k')\left\{1+\frac{(1+k')\,\mathrm{sn}\,u\,\mathrm{cn}\,u}{\mathrm{dn}\,u}\right\},$$

$$= \tfrac{1}{2}(1-k') + \tfrac{1}{2}\frac{k^2\,\mathrm{sn}\,u\,\mathrm{cn}\,u}{\mathrm{dn}\,u},$$

or observing that

$$\frac{d}{du}\log\mathrm{dn}\,u = \frac{-k^2\,\mathrm{sn}\,u\,\mathrm{cn}\,u}{\mathrm{dn}\,u},$$

and integrating so that the value may vanish for $u = -\tfrac{1}{2}K$, we have

$$\Pi(u+\tfrac{1}{2}K, \tfrac{1}{2}K) = \tfrac{1}{2}(1-k')(u+\tfrac{1}{2}K) - \tfrac{1}{2}\log\mathrm{dn}\,u + \tfrac{1}{2}\log\sqrt{k'}.$$

190. *Thirdly* $a = \tfrac{1}{2}K + \tfrac{1}{2}iK'$,

$$\mathrm{sn}(\tfrac{1}{2}K+\tfrac{1}{2}iK'),\ \mathrm{cn}(\tfrac{1}{2}K+\tfrac{1}{2}iK'),\ \mathrm{dn}(\tfrac{1}{2}K+\tfrac{1}{2}iK')$$

$$= \sqrt{\frac{k+ik'}{k}},\ \sqrt{\frac{-ik'}{k}},\ \sqrt{-ik'(k+ik')},$$

$$\mathrm{sn}^2(u+\tfrac{1}{2}K+\tfrac{1}{2}iK') = \frac{k+ik'}{k}\,\frac{\mathrm{cn}\,u + (k-ik')\,\mathrm{sn}\,u\,\mathrm{dn}\,u}{\mathrm{cn}\,u + (k+ik')\,\mathrm{sn}\,u\,\mathrm{dn}\,u},$$

and thence

$$\frac{d}{du}\Pi\left(u + \tfrac{1}{2}K + \tfrac{1}{2}iK', \tfrac{1}{2}K + \tfrac{1}{2}iK'\right)$$

$$= \tfrac{1}{2}(k + ik')\left\{1 + \frac{(k - ik')\operatorname{sn} u \operatorname{dn} u}{\operatorname{cn} u}\right\},$$

$$= \tfrac{1}{2}(k + ik') + \tfrac{1}{2}\frac{\operatorname{sn} u \operatorname{dn} u}{\operatorname{cn} u},$$

whence observing that

$$\frac{d}{du}\log \operatorname{cn} u = \frac{-\operatorname{sn} u \operatorname{dn} u}{\operatorname{cn} u},$$

and integrating so that the value may vanish for

$$u = -\tfrac{1}{2}K - \tfrac{1}{2}iK',$$

we have

$$\Pi\left(u + \tfrac{1}{2}K + \tfrac{1}{2}iK', \tfrac{1}{2}K + \tfrac{1}{2}iK'\right) =$$

$$\tfrac{1}{2}(k + ik')\left(u + \tfrac{1}{2}K + \tfrac{1}{2}iK'\right) - \tfrac{1}{2}\log \operatorname{cn} u + \tfrac{1}{2}\log\sqrt{\frac{-ik'}{k}}.$$

The Function Zu. Art. Nos. 191 to 196.

191. We now proceed to consider the function Zu: it has already been, *ante* No. 131, seen that we have here the addition-equation

$$Zu + Zv - Z(u + v) = k^2 \operatorname{sn} u \operatorname{sn} v \operatorname{sn}(u + v).$$

192. Starting from the equation

$$Zu = -\frac{E}{K}u + \int_0 \operatorname{dn}^2 u \, du,$$

and writing herein iu for u and k' for k, we have

$$Z(iu, k') = -\frac{E'}{K'}iu + i\int_0 \operatorname{dn}^2(iu, k')\, du,$$

(which observing that $\operatorname{dn}(iu, k') = \dfrac{\operatorname{dn} u}{\operatorname{cn} u}$ may also be written

$$= -\frac{E'}{K'}iu + i\int_0 \frac{\operatorname{dn}^2 u}{\operatorname{cn}^2 u}\, du\Big);$$

and it is to be shown that we have

$$Zu = \frac{\operatorname{sn} u \operatorname{dn} u}{\operatorname{cn} u} - \frac{\pi u}{2KK'} + iZ(iu, k').$$

I stop to remark that u being indefinitely small this equation is

$$u\left(1 - \frac{E}{K}\right) = u - \frac{\pi u}{2KK'} - u\left(1 - \frac{E'}{K'}\right),$$

which is true in virtue of

$$\frac{E}{K} + \frac{E'}{K'} - 1 = \frac{\pi}{2KK'}.$$

193. To prove the theorem, we verify without difficulty that

$$\frac{d}{du}\frac{\operatorname{sn} u \operatorname{dn} u}{\operatorname{cn} u} = \operatorname{dn}^2 u + \frac{\operatorname{dn}^2 u}{\operatorname{cn}^2 u} - 1;$$

we have therefore

$$\frac{d}{du}\frac{\operatorname{sn} u \operatorname{dn} u}{\operatorname{cn} u} = \operatorname{dn}^2 u + \operatorname{dn}^2(iu, k') - 1,$$

or integrating from $u = 0$,

$$\frac{\operatorname{sn} u \operatorname{dn} u}{\operatorname{cn} u} = \int_0 \operatorname{dn}^2 u\, du + \int_0 \operatorname{dn}^2(iu, k')\, du - u.$$

But the integrals in this formula are

$$= Zu + \frac{E}{K}u \text{ and } -iZ(iu, k') + \frac{E'}{K'}u$$

respectively, and substituting these values and reducing by

$$\frac{E}{K} + \frac{E'}{K'} - 1 = \frac{\pi}{2KK'},$$

we have the required formula

$$Zu = \frac{\operatorname{sn} u \operatorname{dn} u}{\operatorname{cn} u} - \frac{\pi u}{2KK'} + iZ(iu, k').$$

194. Writing in this equation $u = iK'$, and observing that

$$Z(-K', k') = -Z(K', k') = 0,$$

since $Z(K', k')$ is what $Z(K)$ becomes on writing therein k' for k, we have

$$Z(iK') = \frac{\operatorname{sn} iK' \operatorname{dn} iK'}{\operatorname{cn} iK'} - \frac{i\pi}{2K},$$

which is infinite, $= kI$, if I is the infinite value of $\operatorname{sn} iK'$. Writing in the addition-equation $u = v = \frac{1}{2} iK'$, we have

$$2Z(\tfrac{1}{2}iK') = Z(iK') + k^2 \operatorname{sn}^2 \tfrac{1}{2} iK' \sin iK',$$

$$= \frac{\operatorname{sn} iK' \operatorname{dn} iK'}{\operatorname{cn} iK'} + k^2 \operatorname{sn}^2 \tfrac{1}{2} iK' \sin iK' - \frac{i\pi}{2K},$$

or substituting for $\operatorname{sn}^2 \frac{1}{2} iK'$ its value $= \dfrac{1 - \operatorname{cn} iK'}{1 + \operatorname{dn} iK'}$, this is

$$= \sin iK' \left\{\frac{\operatorname{dn} iK'}{\operatorname{cn} iK'} + \frac{k^2(1 - \operatorname{cn} iK')}{1 + \operatorname{dn} iK'}\right\} - \frac{i\pi}{2K},$$

where the first term is

$$\frac{\sin iK' (\operatorname{dn} iK' + k^2 \operatorname{cn} iK' + k'^2)}{\operatorname{cn} iK' (1 + \operatorname{dn} iK')};$$

and substituting herein for $\sin iK'$, $\operatorname{cn} iK'$, $\operatorname{dn} iK'$ their values, $= I$, $-iI$, $-ikI$ respectively, and then making I infinite, the term is $= i(1 + k)$, and we thus obtain

$$Z(\tfrac{1}{2} iK') = \tfrac{1}{2} i (1 + k) - \frac{i\pi}{4K}.$$

It will be recollected that

$$Z(\tfrac{1}{2} K) = \tfrac{1}{2}(1 - k'),$$

and we thence by the addition-equation find

$$Z(\tfrac{1}{2} K + \tfrac{1}{2} iK') = \tfrac{1}{2}(k + ik') - \frac{i\pi}{4K}.$$

195. Starting from

$$Zu = -\frac{E}{K} u + \int_0 \operatorname{dn}^2 u \, du,$$

we have

$$Z(u+a) = -\frac{E}{K}(u+a) + \int_{-a} \mathrm{dn}^2(u+a)\,du,$$

$$= -\frac{E}{K}(u+a) + \int_0 \mathrm{dn}^2(u+a)\,du + \int_0^a \mathrm{dn}^2 u\,du,$$

$$= -\frac{E}{K}u + \int_0 \mathrm{dn}^2(u+a)\,du + Za,$$

that is

$$\int_0 \mathrm{dn}^2(u+a)\,du = \frac{E}{K}u + Z(u+a) - Za.$$

And similarly, observing that $Z(-a) = -Za$, we have

$$\int_0 \mathrm{dn}^2(u-a)\,du = \frac{E}{K}u + Z(u-a) + Za,$$

whence

$$\int_0 \mathrm{dn}^2(u+a)\,du - \int_0 \mathrm{dn}^2(u-a)\,du = Z(u+a) - Z(u-a) - 2Za.$$

196. We find without difficulty

$$\frac{d}{du}\frac{\mathrm{cn}\,u\,\mathrm{dn}\,u}{\mathrm{sn}\,u} = -\frac{\mathrm{cn}^2 u}{\mathrm{sn}^2 u} - \mathrm{dn}^2 u,$$

$$= \mathrm{dn}^2(u+iK') - \mathrm{dn}^2 u,$$

and thence

$$\frac{\mathrm{cn}\,u\,\mathrm{dn}\,u}{\mathrm{sn}\,u} = C + \int \mathrm{dn}^2(u+iK')\,du - \int \mathrm{dn}^2 u\,du,$$

$$= C + Z(u+iK') - Zu.$$

To determine the constant, write $u = -\tfrac{1}{2}iK'$, we have

$$-\frac{\mathrm{cn}\,\tfrac{1}{2}iK'\,\mathrm{dn}\,\tfrac{1}{2}iK'}{\mathrm{sn}\,\tfrac{1}{2}iK'} = C + 2Z(\tfrac{1}{2}iK').$$

or substituting for the functions of $\frac{1}{2}iK'$ their values, this is

$$i(1+k) = C + i(1+k) - \frac{i\pi}{2K},$$

and therefore

$$C = \frac{i\pi}{2K}:$$

hence the equation is

$$\frac{\operatorname{cn} u \operatorname{dn} u}{\operatorname{sn} u} = \frac{i\pi}{2K} + Z(u+iK') - Zu.$$

The Function Θu. Art. Nos. 197 to 199.

197. The definition has been given at the beginning of the present Chapter.

Θu is obviously an even function, $\Theta(-u) = \Theta u$; and we have $\Theta 0 = \sqrt{\dfrac{2k'K}{\pi}}$.

We have

$$\Theta(iu) = \sqrt{\frac{2k'K}{\pi}}\, e^{i\int_0 Z(iu)\,du},$$

and therefore also

$$\Theta(iu, k') = \sqrt{\frac{2kK'}{\pi}}\, e^{i\int_0 Z(iu, k')\,du}.$$

198. From the equation

$$Zu = \frac{\operatorname{sn} u \operatorname{dn} u}{\operatorname{cn} u} - \frac{\pi u}{2KK'} + iZ(iu, k'),$$

multiplying by du and integrating from $u = 0$ (observing that $\dfrac{\operatorname{sn} u \operatorname{dn} u}{\operatorname{cn} u} = -\dfrac{d}{du}\log \operatorname{cn} u$), we have

$$\int_0 Zu\,du = -\log \operatorname{cn} u - \frac{\pi u^2}{4KK'} + i\int_0 Z(iu, k'),$$

and taking the exponential of each side and expressing the values in terms of Θ we have

$$\Theta u = \sqrt{\frac{k'K}{kK'}}\, e^{-\frac{\pi u^2}{4KK'}} \frac{1}{\operatorname{cn} u}\, \Theta(iu, k')^*.$$

199. From the equation

$$Zu = -\frac{\operatorname{cn} u \operatorname{dn} u}{\operatorname{sn} u} + \frac{i\pi}{2K} + Z(u+iK'),$$

we deduce in like manner

$$\Theta u = C e^{\frac{i\pi u}{2K}} \frac{1}{\operatorname{sn} u}\, \Theta(u+iK'),$$

and, in order to determine the constant, writing herein $u = -\tfrac{1}{2}iK'$, we find

$$C = -\sin \tfrac{1}{2}iK'\, e^{-\frac{\pi K'}{4K}} = -\frac{i}{\sqrt{k}}\, e^{-\frac{\pi K'}{4K}},$$

whence the equation is

$$\Theta u = -\frac{i}{\sqrt{k}}\, e^{-\frac{\pi}{4K}(K'-2iu)} \frac{1}{\operatorname{sn} u}\, \Theta(u+iK'),$$

an equation which will presently be proved in a different manner.

Expression of $\Pi(u, a)$ *in terms of* Θu. Art. No. 200.

200. We have

$$\operatorname{sn}(u+a) + \operatorname{sn}(u-a) = \frac{2 \operatorname{sn} u \operatorname{cn} a \operatorname{dn} a}{1 - k^2 \operatorname{sn}^2 a \operatorname{sn}^2 u},$$

$$\operatorname{sn}(u+a) - \operatorname{sn}(u-a) = \frac{2 \operatorname{sn} a \operatorname{cn} u \operatorname{dn} u}{1 - k^2 \operatorname{sn}^2 a \operatorname{sn}^2 u},$$

* This is in effect Jacobi's formula, *Fund. Nova*, p. 163 (5), viz. interchanging therein k and k', it becomes

$$\frac{\Theta(iu, k')}{\Theta(0, k')} = e^{\frac{\pi u^2}{4KK'}} \operatorname{cn} u \frac{\Theta(u, k)}{\Theta(0, k)},$$

that is

$$\Theta(u, k) = \frac{\Theta(0, k)}{\Theta(0, k')}\, e^{-\frac{\pi u^2}{4KK'}} \frac{1}{\operatorname{cn} u}\, \Theta(iu, k'),$$

or substituting for $\Theta(0, k')$, $\Theta(0, k)$ their values, this is the formula of the text.

and thence

$$\operatorname{sn}^2(u+a)-\operatorname{sn}^2(u-a)=\frac{4\operatorname{sn}a\operatorname{cn}a\operatorname{dn}a\operatorname{sn}u\operatorname{cn}u\operatorname{dn}u}{(1-k^2\operatorname{sn}^2a\operatorname{sn}^2u)^2},$$

$$=2\frac{d}{du}\cdot\frac{\operatorname{sn}a\operatorname{cn}a\operatorname{dn}a\operatorname{sn}^2u}{1-k^2\operatorname{sn}^2a\operatorname{sn}^2u},$$

as is at once verified from the relation $\frac{d}{du}\operatorname{sn}u=\operatorname{cn}u\operatorname{dn}u$.

The equation may be written

$$-\tfrac{1}{2}\operatorname{dn}^2(u+a)+\tfrac{1}{2}\operatorname{dn}^2(u-a)$$

$$=\frac{d}{du}\cdot\frac{k^2\operatorname{sn}a\operatorname{cn}a\operatorname{dn}a\operatorname{sn}^2u}{1-k^2\operatorname{sn}^2a\operatorname{sn}^2u},=\frac{d^2}{du^2}\Pi(u,a),$$

whence multiplying by du and integrating from $u=0$,

$$-\tfrac{1}{2}\int_0\operatorname{dn}^2(u+a)\,du+\tfrac{1}{2}\int_0\operatorname{dn}^2(u-a)\,du=\frac{d}{du}\Pi(u,a),$$

or, what is the same thing,

$$\frac{d}{du}\Pi(u,a)=Za+\tfrac{1}{2}Z(u-a)-\tfrac{1}{2}Z(u+a).$$

Substituting herein for $Z(u-a)$, $Z(u+a)$ their values

$$\frac{\Theta'(u-a)}{\Theta(u-a)},\ \frac{\Theta'(u+a)}{\Theta(u+a)},$$

multiplying by du and integrating from $u=0$, we have

$$\Pi(u,a)=uZa+\tfrac{1}{2}\log\frac{\Theta(u-a)}{\Theta(u+a)},$$

where for Za we may of course substitute its value, $=\frac{\Theta'a}{\Theta a}$.

The Function Θu resumed. Art. Nos. 201 to 206.

201. We have

$$\frac{d}{du}\Pi(u,a)=\frac{k^2\operatorname{sn}a\operatorname{cn}a\operatorname{dn}a\operatorname{sn}^2u}{1-k^2\operatorname{sn}^2a\operatorname{sn}^2u},=-\tfrac{1}{2}\frac{d}{da}\log(1-k^2\operatorname{sn}^2a\operatorname{sn}^2u),$$

that is

$$2\frac{\Theta'a}{\Theta a}+\frac{\Theta'(u-a)}{\Theta(u-a)}-\frac{\Theta'(u+a)}{\Theta(u+a)}=-\frac{d}{da}\log(1-k^2\operatorname{sn}^2a\operatorname{sn}^2u),$$

or what is the same thing,

$$\frac{d}{da}\log\Theta(u-a)+\frac{d}{da}\log\Theta(u+a)$$

$$=2\frac{d}{da}\log\Theta a+\frac{d}{da}\log(1-k^2\,\mathrm{sn}^2 a\,\mathrm{sn}^2 u).$$

Integrating in regard to a we have

$$\Theta(u-a)\,\Theta(u+a)=C\Theta^2 a\,(1-k^2\,\mathrm{sn}^2 a\,\mathrm{sn}^2 u),$$

where of course the constant of integration C may be a function of u. To determine it write $a=0$, we have

$$\Theta^2 u=C\Theta^2 0=C\frac{2k'K}{\pi},$$

and then the equation is

$$\Theta(u-a)\,\Theta(u+a)=\frac{\pi}{2k'K}\Theta^2 u\,\Theta^2 a\,(1-k^2\,\mathrm{sn}^2 a\,\mathrm{sn}^2 u).$$

202. Writing the differential formula under the form

$$\frac{k^2\,\mathrm{sn}\,a\,\mathrm{cn}\,a\,\mathrm{dn}\,a\,\mathrm{sn}^2 u}{1-k^2\,\mathrm{sn}^2 a\,\mathrm{sn}^2 u}=Za+\tfrac{1}{2}Z(u-a)-\tfrac{1}{2}Z(u+a),$$

if we herein interchange a, u, this becomes

$$\frac{k^2\,\mathrm{sn}\,u\,\mathrm{cn}\,u\,\mathrm{dn}\,u\,\mathrm{sn}^2 a}{1-k^2\,\mathrm{sn}^2 a\,\mathrm{sn}^2 u}=Zu-\tfrac{1}{2}Z(u-a)-\tfrac{1}{2}Z(u+a),$$

and adding the two together we have

$$Zu+Za-Z(u+a)=k^2\,\mathrm{sn}\,u\,\mathrm{sn}\,a\,\mathrm{sn}\,(u+a),$$

viz. we thus reproduce the addition-formula for the function Z.

203. Starting with

$$\Pi(u, a)=u\,Za-\tfrac{1}{2}\log\frac{\Theta(u+a)}{\Theta(u-a)},$$

and writing herein $u+a$ in place of u, we have

$$\Pi(u+a, a)=(u+a)\,Za-\tfrac{1}{2}\log\frac{\Theta(u+2a)}{\Theta u};$$

we have in the present chapter found the values of $\Pi(u+a, a)$ in the several cases $a=\tfrac{1}{2}iK'$, $a=\tfrac{1}{2}K$, $a=\tfrac{1}{2}K+\tfrac{1}{2}iK'$.

204. *First* $a = \frac{1}{2} iK'$, we have

$$\tfrac{1}{2} i(1+k)(u + \tfrac{1}{2} iK') - \tfrac{1}{2}\log \operatorname{sn} u + \tfrac{1}{2}\log\left(\frac{-i}{\sqrt{k}}\right)$$

$$= Z(\tfrac{1}{2} iK')(u + \tfrac{1}{2} iK') - \tfrac{1}{2}\log\frac{\Theta(u+iK')}{\Theta u},$$

that is

$$\log \operatorname{sn} u = [i(1+k) - 2Z(\tfrac{1}{2} iK')](u + \tfrac{1}{2} iK')$$

$$+ \log\left(\frac{-i}{\sqrt{k}}\right) + \log\frac{\Theta(u+iK')}{\Theta u},$$

which substituting for $Z(\frac{1}{2} iK')$ its value becomes

$$= \frac{i\pi}{2K}(u + \tfrac{1}{2} iK') + \log\left\{\frac{-i}{\sqrt{k}}\,\frac{\Theta(u+iK')}{\Theta u}\right\},$$

or writing the first term under the form $-\frac{\pi}{4K}(K' - 2iu)$, and taking the exponentials of each side,

$$\operatorname{sn} u = e^{-\frac{\pi}{4K}(K'-2iu)} \cdot \frac{-i}{\sqrt{k}}\,\frac{\Theta(u+iK')}{\Theta u}.$$

205. *Secondly* $a = \frac{1}{2} K$, we have

$$\tfrac{1}{2}(1-k')(u + \tfrac{1}{2} K) - \tfrac{1}{2}\log \operatorname{dn} u + \tfrac{1}{2}\log\sqrt{k'}$$

$$= Z(\tfrac{1}{2} K)(u + \tfrac{1}{2} K) - \tfrac{1}{2}\log\frac{\Theta(u+K)}{\Theta u},$$

that is

$$\log \operatorname{dn} u = [1 - k' - 2Z(\tfrac{1}{2} K)](u + \tfrac{1}{2} K)$$

$$+ \log\sqrt{k'} + \log\frac{\Theta(u+K)}{\Theta u},$$

where the term in $u + \frac{1}{2} K$ vanishes by reason of the value of $Z(\frac{1}{2} K)$, and passing to the exponentials we have

$$\operatorname{dn} u = \sqrt{k'}\,\frac{\Theta(u+K)}{\Theta u}.$$

206. *Thirdly* $a = \frac{1}{2} K + \frac{1}{2} iK'$, we have

$$\tfrac{1}{2}(k + ik')(u + \tfrac{1}{2} K + \tfrac{1}{2} iK') - \tfrac{1}{2}\log \operatorname{cn} u + \tfrac{1}{2}\log\sqrt{\frac{-ik'}{k}}$$

$$= Z(\tfrac{1}{2} K + \tfrac{1}{2} iK')(u + \tfrac{1}{2} K + \tfrac{1}{2} iK') - \tfrac{1}{2}\log\frac{\Theta(u+K+iK')}{\Theta u},$$

that is

$$\log \operatorname{cn} u = \{k + ik' - 2Z(\tfrac{1}{2}K + \tfrac{1}{2}iK')\}(u + \tfrac{1}{2}K + \tfrac{1}{2}iK')$$
$$+ \log\sqrt{\frac{-ik'}{k}} + \log\frac{\Theta(u + K + iK')}{\Theta u},$$

or substituting for $Z(\tfrac{1}{2}K + \tfrac{1}{2}iK')$ its value, the first term is $\frac{i\pi}{2K}(u + \tfrac{1}{2}K + \tfrac{1}{2}iK')$, which is $= -\frac{\pi}{4K}(K' - 2iu) + \frac{i\pi}{4}$: hence passing to the exponentials and observing that

$$e^{\frac{i\pi}{4}}\sqrt{\frac{-ik'}{k}} = \sqrt{i}\sqrt{\frac{-ik'}{k}}, = \sqrt{\frac{k'}{k}},$$

we have

$$\operatorname{cn} u = e^{-\frac{\pi}{4K}(K' - 2iu)}\sqrt{\frac{k'}{k}}\,\frac{\Theta(u + K + iK')}{\Theta u}.$$

Recapitulation. Art. No. 207.

207. We now see that the elliptic functions sn u, cn u, dn u, that the elliptic function of the second kind considered as a function of u, and for convenience replaced by Jacobi's Zu, and that the function of the third kind considered under Jacobi's form $\Pi(u, a)$, are all of them expressed in terms of the single function $\Theta(u)$, and the k-functions K, K', viz. that we have

$$\operatorname{sn} u = e^{-\frac{\pi}{4K}(K' - 2iu)} \cdot \frac{-i}{\sqrt{k}}\,\Theta(u + iK'), \qquad (\div)$$

$$\operatorname{cn} u = e^{-\frac{\pi}{4K}(K' - 2iu)}\sqrt{\frac{k'}{k}}\,\Theta(u + K + iK'), \qquad (\div)$$

$$\operatorname{dn} u = \qquad \sqrt{k'}\,\Theta(u + K), \qquad (\div)$$

$$\text{denom.} = \qquad \Theta u,$$

viz. these are fractional functions having the common denominator Θu, and having also Θ-functions in their numerators; and further that

$$Zu = \frac{\Theta' u}{\Theta u}, \quad \Pi(u, a) = u\frac{\Theta' a}{\Theta a} + \tfrac{1}{2}\log\frac{\Theta(u - a)}{\Theta(u + a)};$$

and conversely that Θu is a function derived from sn u by the equation

$$\Theta u = \sqrt{\frac{2k'K}{\pi}}\, e^{\frac{1}{2}\left(1-\frac{E}{K}\right)u^2 - k^2\int_0 du \int_0 du\, \mathrm{sn}^2 u}$$

(involving the k-function E, Legendre's $E_1 k$).

And we have also proved the formula

$$\Theta u = \sqrt{\frac{k'K}{kK'}}\, e^{-\frac{\pi u^2}{4KK'}}\, \frac{1}{\mathrm{cn}\, u}\, \Theta(iu, k'),$$

or as this may also be written

$$\Theta(iu) = \sqrt{\frac{k'K}{kK'}}\, e^{\frac{\pi u^2}{4KK'}}\, \mathrm{cn}(iu, k')\, \Theta(iu, k');$$

and the formula

$$\Theta(u+a)\,\Theta(u-a) = \frac{\pi}{2k'K}\,\Theta^2 u\,\Theta^2 a\,(1 - k^2\,\mathrm{sn}^2 u\,\mathrm{sn}^2 a).$$

The Function Hu. Art. Nos. 208, 209.

208. If introducing for convenience a new function Hu*, we write

$$Hu = -ie^{-\frac{\pi}{4K}(K' - 2iu)}\,\Theta(u + iK'),$$

and therefore also

$$H(u+K) = -ie^{-\frac{\pi}{4K}(K' - 2iu) + \frac{1}{2}i\pi}\,\Theta(u + K + iK'),$$

$$= e^{-\frac{\pi}{4K}(K' - 2iu)}\,\Theta(u + K),$$

* If instead of Jacobi's Θ, H we use the four functions Θu, $\Theta_1 u$, $\Theta_2 u$, $\Theta_3 u$, $= \Theta u$, $\frac{1}{\sqrt{k}} Hu$, $\sqrt{\frac{k'}{k}} H(u+K)$, $\sqrt{k'}\,\Theta(u+K)$ respectively, then $\Theta_1 u$, $\Theta_2 u$, $\Theta_3 u$ are the numerators, and Θu the common denominator, for the three elliptic functions sn, cn, dn u. The four functions Θ_1, Θ_2, Θ_3, Θ have been tabulated under the superintendence of Mr J. W. L. Glaisher, and are in course of publication.

then the formulæ for the elliptic functions become

$$\text{sn}\, u = \frac{1}{\sqrt{k}} Hu, \qquad (+)$$

$$\text{cn}\, u = \sqrt{\frac{k'}{k}}\, H(u+K), \qquad (+)$$

$$\text{dn}\, u = \sqrt{k'}\, \Theta(u+K), \qquad (+)$$

where $\qquad \text{denom.} = \Theta u.$

It hence appears that Hu is an odd function of u, which for u indefinitely small becomes $= \sqrt{\frac{2kk'K}{\pi}}\, u$.

209. Combining with

$$\Theta(u+a)\,\Theta(u-a) = \frac{\pi}{2k'K}\, \Theta^2 u\, \Theta^2 a\, (1 - k^2\, \text{sn}^2 u\, \text{sn}^2 a),$$

the equation

$$\text{sn}(u+a)\,\text{sn}(u-a) = \frac{\text{sn}^2 u - \text{sn}^2 a}{1 - k^2\,\text{sn}^2 u\,\text{sn}^2 a},$$

and attending to the expressions of sn u, sn a in terms of H, Θ, we have

$$H(u+a)\, H(u-a) = \frac{\pi}{2k'K}\, (H^2 u\, \Theta^2 a - H^2 a\, \Theta^2 u).$$

The Function $\Pi(u, a)$ *resumed.* Art. Nos. 210 to 215.

210. We deduce the addition-equation for the function of the third kind $\Pi(u, a)$, viz. we have first

$$\Pi(u, a) + \Pi(v, a) - \Pi(u+v, a)$$

$$= \tfrac{1}{2} \log \frac{\Theta(u-a)\,\Theta(v-a)\,\Theta(u+v+a)}{\Theta(u+a)\,\Theta(v+a)\,\Theta(u+v-a)} \;\; (= \tfrac{1}{2}\log \Omega, \text{ suppose}),$$

where the logarithmic term containing the functions Θ may be in three different ways made to depend on the functions sn.

211. *First* we have

$$\Theta(u-a)\,\Theta(v-a) = \frac{1}{\Theta^2 0}\,\Theta^2\tfrac{1}{2}(u-v)\,\Theta^2\left(\tfrac{1}{2}(u+v)-a\right)$$
$$\left\{1-k^2\,\mathrm{sn}^2\tfrac{1}{2}(u-v)\,\mathrm{sn}^2\left(\tfrac{1}{2}(u+v)-a\right)\right\},$$

$$\Theta(u+a)\,\Theta(v+a) = \frac{1}{\Theta^2 0}\,\Theta^2\tfrac{1}{2}(u-v)\,\Theta^2\left(\tfrac{1}{2}(u+v)+a\right)$$
$$\left\{1-k^2\,\mathrm{sn}^2\tfrac{1}{2}(u-v)\,\mathrm{sn}^2\left(\tfrac{1}{2}(u+v)+a\right)\right\},$$

$$\Theta a\,\Theta(u+v-a) = \frac{1}{\Theta^2 0}\,\Theta^2\tfrac{1}{2}(u+v)\,\Theta^2\left(\tfrac{1}{2}(u+v)-a\right)$$
$$\left\{1-k^2\,\mathrm{sn}^2\tfrac{1}{2}(u+v)\,\mathrm{sn}^2\left(\tfrac{1}{2}(u+v)-a\right)\right\},$$

$$\Theta a\,\Theta(u+v+a) = \frac{1}{\Theta^2 0}\,\Theta^2\tfrac{1}{2}(u+v)\,\Theta^2\left(\tfrac{1}{2}(u+v)+a\right)$$
$$\left\{1-k^2\,\mathrm{sn}^2\tfrac{1}{2}(u+v)\,\mathrm{sn}^2\left(\tfrac{1}{2}(u+v)+a\right)\right\},$$

and taking the product of the first and fourth expressions divided by that of the second and third, we have

$$\Omega = \frac{\{1-k^2\mathrm{sn}^2\tfrac{1}{2}(u-v)\mathrm{sn}^2(\tfrac{1}{2}(u+v)-a)\}\{1-k^2\mathrm{sn}^2\tfrac{1}{2}(u+v)\mathrm{sn}^2(\tfrac{1}{2}(u+v)+a)\}}{\{1-k^2\mathrm{sn}^2\tfrac{1}{2}(u-v)\mathrm{sn}^2(\tfrac{1}{2}(u+v)+a)\}\{1-k^2\mathrm{sn}^2\tfrac{1}{2}(u+v)\mathrm{sn}^2(\tfrac{1}{2}(u+v)-a)\}}.$$

212. *Secondly* we have

$$\Theta^2(u-a)\,\Theta^2(v-a) = \Theta^2 0\,.\,\Theta(u-v)\,\Theta(u+v-2a)$$
$$+\left\{1-k^2\,\mathrm{sn}^2(u-a)\,\mathrm{sn}^2(v-a)\right\},$$
$$\Theta^2(u+a)\,\Theta^2(v+a) = \Theta^2 0\,.\,\Theta(u-v)\,\Theta(u+v+2a)$$
$$+\left\{1-k^2\,\mathrm{sn}^2(u+a)\,\mathrm{sn}^2(v+a)\right\},$$
$$\Theta^2 a\,\Theta^2(u+v-a) = \Theta^2 0\,.\,\Theta(u+v)\,\Theta(u+v-2a)$$
$$+\left\{1-k^2\,\mathrm{sn}^2 a\,\mathrm{sn}^2(u+v-a)\right\},$$
$$\Theta^2 a\,\Theta^2(u+v+a) = \Theta^2 0\,.\,\Theta(u+v)\,\Theta(u+v+2a)$$
$$+\left\{1-k^2\,\mathrm{sn}^2 a\,\mathrm{sn}^2(u+v+a)\right\},$$

and then in like manner we obtain

$$\Omega = \sqrt{\frac{\{1-k^2\,\mathrm{sn}^2(u+a)\,\mathrm{sn}^2(v+a)\}\{1-k^2\,\mathrm{sn}^2 a\,\mathrm{sn}^2(u+v-a)\}}{\{1-k^2\,\mathrm{sn}^2(u-a)\,\mathrm{sn}^2(v-a)\}\{1-k^2\,\mathrm{sn}^2 a\,\mathrm{sn}^2(u+v+a)\}}}.$$

213. But, *thirdly*, from the form originally obtained for the addition-equation, the same quantity should be

$$\Omega = \frac{1-k^2\,\mathrm{sn}\,a\,\mathrm{sn}\,u\,\mathrm{sn}\,v\,\mathrm{sn}\,(u+v-a)}{1+k^2\,\mathrm{sn}\,a\,\mathrm{sn}\,u\,\mathrm{sn}\,v\,\mathrm{sn}\,(u+v+a)}.$$

The transformation is effected as follows:

We have

$$\{1-k^2\,\mathrm{sn}^2\tfrac{1}{2}(u+v)\,\mathrm{sn}^2\tfrac{1}{2}(u-v)\}\,\mathrm{sn}\,u\,\mathrm{sn}\,v = \mathrm{sn}^2\tfrac{1}{2}(u+v)-\mathrm{sn}^2\tfrac{1}{2}(u-v),$$

$$\{1-k^2\,\mathrm{sn}^2\tfrac{1}{2}(u+v)\,\mathrm{sn}^2(\tfrac{1}{2}(u+v)-a)\}\,\mathrm{sn}\,a\,\mathrm{sn}\,(u+v-a) = \mathrm{sn}^2\tfrac{1}{2}(u+v)-\mathrm{sn}^2(\tfrac{1}{2}(u+v)-a),$$

and taking the products of the two sides each multiplied by $-k^2$, and adding a common term on each side, we have

$$\{1-k^2\mathrm{sn}^2\tfrac{1}{2}(u+v)\mathrm{sn}^2\tfrac{1}{2}(u-v)\}\,\{1-k^2\mathrm{sn}^2\tfrac{1}{2}(u+v)\mathrm{sn}^2(\tfrac{1}{2}(u+v)-a)\}\times\{1-k^2\,\mathrm{sn}\,a\,\mathrm{sn}\,u\,\mathrm{sn}\,v\,\mathrm{sn}\,(u+v-a)\}$$

$$=\{1-k^2\mathrm{sn}^2\tfrac{1}{2}(u+v)\mathrm{sn}^2\tfrac{1}{2}(u-v)\}\,\{1-k^2\mathrm{sn}^2\tfrac{1}{2}(u+v)\,\mathrm{sn}^2(\tfrac{1}{2}(u+v)-a)\} - k^2\,\{\mathrm{sn}^2\tfrac{1}{2}(u+v)-\mathrm{sn}^2\tfrac{1}{2}(u-v)\}\,\{\mathrm{sn}^2\tfrac{1}{2}(u+v)-\mathrm{sn}^2(\tfrac{1}{2}(u+v)-a)\},$$

$$=1+k^4\,\mathrm{sn}^4\tfrac{1}{2}(u+v)\,\mathrm{sn}^2\tfrac{1}{2}(u-v)\,\mathrm{sn}^2(\tfrac{1}{2}(u+v)-a) - k^2\,\mathrm{sn}^4\tfrac{1}{2}(u+v)-k^2\,\mathrm{sn}^2\tfrac{1}{2}(u-v)\,\mathrm{sn}^2(\tfrac{1}{2}(u+v)-a),$$

$$=\{1-k^2\mathrm{sn}^4\tfrac{1}{2}(u+v)\}\,\{1-k^2\mathrm{sn}^2\tfrac{1}{2}(u-v)\,\mathrm{sn}^2(\tfrac{1}{2}(u+v)-a)\}^*.$$

Changing the sign of a we have a second like equation, and dividing one by the other, we find the required equation

$$\frac{\{1-k^2\mathrm{sn}^2\tfrac{1}{2}(u+v)\mathrm{sn}^2(\tfrac{1}{2}(u+v)-a)\}\,\{1-k^2\mathrm{sn}^2\tfrac{1}{2}(u-v)\,\mathrm{sn}^2(\tfrac{1}{2}(u+v)+a)\}}{\{1-k^2\mathrm{sn}^2\tfrac{1}{2}(u+v)\,\mathrm{sn}^2(\tfrac{1}{2}(u+v)+a)\}\,\{1-k^2\mathrm{sn}^2\tfrac{1}{2}(u-v)\,\mathrm{sn}^2(\tfrac{1}{2}(u+v)-a)\}} = \frac{1+k^2\,\mathrm{sn}\,a\,\mathrm{sn}\,u\,\mathrm{sn}\,v\,\mathrm{sn}\,(u+v+a)}{1-k^2\,\mathrm{sn}\,a\,\mathrm{sn}\,u\,\mathrm{sn}\,v\,\mathrm{sn}\,(u+v-a)}.$$

214. The conclusion is

$$\Pi(u, a)+\Pi(v, a)-\Pi(u+v, a)=\tfrac{1}{2}\log\Omega,$$

where Ω is expressed in the three forms just obtained.

215. In the equation

$$\Pi(u, a)=uZa+\tfrac{1}{2}\log\frac{\Theta(u-a)}{\Theta(u+a)},$$

* The identity, writing therein

$$u,\ a,\ v \text{ for } \tfrac{1}{2}(u-v),\ \tfrac{1}{2}(u+v),\ \tfrac{1}{2}(u+v)-a,$$

becomes

$$1-k^2\,\mathrm{sn}\,(a+u)\,\mathrm{sn}\,(a-u)\,\mathrm{sn}\,(a+v)\,\mathrm{sn}\,(a-v)=\frac{\{1-k^2\,\mathrm{sn}^4 a\}\,\{1-k^2\,\mathrm{sn}^2 u\,\mathrm{sn}^2 v\}}{\{1-k^2\,\mathrm{sn}^2 a\,\mathrm{sn}^2 u\}\,\{1-k^2\,\mathrm{sn}^2 a\,\mathrm{sn}^2 v\}}.$$

interchanging u and a, we obtain (observing that Θ is an even function)

$$\Pi(a, u) = aZu + \tfrac{1}{2}\log\frac{\Theta(u-a)}{\Theta(u+a)},$$

and thence

$$\Pi(u, a) - \Pi(a, u) = uZa - aZu,$$

which is the theorem for the interchange of amplitude and parameter.

And we hence deduce

$$\Pi(u, a) + \Pi(u, b) - \Pi(u, a+b) =$$
$$\Pi(a, u) + \Pi(b, u) - \Pi(a+b, u) + u\,[Za + Zb - Z(a+b)].$$

Here on the right-hand side by the addition-theorem the first term is $= \tfrac{1}{2}\log\Omega'$, where Ω' is the same function of a, b, u that Ω is of u, v, a: we have thus Ω' in three forms one of which is

$$\Omega' = \frac{1 - k^2 \operatorname{sn} a \operatorname{sn} b \operatorname{sn}(a+b-u) \operatorname{sn} u}{1 + k^2 \operatorname{sn} a \operatorname{sn} b \operatorname{sn}(a+b+u) \operatorname{sn} u},$$

and the second term is, by the addition-theorem for Z,

$$= k^2 \operatorname{sn} a \operatorname{sn} b \operatorname{sn}(a+b)\, u\,;$$

we have therefore

$$\Pi(u, a) + \Pi(u, b) - \Pi(u, a+b)$$
$$= k^2 \operatorname{sn} a \operatorname{sn} b \operatorname{sn}(a+b)\, u + \tfrac{1}{2}\log\Omega',$$

which is the theorem for the addition of parameters.

Multiplication of the Functions Θu, Hu. Art. Nos. 210, 217.

210. From the equation

$$\Theta(u+v)\,\Theta(u-v) = \frac{\Theta^2 u\,\Theta^2 v}{\Theta^2 0}\,(1 - k^2 \operatorname{sn}^2 u \operatorname{sn}^2 v),$$

we deduce

$$\Theta(2u) = \frac{\Theta^4 u}{\Theta^2 0}\,(1 - k^2 \operatorname{sn}^4 u),$$

$$\Theta(3u) = \frac{\Theta^2(2u)\,\Theta u}{\Theta^2 0}\,(1 - k^2 \operatorname{sn}^2 u \operatorname{sn}^2 2u),$$

$$\Theta(4u) = \frac{\Theta^4(2u)}{\Theta^2 0}\,(1 - k^2 \operatorname{sn}^4 2u),$$

and it is hence easy to see that $\Theta nu \div \Theta^{n^2}(u)$ is a rational and integral function of $\operatorname{sn}^2 u$ of the degree $\tfrac{1}{2}n^2$ or $\tfrac{1}{2}(n^2-1)$ (that is n^2 or n^2-1 in $\operatorname{sn} u$) according as n is even or odd. More precisely we may say that $\Theta nu \,.\, \Theta^{n^2-1}0 \div \Theta^{n^2}u$ is such a function, reducing itself to unity for $\operatorname{sn} u = 0$; and it thus appears that considering $\operatorname{sn} nu$, $\operatorname{cn} nu$, $\operatorname{dn} nu$ as expressed in terms of $\operatorname{sn} u$ by the multiplication formulæ, in such wise that for $u = 0$ the denominator is $= 1$, then this denominator will be

$$= \Theta nu \,.\, \Theta^{n^2-1}0 \div \Theta^{n^2}u.$$

217. And it hence of course follows that the three numerators are

$$= \frac{1}{\sqrt{k}} Hnu \,.\, \Theta^{n^2-1}0 \div \Theta^{n^2}u,$$

$$= \sqrt{\frac{k'}{k}}\, H(nu+K)\, \Theta^{n^2-1}0 \div \Theta^{n^2}u,$$

$$= \sqrt{k'}\, \Theta(nu+K)\, \Theta^{n^2-1}0 \div \Theta^{n^2}u,$$

respectively. It will appear in the sequel how we thence obtain the expressions of these numerators and denominator.

CHAPTER VII.

TRANSFORMATION. GENERAL OUTLINE.

218. THE theory of transformation is considered in the first instance in regard to the differential expression $\frac{dx}{\sqrt{X}}$ (which, for the elliptic integrals, has the particular form $\frac{dx}{\sqrt{1-x^2 . 1-k^2x^2}}$), and then to the elliptic functions sn, cn, dn.

Case of a general quartic radical $\sqrt{X}$. Art. Nos. 219 to 222.

219. Consider the differential expression $\frac{dy}{\sqrt{Y}}$, where Y is a given rational and integral quartic function of y. Write herein $y=\frac{U}{V}$ where U and V are rational and integral functions of x, one of them of the order p, the other of the order p or $p-1$: such a fraction is said to be of the order p. It is to be shown that the coefficients of U, V may be so determined as to lead to an equation

$$\frac{Mdy}{\sqrt{Y}}=\frac{dx}{\sqrt{X}},$$

where X is a rational and integral quartic function of x, and M is a constant. We have

$$dy=\frac{1}{V^2}(VU'-V'U)\,dx, \quad \left(U', V'=\frac{dU}{dx}, \frac{dV}{dx}\right),$$

$$Y=\frac{1}{V^4}(V, U)^4,$$

where considering Y as a homogeneous quartic function of $(1, y)$, then $(V, U)^4$ is what this becomes on writing therein V, U in place of $1, y$ respectively: viz. $(V, U)^4$ is a homogeneous quartic function of U, V, and therefore of the order $4p$ in x; $VU' - V'U$, if V, U are of the same order p, would at first sight appear to be of the order $2p-1$, but in this case the coefficient of x^{2p-1} vanishes and the order is really $= 2p-2$; viz. whether the orders of U, V are p, p or $p, p-1$, the order of $VU' - V'U$ is $= 2p-2$. The foregoing values give

$$\frac{dy}{\sqrt{Y}} = \frac{(VU' - V'U)\,dx}{\sqrt{(V, U)^4}}.$$

220. It is at once seen that if $(V, U)^4$ has a square factor $(x-a)^2$ then $x-a$ divides $VU' - V'U$. Similarly if $(V, U)^4$ has $2p-2$ such factors, or if it is $= T^2X$, where T^2 is of the order $4p-4$ and therefore X of the order 4, then the product T of the roots of the square factors divides $VU' - V'U$, and since $VU' - V'U$ and T are each of the order $2p-2$ the quotient $(VU' - V'U) \div T$ must be an absolute constant M^{-1}. But in this case we have

$$\frac{Mdy}{\sqrt{Y}} = \frac{dx}{\sqrt{X}},$$

an equation of the required form.

221. Regarding U, V as being each of them of the order p, the expression $\frac{U}{V}$ contains $2p+1$ constants, and in determining U, V so as to satisfy the condition $(V, U)^4 = T^2X$ we determine $2p-2$ of these: there thus remain three arbitrary constants: this is as it should be, for if the required condition is satisfied by any particular values U, V, it will also be satisfied by the new values obtained by writing in the fraction $\frac{U}{V}$, in place of x, the function $\frac{\alpha+\beta x}{1+\delta x}$ with three arbitrary constants. We may by such linear transformation make either U or V to be of the order $p-1$, or if we please begin by assuming this

to be so. But we cannot have either U or V of an order inferior to $p-1$; for if this were the case $VU' - V'U$ would be of an order inferior to $2p-2$, while in fact it divides by T which is of the order $2p-2$.

Considering Y as a given quartic function of y, the function X is obtained as an arbitrary linear transformation of a determinate quartic function of x: or what is the same thing, it is a quartic function containing a single parameter which cannot be assumed at pleasure, but is a determinate function of the coefficients of Y, different according to the different values of the number p: which number is termed the order of the transformation.

222. It is to be observed that we cannot have any other really distinct transformation of the differential expression $\frac{dy}{\sqrt{Y}}$ into the form $\frac{M^{-1}dx}{\sqrt{X}}$ with the same radical $\sqrt{X}$ and a constant value of M: for suppose that such transformation existed; say by writing $y =$ Function (z) we could obtain $\frac{dy}{\sqrt{Y}} = \frac{N^{-1}dz}{\sqrt{Z}}$ where Z is the same quartic function of z that X is of x and N is a constant: then $\frac{dy}{\sqrt{Y}} = \frac{M^{-1}dx}{\sqrt{X}} = \frac{N^{-1}dz}{\sqrt{Z}}$, that is $\frac{Ndx}{\sqrt{X}} = \frac{Mdz}{\sqrt{Z}}$; such an equation is integrable algebraically when M, N are commensurable, that is proportional to integer numbers m, n; and from the form of the integral we infer that the equation is not integrable algebraically unless M, N are commensurable: hence N, M must be commensurable or the last-mentioned equation must be of the form $\frac{ndx}{\sqrt{X}} = \frac{mdz}{\sqrt{Z}}$; and we have thus a known algebraical relation between the quantities x, z such that by means of it we can pass from one to the other of the transformations $y = \frac{U}{V}$, $y =$ Funct. (z): the two transformations would on this account be regarded as not essentially distinct the one from the other.

The standard form $\dfrac{dx}{\sqrt{1-x^2 . 1-k^2x^2}}$. Art. No. 223.

223. The theory applies in particular to the case of a differential expression of the form

$$\frac{dy}{\sqrt{1-y^2 . 1-\lambda^2y^2}},$$

viz. this by a transformation of the form $y=\dfrac{U}{V}$, of the order n, can be converted into one of a like form in regard to x, that is we obtain a relation

$$\frac{Mdy}{\sqrt{1-y^2 . 1-\lambda^2y^2}}=\frac{dx}{\sqrt{1-x^2 . 1-k^2x^2}},$$

where, k or λ being given, the other of them and also the value of the multiplier M are each determined, not uniquely but by means of an equation called the modular equation, between k and λ: more precisely, if k or λ be given, the other of them may be taken to be any particular root of the modular equation, and then the coefficients of U, V, and the multiplier M, are determinate functions of k, λ.

Distinction of cases according to the form of n.
Art. Nos. 224 and 225.

224. In the case where n is a composite number $=qr$, the modular equation breaks up, and the transformation in fact decomposes into distinct transformations. That this *may* be the case is clear *à priori*, viz. if we have $z=\dfrac{U_1}{V_1}$ a rational function of x of the order q, giving rise to a relation

$$\frac{M_1dz}{\sqrt{1-z^2 . 1-l^2z^2}}=\frac{dx}{\sqrt{1-x^2 . 1-k^2x^2}},$$

and $y=\dfrac{U_2}{V_2}$ a rational function of z of the order r, giving rise to a relation

$$\frac{M_2dy}{\sqrt{1-y^2 . 1-\lambda^2y^2}}=\frac{dz}{\sqrt{1-z^2 . 1-l^2z^2}},$$

then for z substituting its value in terms of x, we have clearly $y = \frac{U}{V}$ a rational function of x of the order qr, giving

$$\frac{M_1 M_2 dy}{\sqrt{1-y^2 . 1-\lambda^2 y^2}} = \frac{dx}{\sqrt{1-x^2 . 1-k^2 x^2}};$$

but to show that the case is of necessity so would require further investigation, and the question is not entered upon in the present work.

Assuming the property in question, it appears that the transformations belonging to the several prime numbers need alone be considered; viz. the cases $n = 2$ and n an odd prime $= p$. The case $n = 2$ presents certain peculiarities.

225. $n = 2$. There are in this case two distinct rational transformations, one of them of the form $y = \frac{bx}{c + dx^2}$ (viz. here y vanishes with x), for which the new modulus is $\lambda, = \frac{1-k'}{1+k'}$, and the other of them of the form $y = \frac{a + bx^2}{c + dx^2}$, for which the new modulus is $\gamma, = \frac{2\sqrt{k}}{1+k}$: these will be considered.

It is to be observed that for the case in question $n = 2$, λ and γ correspond respectively to the real moduli λ and λ_1 belonging to the case n, an odd prime, as presently mentioned: viz. we have the equations $\frac{1}{2}\frac{\Lambda'}{\Lambda} = \frac{K'}{K} = 2\frac{\Gamma'}{\Gamma}$ precisely corresponding to the equations $\frac{1}{n}\frac{\Lambda'}{\Lambda} = \frac{K'}{K} = n\frac{\Lambda_1'}{\Lambda_1}$ afterwards mentioned. But in the case of n an odd prime, λ, λ_1 are roots of one and the same irreducible equation: moreover (as afterwards appears) $y, = \text{sn}\left(\frac{u}{M}, \lambda\right)$ and $y, = \text{sn}\left(\frac{u}{M_1}, \lambda_1\right)$ are each given in terms of $x, = \text{sn}\, u$, by a rational transformation of the form $y = \frac{U}{V}$ where y vanishes with x: whereas in the present case

$n=2$, the corresponding functions

$$y_1 = \text{sn}\,(\overline{1+k'}\,u, \lambda),\quad y_1 = \text{sn}\,(\overline{1+k}\,u, \gamma)$$

are (as will be seen) given in terms of $x, = \text{sn}\,u$, the former by an irrational, the latter by a rational transformation, y in each of them vanishing with x.

Instead of at once proceeding to the case of n an odd prime, we take in the first instance, n any odd number whatever.

n an odd number: further development of the theory.
Art. Nos. 226 to 231.

226. We have here the formula

$$y = \frac{x\,(1,\ x^2)^{\frac{1}{2}(n-1)}}{(1,\ x^2)^{\frac{1}{2}(n-1)}};$$

viz. the numerator is an odd function of the order n, and the denominator an even function of the order $n-1$. We may proceed somewhat further in the determination of the form: for this purpose take P, Q even functions of x, such that $P+Qx$ is of the degree $\frac{1}{2}(n-1)$: for instance

$n=3,\ P+Qx = \alpha+\beta x,$ ord. $P=0$, ord. $Q=0$,
$n=5,\ P+Qx = \alpha+\beta x+\gamma x^2,$ ord. $P=2$, ord. $Q=0$,
$n=7,\ P+Qx = \alpha+\beta x+\gamma x^2+\delta x^3,$ ord. $P=2$, ord. $Q=2$,
$n=9,\ P+Qx = \alpha+\beta x+\gamma x^2+\delta x^3+\epsilon x^4,$ ord. $P=4$, ord. $Q=2$,

and so in general; viz. $n=4p-1$, the orders of P and Q are each $=2p-2$, but $n=4p+1$, order of P is $=2p$ and that of Q is $=2p-2$.

227. This being so, assuming

$$\frac{1-y}{1+y} = \frac{(P-Qx)^2}{(P+Qx)^2}\,\frac{1-x}{1+x},$$

we see that

$$y_1 = \frac{x\,(P^2+2PQ+Q^2x^2)}{P^2+2PQx^2+Q^2x^2},$$

is a function of the above-mentioned form; and not only so; but forming the equations

$$1-y=(P-Qx)^2(1-x), \qquad (\div)$$

$$1+y=(P+Qx)^2(1+x), \qquad (\div)$$

where

$$\text{denom.} = P^2+2PQx^2+Q^2x^2,$$

we see that $1-y$ and $1+y$ have each of them the required property of having in the numerators a square factor of the proper order.

228. It is next to be observed that the functions P, Q may be so determined that the expression for y remains unaltered when we simultaneously change x into $\frac{1}{kx}$, and y into $\frac{1}{\lambda y}$.

To see how this is, write for shortness

$$y=\frac{x\,N(1,\, x^2)}{D(1,\, x^2)},$$

N, D being as above functions each of the order $\frac{1}{2}(n-1)$ in x^2. We have

$$N\left(1,\, \frac{1}{k^2x^2}\right)=\left(\frac{1}{kx}\right)^{n-1}N(k^2x^2,\, 1),$$

and considering the coefficients, say of $N(1,\, x^2)$, as given, we can at once determine those of $D(1,\, x^2)$ in such manner that, Ω being a constant, we have identically

$$N(k^2x^2,\, 1)=\Omega D(1,\, x^2).$$

In fact the coefficients of D will be those of N taken in the reverse order and multiplied each by the proper power of k. This being so, we have

$$N\left(1,\, \frac{1}{k^2x^2}\right)=\Omega\left(\frac{1}{kx}\right)^{n-1}D(1,\, x^2);$$

and this identical equation, writing $\frac{1}{kx}$ for x, becomes

$$N(1,\, x^2)=\Omega x^{n-1}D\left(1,\, \frac{1}{k^2x^2}\right),$$

whence identically

$$N(1,\, x^2)\,N\left(1,\, \frac{1}{k^2x^2}\right)=\frac{\Omega^2}{k^{n-1}}D(1,\, x^2)\,D\left(1,\, \frac{1}{k^2x^2}\right).$$

Suppose that writing $\frac{1}{kx}$ for x, y is changed into $\bar{y}$, then

$$\bar{y} = \frac{\frac{1}{kx} N\left(1, \frac{1}{k^2x^2}\right)}{D\left(1, \frac{1}{k^2x^2}\right)},$$

or multiplying by y and reducing by means of the result just obtained, we have

$$y\bar{y} = \frac{\Omega^2}{k^n},$$

viz. writing $\frac{\Omega^2}{k^n} = \frac{1}{\lambda}$ we have $\bar{y} = \frac{1}{\lambda y}$; and thus we may simultaneously change x, y into $\frac{1}{kx}$, $\frac{1}{\lambda y}$, the theorem in question.

229. Or, in a somewhat different form, the theorem is at once seen to hold good provided we have

$$y = \frac{\frac{x}{M}\left(1 - \frac{x^2}{a^2}\right)\left(1 - \frac{x^2}{b^2}\right)\dots}{(1 - k^2a^2x^2)(1 - k^2b^2x^2)\dots},$$

for then, making the change in question it becomes

$$\frac{1}{\lambda y} = \frac{1}{Mk^n (ab\dots)^4} \frac{(1 - k^2a^2x^2)(1 - k^2b^2x^2)\dots}{\left(1 - \frac{x^2}{a^2}\right)\left(1 - \frac{x^2}{b^2}\right)\dots},$$

which is in fact the original equation provided only

$$\lambda = M^2k^n (ab\dots)^4.*$$

We thus in effect determine λ as a function of k (viz. these are connected by an equation called the modular equation), and then the coefficients of P, Q are determined in terms of k, λ.

230. The required condition being satisfied, we may in the formulæ which give $1 - y$, $1 + y$ make the same change; and it is easy to see that the resulting formulæ will be of the form

$$1 - \lambda y = (P' - Q'x)^2 (1 - kx), \qquad (+)$$

$$1 + \lambda y = (P' + Q'x)^2 (1 + kx), \qquad (+)$$

* Comparing with the former equation $\lambda = \frac{k^n}{\Omega^2}$, we have $\frac{1}{\Omega} = M(ab\dots)^2$.

the denominator being of course the same as before: hence the required condition as to the square factor is also satisfied by each of the functions $1-\lambda y$, $1+\lambda y$; and the integral relation between y, x leads thus to the required differential equation

$$\frac{Mdy}{\sqrt{1-y^2 . 1-\lambda^2 y^2}} = \frac{dx}{\sqrt{1-x^2 . 1-k^2 x^2}}.$$

231. Supposing that n is not a prime number it will be the product of two or more odd primes, and the transformation will break up into distinct transformations each of which may be separately considered. We therefore now assume n an odd prime: the modular equation is in this case an irreducible equation of the order $n+1$, so that k being given we have $n+1$ different values of λ; and corresponding to each of them we have a distinct formula of transformation. This modular equation is conveniently expressed as an equation between the two quantities $u=\sqrt[4]{k}$, and $v=\sqrt[4]{\lambda}$, viz. it is an equation of the form $(u, v)=0$ where (u, v) is a rational function of the degree $n+1$ as regards each of the quantities (u, v) separately. It is to be added that k^2 being as usual positive and less than 1, there are two and only two real values of λ^2 (which values are also positive and less than 1): and corresponding to them there are two *real* transformations: but this is a property which may in the first instance be disregarded.

Application to the Elliptic Functions. Art. No. 232.

232. We have in what precedes a purely algebraical theory of transformation: in particular, in the case where the order n is an odd number, if in the formulæ we write $y=\sin\chi$, $x=\sin\phi$, the differential equation becomes $\frac{Md\chi}{\Delta(\lambda, \chi)} = \frac{d\phi}{\Delta(k, \phi)}$; and further assuming $\sin\chi = \text{sn}(v, \lambda)$, $\sin\phi = \text{sn}(u, k)$, then it becomes $Mdv = du$, giving (since u and v vanish together) $v=\frac{u}{M}$; whence $x=\text{sn}(u, k)$, $y=\text{sn}\left(\frac{u}{M}, \lambda\right)$: and the theory

is an algebraic theory of transformation, serving to express $\operatorname{sn}\left(\frac{u}{M}, \lambda\right)$ in terms of sn (u, k).

The theory may be completed algebraically without much difficulty in the cases $n = 3, 5, 7$; but there is great difficulty in doing this generally for larger values of n: and it is in fact completed by Jacobi, not algebraically but transcendentally, by expressing λ and the coefficients of the transformation by means of the sn, cn and dn of $\frac{mK + m'iK'}{n}$ (m and m' integers), or say by means of the functions dependent on the n-division of the complete functions K, K'.

*n an odd-prime, the ulterior theory**. Art. Nos. 233 to 235.

233. In particular when n is an odd-prime, there are as already mentioned two real transformations; a *first* transformation from k to a smaller modulus λ, involving the functions of $\frac{K}{n}$; and a *second* transformation from k to a larger modulus λ_1 involving the functions of $\frac{iK'}{n}$. And in these two cases (taking $K, \Lambda, \Lambda_1, K', \Lambda', \Lambda_1'$ for the complete functions to the moduli $k, \lambda, \lambda_1, k', \lambda', \lambda_1'$ respectively) the modular equation is replaced by the equations $\frac{\Lambda'}{\Lambda} = n\frac{K'}{K}$, $\frac{K'}{K} = n\frac{\Lambda_1'}{\Lambda_1}$ respectively: viz. these transcendental equations contain the relations between the original modulus k and the new moduli λ and λ_1 respectively.

* Observe that λ, heretofore used to denote any one whatever of the $n+1$ roots of the modular equation, is in what immediately follows used to denote a particular root, and λ_1 another particular root, the roots belonging to the first and second real transformations respectively. In Nos. 237 *et seq.* λ is again used at the beginning to denote any root, and (λ) a determinate root corresponding thereto, these are taken to be first the particular roots (λ, λ_1), and secondly the particular roots (λ_1, λ). It would, abstractedly, be advantageous to reserve λ as the symbol of any root whatever, using λ_1, λ_2 for the particular roots: but this would have occasioned a very frequent alteration of Jacobi's notation.

234. The equations just referred to are obtained from the following:

$$\Lambda = \frac{K}{nM}, \quad \Lambda' = \frac{K'}{M},$$

$$\Lambda_1 = \frac{K}{M_1}, \quad \Lambda_1' = \frac{K'}{nM_1},$$

which present themselves in the theory. As regards these equations it may be observed here as follows:

235. The *first* transformation is a relation between $\operatorname{sn}\left(\frac{u}{M}, \lambda\right)$, $\operatorname{sn}(u, k)$, and it leads to the equation $\Lambda = \frac{K}{nM}$. Effecting on the transformation-equation Jacobi's imaginary substitution, we obtain from it a *complementary first* transformation, giving $\operatorname{sn}\left(\frac{u}{M}, \lambda'\right)$ in terms of $\operatorname{sn}(u, k')$, and this leads to the equation $\Lambda' = \frac{K'}{M}$.

Similarly the *second* transformation is a relation between $\operatorname{sn}\left(\frac{u}{M_1}, \lambda_1\right)$, $\operatorname{sn}(u, k)$, and it leads to the equation $\Lambda_1 = \frac{K}{M_1}$. Effecting on the transformation-equation Jacobi's imaginary substitution, we obtain from it a *complementary second* transformation, giving $\operatorname{sn}\left(\frac{u}{M_1}, \lambda'_1\right)$ in terms of $\operatorname{sn}(u, k')$, and this leads to the equation $\Lambda_1' = \frac{K'}{nM_1}$, or recapitulating,

first transformation gives $\Lambda = \frac{K}{nM}$,

complementary first „ $\Lambda' = \frac{K'}{M}$,

second „ $\Lambda_1 = \frac{K}{M_1}$,

complementary second „ $\Lambda_1' = \frac{K'}{nM_1}$,

the chief object of the complementary transformations being in fact the deduction of these second and fourth equations.

Connexion with Multiplication. Art. Nos. 236 to 241.

236. The theory of transformation is connected in a very remarkable manner with that of multiplication. This is the case as well for an even as an order number n, and indeed the connexion will be exhibited in the case, $n=2$, of the quadric transformation, but here one of the transformations is irrational: and it is convenient to restrict the attention to the case n an odd number, where the transformations are both rational; or rather (this being the only case which has been completely developed) we may at once take n to be an odd-prime.

237. This being so, starting with the transformation-equation $y=\frac{U}{V}$ of the order n, which gives

$$\frac{Mdy}{\sqrt{1-y^2 \,.\, 1-\lambda^2y^2}}=\frac{dx}{\sqrt{1-x^2 \,.\, 1-k^2x^2}},$$

we may imagine a new variable z connected with y by a transformation-equation $z=\frac{P}{Q}$ of the same order n (P, Q rational and integral functions of y) giving

$$\frac{Ndz}{\sqrt{1-z^2 \,.\, 1-(\lambda)^2z^2}}=\frac{dy}{\sqrt{1-y^2 \,.\, 1-\lambda^2y^2}},$$

where (λ) is not of necessity the same function of λ that λ is of k, but a like function; viz. λ, k are connected by the modular equation, and changing herein k into λ and λ into (λ) we have the relation between λ, (λ). And we have then z a fractional function of x such that

$$\frac{MNdz}{\sqrt{1-z^2 \,.\, 1-(\lambda)^2z^2}}=\frac{dx}{\sqrt{1-x^2 \,.\, 1-k^2x^2}}.$$

238. It is a property of the modular equation that we may have $(\lambda)=k$, and further that when this is so $MN=\frac{1}{n}$: the last-mentioned equation then is

$$\frac{dz}{\sqrt{1-z^2 \,.\, 1-k^2z^2}}=\frac{ndx}{\sqrt{1-x^2 \,.\, 1-k^2x^2}},$$

viz. u being as before taken $=\operatorname{sn}(u, k)$, we have $z=\operatorname{sn}(nu, k)$; and the relation between z, x then gives $\operatorname{sn}(nu, k)$ as a function of $\operatorname{sn}(u, k)$, viz. the expression is a fraction, the numerator being an odd function of the order n^2 and the denominator an even function of the order n^2-1; this is in fact the expression of $\operatorname{sn}(nu, k)$ in terms of $\operatorname{sn}(u, k)$ given by the multiplication-equation. Observe that for obtaining in this manner the transformation x to z (or $\operatorname{sn}(u, k)$ to $\operatorname{sn}(nu, k)$), the transformation x to y may be any one at pleasure of the different transformations, but that regarding it as given we must combine with it a determinate transformation y to z, the resulting transformation x to z being of course independent of the selected x to y transformation: there are thus as many ways of obtaining the final x to z transformation as there are transformations x to y. In the case n an odd-prime, this may be considered more in detail.

230. Selecting the root λ of the modular equation we have a real transformation (Jacobi's *first* transformation) $y=\frac{U}{V}$ giving (M real)

$$\frac{Mdy}{\sqrt{1-y^2 . 1-\lambda^2y^2}}=\frac{dx}{\sqrt{1-x^2 . 1-k^2x^2}},$$

and selecting the root λ_1 of the modular equation we have a real transformation (Jacobi's *second* transformation) $y=\frac{U_1}{V_1}$, giving (M_1 real)

$$\frac{M_1dy}{\sqrt{1-y^2 . 1-\lambda_1^2y^2}}=\frac{dx}{\sqrt{1-x^2 . 1-k^2x^2}}.$$

Now λ is in fact the same function of k that k is of λ_1: this at once appears from the before-mentioned relations

$$\frac{\Lambda'}{\Lambda}=n\frac{K'}{K},\quad \frac{K'}{K}=n\frac{\Lambda_1'}{\Lambda_1}.$$

Hence taking z such a function of y, λ as $\frac{U_1}{V_1}$ is of x, k, the differential relation between z, y is

$$\frac{Ndz}{\sqrt{1-z^2 . 1-k^2z^2}}=\frac{dy}{\sqrt{1-y^2 . 1-\lambda^2y^2}},$$

and consequently, MN being $=\frac{1}{n}$, we have

$$\frac{ds}{\sqrt{1-s^2 \,.\, 1-k^2s^2}}=\frac{ndz}{\sqrt{1-z^2 \,.\, 1-k^2z^2}}.$$

240. Or again, taking s such a function of y, λ_1 as $\frac{U}{V}$ is of x, k, the differential equation between s, y is

$$\frac{N_1 ds}{\sqrt{1-s^2 \,.\, 1-k^2s^2}}=\frac{dy}{\sqrt{1-y^2 \,.\, 1-\lambda_1^2y^2}},$$

and consequently, M_1N_1 being $=\frac{1}{n}$, we have in this case also

$$\frac{ds}{\sqrt{1-s^2 \,.\, 1-k^2s^2}}=\frac{ndz}{\sqrt{1-z^2 \,.\, 1-k^2z^2}},$$

so that in each case, x being $=\text{sn}\,(u, k)$, we obtain the same value $s=\text{sn}\,(nu, k)$: viz. in the first case we pass by a *first* and then a *second* transformation from k through λ to k; and in the second case by a *second* and then a *first* transformation from k through λ_1 to k.

241. As regards the equations $MN=\frac{1}{n}$, $M_1N_1=\frac{1}{n}$, these follow from the before-mentioned equations

$$M=\frac{K}{n\Lambda},\quad M_1=\frac{K}{\Lambda_1},$$

viz. N being what M_1 becomes on changing therein k, λ_1 into λ, k, and N_1 what M becomes on changing k, λ into λ_1, k, we derive from these

$$N=\frac{\Lambda}{K},\quad N_1=\frac{\Lambda_1}{nK},$$

and thence the equations in question.

Jacobi in connexion with the equations $\frac{\Lambda'}{\Lambda}=n\frac{K'}{K}$ and $\frac{\Lambda_1'}{\Lambda_1}=\frac{1}{n}\frac{K'}{K}$ remarks, *Fundamenta Nova*, p. 59, that if n be a composite number $=n'n''$, then, in the transformation of the order n, there is corresponding to each real root of the modular equation a relation of the form $\frac{\Lambda'}{\Lambda} = \frac{n'}{n} \frac{K'}{K}$: whence in particular if n be a square number, the equation is $\frac{\Lambda'}{\Lambda} = \frac{K'}{K}$, viz. we then have $\lambda = k$, showing that in the case where n is a square number there is among the transformations of the order n one which gives the multiplication by $\sqrt{n}$.

He further remarks, p. 75, that λ being *any root whatever* of the modular equation there exist equations of the form

$$a\Lambda + i\beta\Lambda' = \frac{aK + ibK'}{nM},$$

$$a'\Lambda' + i\beta'\Lambda = \frac{a'K' + ib'K}{nM},$$

where a, a', a, a' are odd numbers, b, b', β, β' even numbers, such that $aa' + bb' = 1$, $aa' + \beta\beta' = 1$: and (same page in a foot-note) as follows: "Accuratior numerorum a, a', b, b', &c. determinatio pro singulis ejusdem ordinis transformationibus gravibus laborare difficultatibus videtur. Immo hæc determinatio, nisi egregie fallimur, maxime à limitibus pendet, inter quos modulus k versatur, ita ut pro limitibus diversis plane alia evadat. Id quod quam intricatam reddat quæstionem, expertus cognoscet. Ante omnia autem accuratius in naturam modulorum imaginariorum inquirendum esse videtur, quæ adhuc tota jacet quæstio." That some such equations exist may be inferred without difficulty from the general formulæ of transformation, but the strict proof, and certainly the determination in question, would depend upon investigations out of the field of the *Fundamenta Nova*. The property is used by Jacobi to show that the proof which he gives of the equation $M^2 = \frac{1}{n} \frac{\lambda\lambda'^2}{kk'^2} \frac{dk}{d\lambda}$, where λ denotes in the first instance the real root, applies to the case of any root whatover.

CHAPTER VIII.

THE QUADRIC TRANSFORMATION, $n=2$; AND THE ODD-PRIME TRANSFORMATIONS $n=3, 5, 7$. PROPERTIES OF THE MODULAR EQUATION AND THE MULTIPLIER.

242. THE case $n=2$, although very analogous to the case n an odd prime, presents, as remarked in the preceding Chapter, some essential differences; there are analytically distinct transformations relating to the two new moduli λ and γ respectively, viz. these are not roots of one and the same irreducible modular equation: and it is an irrational transformation which in some sort corresponds to one of the real transformations in the other case. There is an *à priori* necessity for this: viz. as $\operatorname{sn} 2u$ is not a rational function of $\operatorname{sn} u$, we cannot have here two rational transformations leading to the duplication: the duplication must arise from the combination of a rational and an irrational transformation. It should be noticed that the case may be studied quite independently of, and in fact previous to, the general theory explained in the preceding Chapter.

The Quadric Transformation. Art. Nos. 243 to 258.

243. It has been shown geometrically that, considering a new modulus λ connected with k by the equation $\lambda=\dfrac{1-k'}{1+k'}$, and establishing between ϕ, θ the relation $\lambda\sin\theta=\sin(2\phi-\theta)$, or, what is the same thing,

$$\sin\theta=\frac{\frac{1}{2}(1+k')\sin 2\phi}{\sqrt{1-k^2\sin^2\phi}},$$

we have between ϕ, θ the differential equation

$$\frac{(1+k')\,d\phi}{\Delta(k,\phi)} = \frac{d\theta}{\Delta(\lambda,\theta)}.$$

Writing herein $\sin\phi = x$, $\sin\theta = y$, the relation between y, x is

$$y = \frac{(1+k')\,x\sqrt{1-x^2}}{\sqrt{1-k^2x^2}},$$

this is in fact the first form of quadric transformation, and (as is about to be shown) it is connected with a second form $y = \dfrac{(1+k)\,x}{1+kx^2}$.

Modular relations.

244. From the original modulus k we derive two moduli γ, λ; these form a decreasing series γ, k, λ, the relations between them being

$$\gamma = \frac{2\sqrt{k}}{1+k}, \quad k = \frac{2\sqrt{\lambda}}{1+\lambda},$$

$$\gamma' = \frac{1-k}{1+k}, \quad k' = \frac{1-\lambda}{1+\lambda},$$

$$k = \frac{1-\gamma'}{1+\gamma'}, \quad \lambda = \frac{1-k'}{1+k'};$$

and the corresponding complete functions Γ, Γ', K, K′, Λ, Λ', are connected by the equations

$$(1+\lambda)\,\Lambda = \mathrm{K} = \frac{1}{1+k}\,\Gamma,$$

$$\tfrac{1}{2}(1+\lambda)\,\Lambda' = \mathrm{K}' = \frac{2}{1+k}\,\Gamma',$$

whence also
$$\tfrac{1}{2}\frac{\Lambda'}{\Lambda} = \frac{\mathrm{K}'}{\mathrm{K}} = 2\frac{\Gamma'}{\Gamma}.$$

First and Second Transformations.

245. We pass by a quadric transformation from the differential expression $\dfrac{dx}{\sqrt{1-x^2\,.\,1-k^2x^2}}$ to $\dfrac{dy}{\sqrt{1-y^2\,.\,1-\lambda^2y^2}}$ or

$\dfrac{dy}{\sqrt{1-y^2 . 1-\gamma^2 y^2}}$, viz. in the former case the transformation is from (x, k) to (y, λ), in the latter case from (x, k) to (y, γ).

The *first* form is

$$y = \frac{(1+k')\,x\sqrt{1-x^2}}{\sqrt{1-k^2x^2}}.$$

Here, taking throughout, denom. $= 1-k^2x^2$, we have

$$1-\ \ y^2 = \{1-(1+k')\,x^2\}^2 \qquad (+),$$

$$1-\lambda^2 y^2 = \{1-(1-k')\,x^2\}^2 \qquad (+),$$

$$\sqrt{1-y^2 . 1-\lambda^2 y^2} = 1-2x^2+k^2x^4 \qquad (+),$$

$$(1+\lambda)\,dy = \frac{2\,(1-2x^2+k^2x^4)\,dx}{\sqrt{1-x^2 . 1-k^2x^2}} \qquad (+),$$

and therefore

$$\frac{(1+\lambda)\,dy}{\sqrt{1-y^2 . 1-\lambda^2 y^2}} = \frac{2dx}{\sqrt{1-x^2 . 1-k^2x^2}}.$$

As x passes from 0 to $\dfrac{1}{\sqrt{1+k'}}$, y passes from 0 to 1, and as x continues to increase to 1, y diminishes from 1 to 0; we thus obtain the relation $2 . (1+\lambda)\,\Lambda = 2K$, that is $(1+\lambda)\,\Lambda = K$, which is one of the above-mentioned integral relations.

240. The *second* form is

$$y = \frac{(1+k)\,x}{1+kx^2}.$$

Taking here, denom. $= 1+kx^2$, we have

$$1-\ \ y = (1-x)\,(1-kx) \qquad (+),$$

$$1+\ \ y = (1+x)\,(1+kx) \qquad (+),$$

$$1-\gamma y = (1-x\sqrt{k})^2 \qquad (+),$$

$$1+\gamma y = (1+x\sqrt{k})^2 \qquad (+),$$

consequently

$$\sqrt{1-y^2 . 1-\gamma^2 y^2} = (1-kx^2)\sqrt{1-x^2 . 1-k^2x^2} \qquad (+),$$

and

$$dy = (1+k)\,(1-kx^2)\,dx \qquad (+).$$

in which two formulæ

$$\text{denom.} = (1 + kx^2)^2,$$

and we have therefore

$$\frac{dy}{\sqrt{1 - y^2 . 1 - \lambda^2 y^2}} = \frac{(1 + k)\, dx}{\sqrt{1 - x^2 . 1 - k^2 x^2}}.$$

Here x and y increase simultaneously from 0 to 1: hence taking the integrals between these limits we have another of the above-mentioned integral relations, $\Lambda = (1 + k)\,K$.

Complementary Transformations.

247. If in the *first* form we effect Jacobi's imaginary transformation, that is write $x = \dfrac{iX}{\sqrt{1 - X^2}}$, and $y = \dfrac{iY}{\sqrt{1 - Y^2}}$, then

$$\frac{dy}{\sqrt{1 - y^2 . 1 - \lambda^2 y^2}} = \frac{i\,dY}{\sqrt{1 - Y^2 . 1 - \lambda'^2 Y^2}},$$

and

$$\frac{dx}{\sqrt{1 - x^2 . 1 - k^2 x^2}} = \frac{i\,dX}{\sqrt{1 - X^2 . 1 - k'^2 X^2}};$$

and the differential relation is therefore changed into

$$\frac{(1 + \lambda)\, dY}{\sqrt{1 - Y^2 . 1 - \lambda'^2 Y^2}} = \frac{2dX}{\sqrt{1 - X^2 . 1 - k'^2 X^2}};$$

the integral equation is changed into

$$\frac{Y}{\sqrt{1 - Y^2}} = \frac{(1 + k')\, X}{\sqrt{1 - X^2 . 1 - k'^2 X^2}},$$

viz. this is

$$Y = \frac{(1 + k')\, X}{1 + k' X^2},$$

which integral form gives therefore the last-mentioned differential relation: observe that this integral form is what the *second* form becomes on writing therein X, Y for x, y, and for k the complementary modulus k'.

Moreover since X, Y increase simultaneously from 0 to 1, the differential equation leads to $(1 + \lambda)\,\Lambda' = 2K'$, which is another of the above-mentioned integral relations.

248. Similarly, if in the *second* form we effect Jacobi's imaginary transformation, then the differential equation is changed into

$$\frac{dY}{\sqrt{1-Y^2 \cdot 1-\gamma'^2 Y^2}} = \frac{(1+k)\,dX}{\sqrt{1-X^2 \cdot 1-k'^2X^2}};$$

the integral relation between x, y is changed into

$$\frac{Y}{\sqrt{1-Y^2}} = \frac{(1+k)\,X\sqrt{1-X^2}}{1-(1+k)\,X^2},$$

leading to

$$Y = \frac{(1+k)\,X\sqrt{1-X^2}}{\sqrt{1-k'^2X^2}},$$

which integral form gives rise therefore to the last-mentioned differential relation: observe that this integral form is what the *first* form becomes on writing X, Y for x, y, and for k the complementary modulus k'. Moreover as X passes from 0 to $\frac{1}{\sqrt{1+k}}$, Y passes from 0 to 1, and as X, continuing to increase, passes to 1, Y passes from 1 to 0: the differential equation gives therefore $2\Gamma' = (1+k)\,K'$, which completes the set of integral relations.

The Duplication Theory.

249. We may in two different ways combine the two transformations, and thus in two different ways obtain a "Duplication by two quadric transformations."

First duplication (through λ). Writing

$$z = \frac{(1+\lambda)\,y}{1+\lambda y^2}, \quad y = \frac{(1+k')\,x\sqrt{1-x^2}}{\sqrt{1-k'^2x^2}},$$

we have by what precedes

$$\frac{dy}{\sqrt{1-y^2 \cdot 1-\lambda^2y^2}} = \frac{2}{1+\lambda}\,\frac{dx}{\sqrt{1-x^2 \cdot 1-k^2x^2}} = \frac{1}{1+\lambda}\,\frac{dz}{\sqrt{1-z^2 \cdot 1-k^2z^2}},$$

and therefore

$$\frac{dz}{\sqrt{1-z^2 \cdot 1-k^2z^2}} = \frac{2dx}{\sqrt{1-x^2 \cdot 1-k^2x^2}},$$

where, from the assumed integral equations,

$$z=\frac{2x\sqrt{1-x^2}\,.\sqrt{1-k^2x^2}}{1-k^2x^4}.$$

250. *Second* duplication (through γ). Writing

$$z=\frac{(1+\gamma)\,y\,\sqrt{1-y^2}}{\sqrt{1-\gamma^2y^2}},\qquad y=\frac{(1+k)\,x}{1+kx^2},$$

we have by what precedes

$$\frac{(1+k)\,dz}{\sqrt{1-z^2\,.\,1-k^2z^2}}=\frac{2dy}{\sqrt{1-y^2\,.\,1-\gamma^2y^2}},$$

$$\frac{dy}{\sqrt{1-y^2\,.\,1-\gamma^2y^2}}=\frac{(1+k)\,dx}{\sqrt{1-x^2\,.\,1-k^2x^2}},$$

and therefore

$$\frac{dz}{\sqrt{1-z^2\,.\,1-k^2z^2}}=\frac{2dx}{\sqrt{1-x^2\,.\,1-k^2x^2}},$$

and the two integral equations give, as in the first duplication,

$$z=\frac{2x\sqrt{1-x^2}\sqrt{1-k^2x^2}}{1-k^2x^4}.$$

251. In the first duplication, assuming $x=\text{sn}(u,k)$, $y=\text{sn}(v,\lambda)$, $z=\text{sn}(w,k)$, and observing that u, v, w vanish together, we obtain $v=(1+k')\,u$, $w=2u$, and the formulæ are

$$x=\text{sn}(u,k),$$

$$y=\text{sn}(\overline{1+k'}\,u,\lambda),\ =\frac{(1+k')\,\text{sn}(u,k)\,\text{cn}(u,k)}{\text{dn}(u,k)},$$

$$z=\text{sn}(2u,k),\qquad =\frac{(1+\lambda)\,\text{sn}(\overline{1+k'}\,u,\lambda)}{1+\lambda\,\text{sn}^2(\overline{1+k'}\,u,\lambda},$$

and similarly in the second duplication the formulæ are

$$x = \mathrm{sn}\,(u,\ k),$$

$$y = \mathrm{sn}\,(\overline{1+k}u,\ \gamma) = \frac{(1+k)\,\mathrm{sn}\,(u,\ k)}{1+k\,\mathrm{sn}^2(u,\ k)}\,,$$

$$z = \mathrm{sn}\,(2u,\ k) \qquad = \frac{(1+\gamma')\,\mathrm{sn}\,(\overline{1+k}u, \gamma)\,\mathrm{cn}\,(\overline{1+k}u, \gamma)}{\mathrm{dn}\,(\overline{1+k}u,\ \gamma)}\,.$$

Transformations of the Elliptic Functions sn, cn, dn.

252. Take the first and second y-formulæ as they stand. In the first z-formula change k, λ into λ', k', and for u write $\frac{1}{2}(1+k')\,u$. In the second z-formula change k, γ into γ', k', and for u write $\frac{1}{2}(1+k)\,u$, observing that $\frac{1}{2}(1+\gamma')(1+k) = 1$. We thus obtain the formulæ:

$$\mathrm{sn}\,(\overline{1+k'}u,\ \lambda) = \frac{(1+k')\,\mathrm{sn}\,(u,\ k)\,\mathrm{cn}\,(u,\ k)}{\mathrm{dn}\,(u,\ k)} \quad \text{(from first } y\text{-formula)},$$

$$\mathrm{sn}\,(\overline{1+k}\,u,\ \gamma) = \frac{(1+k)\,\mathrm{sn}\,(u,\ k)}{1+k\,\mathrm{sn}^2(u,\ k)} \quad \text{(from second } y\text{-formula)},$$

$$\mathrm{sn}\,(\overline{1+k'}u,\ \lambda') = \frac{(1+k')\,\mathrm{sn}\,(u,\ k')}{1+k'\,\mathrm{sn}^2(u,\ k')} \quad \text{(from first } z\text{-formula)},$$

$$\mathrm{sn}\,(\overline{1+k}\,u,\ \gamma') = \frac{(1+k)\,\mathrm{sn}\,(u,\ k')\,\mathrm{cn}\,(u,\ k')}{\mathrm{dn}\,(u,\ k')} \quad \text{(from second } z\text{-formula)},$$

and we may complete the system by adding the values of the functions cn, dn.

253. We have thus the formulæ:

	sn =	cn =	dn =	
$(\overline{1+k'}u,\ \lambda)$	$(1+k')\,\mathrm{sn}\,u\,\mathrm{cn}\,u$	$1-(1+k')\,\mathrm{sn}^2u$	$1-(1-k')\,\mathrm{sn}^2u$	$\div\,\mathrm{dn}\,u$
$(\overline{1+k}u,\ \gamma)$	$(1+k)\,\mathrm{sn}\,u$	$\mathrm{cn}\,u\,\mathrm{dn}\,u$	$1-k\,\mathrm{sn}^2u$	$\div(1+k\,\mathrm{sn}^2u)$
$(\overline{1+k'}u,\ \lambda')$	$(1+k')\,\mathrm{sn}_1u$	$\mathrm{cn}_1u\,\mathrm{dn}_1u$	$1-k'\,\mathrm{sn}_1^2u$	$\div(1+k'\,\mathrm{sn}_1^2u)$
$(\overline{1+k}u,\ \gamma')$	$(1+k)\,\mathrm{sn}_1u\,\mathrm{cn}_1u$	$1-(1+k)\,\mathrm{sn}_1^2u$	$1-(1-k)\,\mathrm{sn}_1^2u$	$\div\,\mathrm{dn}_1u$

where in the first and second lines sn u, &c. denote (as usual) sn(u, k), &c.: and in the third and fourth lines $\text{sn}_1 u$, &c. denote sn(u, k'), &c.

Third and Fourth Transformations.

234. In what precedes we have a complete theory, or say "the standard theory," of the quadric transformation, but we may add a third and fourth form.

The *third* form is:

$$y = \frac{1+\lambda-2x^2}{1+\lambda-2\lambda x^2}.$$

Here, denom. $= 1+\lambda-2\lambda x^2$, we have

$$\begin{aligned} 1-y &= 2(1-\lambda)x^2 & (+),\\ 1+y &= 2(1+\lambda)(1-x^2) & (+),\\ 1-\lambda y &= \lambda'^2 & (+),\\ 1+\lambda y &= (1+\lambda)^2(1-k^2x^2) & (+), \end{aligned}$$

and thence, denom. $= (1+\lambda-2\lambda x^2)^2$, we have

$$\begin{aligned} \sqrt{1-y^2 . 1-\lambda^2y^2} &= 2\lambda'^2(1+\lambda)\,x\sqrt{1-x^2 . 1-k^2x^2},\\ dy &= -4\lambda'^2 x\,dx, \end{aligned}$$

and consequently

$$\frac{(1+\lambda)\,dy}{\sqrt{1-y^2 . 1-\lambda^2y^2}} = -\frac{2dx}{\sqrt{1-x^2 . 1-k^2x^2}}.$$

235. To connect with the standard form, observe that writing $x = \text{sn}(u, k)$, $y = \text{sn}(v, \lambda)$, we have $dv = -\dfrac{2}{1+\lambda}du$, $= -(1+k')\,du$, that is, $v = C-(1+k')\,u$, or (since for $x=0$ we have $y=1$, that is for $u=0$ we have $v=\Lambda$) the value is $v = \Lambda-(1+k')\,u$, and the integral equation is

$$\text{sn}(\Lambda-\overline{1+k'}\,u, \lambda) = \frac{1+\lambda-2\,\text{sn}^2(u, k)}{1+\lambda-2\lambda\,\text{sn}^2(u, k)},$$

or, what is the same thing,

$$= \frac{1-(1+k')\,\text{sn}^2(u, k)}{1-(1-k')\,\text{sn}^2(u, k)},$$

but the left-hand side is $= \text{cn}\,(\overline{1+k'}\,u, \lambda) + \text{dn}\,(\overline{1+k'}\,u, \lambda)$, and substituting herein the values of the two terms from the table No. 253 the formula is verified.

256. The *Fourth* form is

$$y = \frac{1+kx^2}{2\sqrt{k}\,.\,x}.$$

Here

$$\begin{aligned} 1-\ y &= -\ (1-x\sqrt{k})^2 && (+),\\ 1+\ y &= \ \ (1+x\sqrt{k})^2 && (+),\\ 1-\gamma y &= -\gamma\,(1-x)\,(1-kx) && (+),\\ 1+\gamma y &= \ \ \gamma\,(1+x)\,(1+kx) && (+), \end{aligned}$$

where denom. $= 2\sqrt{k}x$,

and hence

$$\begin{aligned} \sqrt{1-y^2\,.\,1-\gamma^2y^2} &= -\gamma\,(1-kx^2)\,\sqrt{1-x^2\,.\,1-k^2x^2} && (+),\\ dy &= -2\sqrt{k}\,(1-kx^2)\,dx && (+), \end{aligned}$$

where denom. $= 4kx^2$.

Consequently

$$\frac{dy}{\sqrt{1-y^2\,.\,1-\gamma^2y^2}} = \frac{(1+k)\,dx}{\sqrt{1-x^2\,.\,1-k^2x^2}}.$$

257. To connect with the standard form, putting $x = \text{sn}\,(u, k)$, $y = \text{sn}\,(v, \gamma)$, we find $v = C + (1+k)\,u$, and then, since $x = 1$ gives $y = \dfrac{1+k}{2\sqrt{k}}, = \dfrac{1}{\gamma}$, we have $\Gamma + i\Gamma' = C + (1+k)\,\text{K}$, or since $(1+k)\,\text{K} = \Gamma$, this gives $C = i\Gamma'$, and therefore $v = i\Gamma' + \overline{1+k}\,u$ and $y = \text{sn}\,(i\Gamma' + \overline{1+k}\,u, \gamma)$: wherefore the equation is

$$\text{sn}\,(i\Gamma' + \overline{1+k}\,u, \gamma) = \frac{1+k\,\text{sn}^2(u, k)}{2\sqrt{k}\,\text{sn}\,(u, k)}.$$

The left-hand side is

$$\frac{1}{\gamma\,\text{sn}\,(\overline{1+k}\,u, \gamma)}, = \frac{1+k\,\text{sn}^2(u, k)}{\gamma\,(1+k)\,\text{sn}\,(u, k)},$$

or, what is the same thing, $=\dfrac{1+k\,\text{sn}^2(u,k)}{2\sqrt{k}\,\text{sn}(u,k)}$, which is right.

258. Making in the *third* form Jacobi's imaginary transformation $x=\dfrac{iX}{\sqrt{1-X^2}}$, $y=\dfrac{iY}{\sqrt{1-Y^2}}$, it becomes $\dfrac{iY}{\sqrt{1-Y^2}}=\dfrac{1+k'X^2}{1-k'X^2}$, giving $Y=\dfrac{1+k'X^2}{2\sqrt{k'}X}$, viz. this is the *fourth* form, writing therein X, Y for x, y, and for k the complementary modulus k'.

And similarly making in the *fourth* form Jacobi's imaginary transformation, it becomes $\dfrac{-Y}{\sqrt{1-Y^2}}=\dfrac{1-1+kX^2}{2\sqrt{k}X\sqrt{1-X^2}}$, giving $Y=\dfrac{1+\lambda'-2X^2}{1+\lambda'-2\lambda'X^2}$, viz. this is what the *third* form becomes on substituting therein X, Y for x, y, and for λ the complementary modulus λ'.

259. The cases $n=3, 5, 7$ are worked out in accordance with the general algebraical theory explained in the preceding Chapter. In the case $n=3$, it is to be observed, that the process introduces a single indeterminate quantity a, in terms of which the moduli k, λ are expressed; the resulting form, containing only this parameter, is an interesting and valuable one, but it is nevertheless proper to obtain the modular equation, and express the formula in terms of the two quantities u, v connected by this modular equation. I have in regard to this same case $n=3$ gone into some details to connect the formulæ with the transcendental ones depending on the trisection of the complete functions, as obtained from the general theory for the case of an odd-prime.

The Cubic Transformation. Art. Nos. 260 to 262.

260. We write

$$\frac{1-y}{1+y}=\left(\frac{1-ax}{1+ax}\right)^2\frac{1-x}{1+x},$$

giving

$$y=\frac{x\{2a+1+a^2x^2\}}{1+a(a+2)x^2},$$

and then the conditions in order to the change x, y into $\frac{1}{kx}$, $\frac{1}{\lambda y}$, are

$$a^2 = \Omega,$$

$$k^2(2a+1) = \Omega a(a+2),$$

$$\lambda = \frac{k^3}{\Omega^2}.$$

It is moreover clear that $\frac{1}{M} = 2a+1$.

261. We have at once everything expressed in terms of a, viz. we have first $\Omega = a^2$, and thence

$$k^2 = \frac{a^3(2+a)}{2a+1},\quad \lambda^2 = a\left(\frac{2+a}{2a+1}\right)^3,$$

and then

$$1-y = (1-ax)^2(1-x), \qquad (\div),$$

$$1+y = (1+ax)^2(1+x), \qquad (\div),$$

$$1-\lambda y = \left(1-\frac{k}{a}x\right)^2(1-kx), \qquad (\div),$$

$$1+\lambda y = \left(1+\frac{k}{a}x\right)^2(1+kx), \qquad (\div),$$

where $\quad$ denom. $= 1 + a(a+2)x^2$,

and thence

$$\frac{dy}{\sqrt{1-y^2 . 1-\lambda^2 y^2}} = \frac{(2a+1)\,dx}{\sqrt{1-x^2 . 1-k^2x^2}},$$

the factor $2a+1$ being obtained directly from the consideration that, x and y being small, $y = (2a+1)x$. The modular equation is here replaced by the two equations

$$k^2 = \frac{a^3(2+a)}{2a+1},\quad \lambda^2 = a\left(\frac{2+a}{2a+1}\right)^3,$$

which in fact determine λ in terms of k. We obtain

$$k'^2 = \frac{(1-a)(1+a)^3}{2a+1},\quad \lambda'^2 = \frac{(1+a)(1-a)^3}{(2a+1)^3}.$$

and thence

$$\sqrt{k\lambda} = \frac{\alpha(2+\alpha)}{2\alpha+1}, \quad \sqrt{k'\lambda'} = \frac{1-\alpha^2}{2\alpha+1},$$

hence $\sqrt{k\lambda} + \sqrt{k'\lambda'} = 1$, which is a form of the modular equation.

We have $\frac{k^3}{\lambda} = \alpha^4$, that is writing $\sqrt[4]{k} = u$, $\sqrt[4]{\lambda} = v$, we have $\alpha = \frac{u^3}{v}$. Moreover $\sqrt{k\lambda} = \frac{\alpha(\alpha+2)}{2\alpha+1}$, that is $u^2v^2 = \frac{\alpha(\alpha+2)}{2\alpha+1}$, or (substituting herein for α its value),

$$u^2v^2 = \frac{u^3(u^3+2v)}{v(2u^3+v)},$$

or $$u(u^3+2v) = v^3(2u^3+v),$$

that is $$u^4 - v^4 + 2uv(1-u^2v^2) = 0,$$

which is the modular equation, expressed as an equation between $u = \sqrt[4]{k}$, and $v = \sqrt[4]{\lambda}$.

262. Introducing into the equations u, v in place of α we have

$$y = [(v+2u^3)vx + u^6x^3] \quad (\div),$$

$$1+y = (v+u^3x)^2(1+x) \quad (\div),$$

$$1-y = (v-u^3x)^2(1-x) \quad (\div),$$

$$1+v^4y = v^2(1+uvx)^2(1+u^4x) \quad (\div),$$

$$1-v^4y = v^2(1-uvx)^2(1-u^4x) \quad (\div),$$

where the denominator is in the first instance obtained in the form $v^2 + u^3(u^3+2v)x^2$; or, altering this by means of the modular equation, we have

$$\text{denom.} = v^2[1 + vu^2(v+2u^3)x^2];$$

and then

$$\frac{v\,dy}{\sqrt{1-y^2 \cdot 1-v^8y^2}} = \frac{(v+2u^3)\,dx}{\sqrt{1-x^2 \cdot 1-u^8x^2}}.$$

The Quintic Transformation. Art. Nos. 263 to 267.

263. We write

$$\frac{1-y}{1+y}=\frac{(1-x)(1-\alpha x+\beta x^2)^2}{(1+x)(1+\alpha x+\beta x^2)^2},$$

giving
$$y=\frac{x\left[(2\alpha+1)+(2\alpha\beta+2\beta+\alpha^2)x^2+\beta^2x^4\right]}{1+(2\beta+2\alpha+\alpha^2)x^2+(\beta^2+2\alpha\beta)x^4}.$$

And then the conditions in order to the change x into $\frac{1}{kx}$, y into $\frac{1}{\lambda y}$, are

$$\beta^2=\Omega,$$
$$k^2(2\alpha\beta+2\beta+\alpha^2)=\Omega(2\alpha+2\beta+\alpha^2),$$
$$k^4(2\alpha+1)\qquad=\Omega(\beta^2+2\alpha\beta),$$

where $\Omega^2=\frac{k^5}{\lambda}$. It is moreover clear that $\frac{1}{M}=2\alpha+1$.

264. Assuming $k=u^4$, $\lambda=v^4$, we have $\Omega^2=\frac{u^{20}}{v^4}$, and thence $\beta=\sqrt{\Omega}=\frac{u^5}{v}$. Substituting these values the last equation becomes $(2\alpha+1)uv^4=u^5+2\alpha v$, that is

$$2\alpha v(1-uv^3)=u(v^4-u^4),\text{ or }2\alpha=\frac{u(v^4-u^4)}{v(1-uv^3)}.$$

The second equation becomes

$$\begin{aligned}(v^2-u^2)(2\beta+\alpha^2)&=u^2(1-u^3v)2\alpha,\\&=\frac{u^3}{v}\,\frac{(v^4-u^4)(1-u^3v)}{1-uv^3},\end{aligned}$$

that is
$$2\beta+\alpha^2=\frac{u^3}{v}\,\frac{(v^2+u^2)(1-u^3v)}{1-uv^3},$$

whence
$$\begin{aligned}\alpha^2&=\frac{u^3}{v}\left\{\frac{(v^2+u^2)(1-u^3v)}{1-uv^3}-2u^2\right\},\\&=\frac{u^3}{v}\,\frac{(v^2-u^2)(1+u^3v)}{1-uv^3}.\end{aligned}$$

And dividing this value by the value first obtained for 2α, we have

$$2\alpha = \frac{4u^2(1+u^3v)}{v^2+u^2}, \qquad = \frac{u\,(v^4-u^4)}{v(1-uv^3)},$$

whence $\quad -(v^2+u^2)(v^4-u^4)+4uv(1-uv^3)(1+u^3v)=0,$

or, what is the same thing,

$$u^6-v^6+5u^2v^2(u^2-v^2)+4uv\,(1-u^4v^4)=0,$$

the modular equation.

265. We then have

$$2\alpha+1=\frac{v-u^5}{v\,(1-uv^3)},$$

$$2\alpha\beta+2\beta+\alpha^2=\frac{u^3(v^2+u^2)(v-u^5)}{v^3\,(1-uv^3)}, \quad 2\beta+2\alpha+\alpha^2=\frac{u\,(v^2+u^2)(v-u^5)}{1-uv^3},$$

$$\beta^2=\frac{u^{10}}{v^5}, \qquad\qquad \beta^2+2\alpha\beta=\frac{vu^4\,(v-u^5)}{1-uv^3},$$

and hence

$$y=\frac{v\,(v-u^5)\,x+u^3(v^2+u^2)(v-u^5)\,x^3+u^{10}(1-uv^3)\,x^5}{v^3(1-uv^3)+uv^2(v^2+u^2)(v-u^5)\,x^2+v^4u^6(v-u^5)\,x^4},$$

or if we please

$$\frac{1-y}{1+y}=\frac{1-x}{1+x}\cdot\left(\frac{1-\dfrac{u\,(v^4-u^4)}{2(1-uv^3)}\,x+\dfrac{u^3}{v}\,x^2}{1+\dfrac{u\,(v^4-u^4)}{2(1-uv^3)}\,x+\dfrac{u^3}{v}\,x^2}\right)^2,$$

leading to

$$\frac{v\,(1-uv^3)\,dy}{\sqrt{1-y^2 . 1-v^8y^2}}=\frac{(v-u^5)\,dx}{\sqrt{1-x^2 . 1-u^8x^2}}.$$

266. If from the original equations we eliminate k, Ω, we obtain

$$(\alpha^2+2\alpha\beta+2\beta)^2(2\alpha+\beta)-(\alpha^2+2\alpha+2\beta)^2(2\alpha+1)\,\beta=0,$$

viz. this is

$$2\alpha^2(1-\beta)\,[\alpha^2-2\beta\,(1+\alpha+\beta)]=0.$$

But $\alpha=0$ gives simply $y=x$; $1-\beta=0$ corresponds to $k=\lambda=1$, and does not give a transformation; rejecting these factors, we have

$$\alpha^2-2\beta(1+\alpha+\beta)=0,$$

viz. if α, β are connected by this equation, and

$$\frac{1-y}{1+y}=\frac{1-x}{1+x}\left(\frac{1-\alpha x+\beta x^2}{1+\alpha x+\beta x^2}\right)^2,$$

then there exist values of M, k, λ, such that

$$\frac{M\,dy}{\sqrt{1-y^2 . 1-\lambda^2y^2}}=\frac{dx}{\sqrt{1-x^2 . 1-k^2x^2}},$$

viz. we have $\dfrac{1}{M}=2\alpha+1$, $k^2=\dfrac{\beta^3(2\alpha+2\beta+\alpha^2)}{(2\alpha\beta+2\beta+\alpha^2)}$, or, what is the same thing, $k^4=\dfrac{\beta^3(\beta^2+2\alpha\beta)}{2\alpha+1}$ and $\lambda^2=\dfrac{k^{10}}{\beta^8}$: this is of course only another form of the theorem.

267. It is worth while to consider the case $\beta=1$: as already mentioned this gives $k^2=\lambda^2=1$: we have

$$\frac{1-y}{1+y}=\frac{1-x}{1+x}\left(\frac{1-\alpha x+x^2}{1+\alpha x+x^2}\right)^2,$$

giving
$$y=\frac{x\,[2\alpha+1+(\alpha^2+2\alpha+2)\,x^2+x^4]}{1+(\alpha^2+2\alpha+2)\,x^2+(2\alpha+1)\,x^4},$$

and calling the denominator D, we have thence

$$1-y^2=\frac{1}{D^2}(1-x^2)\,[1+(2-\alpha^2)\,x^2+x^4]^2.$$

Moreover $dy=\dfrac{1}{D^2}(1,\ x^2)^4\,dx$, but the numerator $(1,\ x^2)^4$ contains, not the square, but only the first power of $1+(2-\alpha^2)x^2+x^4$; we in fact find

$$dy=\frac{1}{D^2}[2\alpha+1+(-\alpha^2-4\alpha+2)x^2+(2\alpha+1)x^4]\,[1+(2-\alpha^2)x^2+x^4]\,dx,$$

and consequently

$$\frac{dy}{1-y^2}=\frac{2\alpha+1+(-\alpha^2-4\alpha+2)\,x^2+(2\alpha+1)\,x^4}{1+(2-\alpha^2)\,x^2+x^4}\cdot\frac{dx}{1-x^2}.$$

viz. the factor which multiplies $\dfrac{dx}{1-x^2}$ is not a mere constant; and we have thus no quintic transformation.

The Septic Transformation. Art. Nos. 268 and 269.

268. We write

$$\frac{1-y}{1+y} = \frac{1-x}{1+x}\left(\frac{1-\alpha x+\beta x^2-\gamma x^3}{1+\alpha x+\beta x^2+\gamma x^3}\right)^2,$$

and thence the conditions in order to the change x, y into $\dfrac{1}{kx}$, $\dfrac{1}{\lambda y}$ are

$$\gamma^2 = \Omega,$$
$$k^2(\beta^2+2\beta\gamma+2\alpha\gamma) = \Omega(2\beta+2\gamma+\alpha^2),$$
$$k^4(2\beta+2\alpha\beta+2\gamma+\alpha^2) = \Omega(\beta^2+2\alpha\beta+2\gamma+2\alpha\gamma),$$
$$k^6(1+2\alpha) = \Omega(\gamma^2+2\beta\gamma),$$

where $\Omega^2 = \dfrac{k^7}{\lambda}$. Writing as before $k = u^4$, $\lambda = v^4$, we have $\Omega = \dfrac{u^{14}}{v^2}$, and thence $\gamma, = \sqrt{\Omega}, = \dfrac{u^7}{v}$. Moreover, by taking x and y each indefinitely small we obtain at once $1+2\alpha = \dfrac{1}{M}$, and substituting these results in the last of the four equations we find $2\beta = u^3v^3\left(\dfrac{1}{M}-\dfrac{u^4}{v^4}\right)$: and the second and third equations become

$$v^2(\beta^2+2\beta\gamma+2\alpha\gamma) = u^6(2\beta+2\gamma+\alpha^2),$$
$$u^2v^2(2\beta+2\alpha\beta+2\gamma+\alpha^2) = \beta^2+2\alpha\beta+2\gamma+2\alpha\gamma,$$

in which equations α, β, γ are to be considered as given functions of u, v, M: the equations therefore determine the relation between u and v (the modular equation); and they also determine the multiplier M as a function of u, v.

269. The final results are simple: but it is by no means easy to deduce them from the equations, or even to verify them, when known: we have

$$(1-u^8)(1-v^8) = (1-uv)^8.$$

or, as this may also be written,

$$(v-u^7)(u-v^7)+7uv(1-uv)^2(1-uv+u^2v^2)^2=0,$$

for the modular equation: and then M is given in either of the two forms

$$\frac{1}{M}=-\frac{7u(1-uv)(1-uv+u^2v^2)}{u-v^7},\quad M=\frac{v(1-uv)(1-uv+u^2v^2)}{v-u^7},$$

values which are identical in virtue of the last-mentioned form of the modular equation. And then as above

$$2\alpha=\frac{1}{M}-1,\quad 2\beta=u^2v^2\left(\frac{1}{M}-\frac{u^4}{v^4}\right),\quad \gamma=\frac{u^7}{v},$$

which are the values of the coefficients α, β, γ.

Forms of the Modular Equation in the Cubic and Quintic Transformations. Art. Nos. 270 to 273.

270. In the cubic transformation, the modular equation is originally given as an equation of the fourth order between (u, v): but we thence easily derive equations of the same order, 4, between (u^2, v^2) (u^4, v^4), and (u^8, v^8): the forms are

I.

	1	u	u^2	u^3	u^4
1					$+1$
v		$+2$			
v^2					
v^3				-2	
v^4	-1				

$=0,$

II.

	1	u^2	u^4	u^6	u^8
1					$+1$
v^2		-4			
v^4			$+6$		
v^6				-4	
v^8	$+1$				

$=0,$

III.

	1	u^4	u^8	u^{12}	u^{16}
1					+1
v^4		−16		+12	
v^8			+6		
v^{12}		+12		−16	
v^{16}	+1				

= 0,

IV.

	1	u^8	u^{16}	u^{24}	u^{32}
1					+1
v^8		−256	+384	−132	
v^{16}		+384	−762	+384	
v^{24}		−132	+384	−256	
v^{32}	+1				

= 0.

271. Here I. is the original form $u^4 - v^4 + 2uv(1 - u^2v^2) = 0$.

II. may be written $(1 - u^8)(1 - v^8) = (1 - u^2v^2)^4$. Jacobi obtains this, *Fund. Nova*, p. 68, as follows: we have

$$(1 - u^4)(1 + v^4) = 1 - u^4v^4 + 2uv(1 - u^2v^2)$$
$$= (1 - u^2v^2)(1 + uv)^2, = (1 - uv)(1 + uv)^3,$$
$$(1 + u^4)(1 - v^4) = 1 - u^4v^4 - 2uv(1 - u^2v^2)$$
$$= (1 - u^2v^2)(1 - uv)^2, = (1 + uv)(1 - uv)^3,$$

whence the form. Writing

$$k^2 = u^8,\ k'^2 = 1 - u^8,\ \lambda^2 = v^8,\ \lambda'^2 = 1 - v^8,$$

the equation is $$k'^2\lambda'^2 = (1 - \sqrt{k\lambda})^4,$$

or, what is the same thing,

$$\sqrt{k\lambda} + \sqrt{k'\lambda'} = 1,$$

the irrational form obtained *ante*, No. 261.

III. may be written $(u^4 - v^4)^4 - 16u^4v^4(1 - u^8)(1 - v^8) = 0$: which form can be at once derived from II. under the form $(1 - u^8)(1 - v^8) = (1 - u^2v^2)^4$, by writing therein

$$1 - u^2v^2 = -(u^4 - v^4) \div 2uv.$$

IV. may be written

$$(u^8 - v^8)^4 = 128\, u^4v^4 (1 - u^8)(1 - v^8)(2 - u^8 - v^8 + 2u^4v^4):$$

or say

$$(k^2 - \lambda^2)^4 = 128\, k^2\lambda^2 (1 - k^2)(1 - \lambda^2)(2 - k^2 - \lambda^2 + 2k^2\lambda^2),$$

Fund. Nova, p. 67, viz. this is the modular equation expressed rationally in terms of k^2, λ^2. Writing, with Jacobi, $q = 1 - 2k^2$, $l = 1 - 2\lambda^2$, it becomes

$$(q - l)^4 = 64 (1 - q^2)(1 - l^2)(3 + ql).$$

272. In the quintic transformation the modular equation is originally given as an equation of the order 6 between u, v: this may be expressed as an equation of the same order 6 between (u^2, v^2), (u^4, v^4), (u^8, v^8), viz. the four forms are

I.

	1	u	u^2	u^3	u^4	u^5	u^6	
1							$+1$	$=0,$
v		$+4$						
v^2					$+5$			
v^3								
v^4			-5					
v^5						-4		
v^6	-1							

II.

	1	u^2	u^4	u^6	u^8	u^{10}	u^{12}	
1							$+1$	$=0,$
v^2		-16				$+10$		
v^4					$+15$			
v^6				-20				
v^8			$+15$					
v^{10}		$+10$				-16		
v^{12}	$+1$							

III.

	1	u^4	u^8	u^{12}	u^{16}	u^{20}	u^{24}	
1							+1	= 0,
v^4		− 256		+ 320		− 70		
v^8			− 640		+ 655			
v^{12}		+ 320		− 660		+ 320		
v^{16}			+ 655		− 640			
v^{20}		− 70		+ 320		− 256		
v^{24}	+1							

IV.

	1	u^8	u^{16}	u^{24}	u^{32}	u^{40}	u^{48}	
1							+1	= 0.
v^8		− 65536	+ 163840	− 138240	+ 43520	− 8590		
v^{16}		+ 163840	− 138120	− 207360	+ 183135	+ 43520		
v^{24}		− 138240	− 207360	+ 691160	− 207360	− 138240		
v^{32}		+ 43520	+ 183135	− 207360	− 183120	+ 163840		
v^{40}		− 8590	+ 43520	− 138240	+ 163840	− 65536		
v^{48}	+1							

273. Here I. is the original form

$$u^4 - v^4 + 5u^2v^2(u^2 - v^2) + 4uv(1 - u^2v^2) = 0.$$

II. may be written $(u^8 - v^8)^2 - 16u^2v^2(1 - u^8)(1 - v^8) = 0$. This Jacobi obtains, *Fund. Nova*, p. 69, directly as follows: writing the modular equation in the form

$$(u^2 - v^2)(u^2 + 6u^2v^2 + v^2) = -4uv(1 - u^2v^2),$$

from this we deduce

$$(u^2 - v^2)(u + v)^4, = (u - v)(u + v)^6, = -4uv(1 - u^4)(1 + v^4),$$

$$(u^2 - v^2)(u - v)^4, = (u - v)^6(u + v), = -4uv(1 + u^4)(1 - v^4),$$

and thence the form in question.

The form IV. may be transformed into:

$$(u^8 - v^8)^4 = 512\, u^8 v^8 \text{ into}$$

	1	u^8	u^{16}	u^{24}	u^{32}
1	+ 128	− 690	+ 270	− 85	+ 7
v^8	− 690	+ 260	+ 405	− 260	− 85
v^{16}	+ 270	+ 405	− 1850	+ 405	+ 270
v^{24}	− 85	− 260	+ 405	+ 260	− 690
v^{32}	+ 7	− 85	+ 270	− 820	+ 128

and thence into:

$$(u^8 - v^8)^4 = 512\, u^8 v^8 (1 - u^8)(1 - v^8) \text{ into}$$

	1	u^8	u^{16}	u^{24}
1	+ 128	− 192	+ 78	− 7
v^8	− 192	− 252	+ 423	+ 78
v^{16}	+ 78	+ 423	− 252	− 192
v^{24}	− 7	+ 78	− 192	+ 128

which is the modular equation expressed rationally in terms of u^8, v^8, $= k^2$, λ^2. If we herein write $q = 1 - 2k^2$, $l = 1 - 2\lambda^2$, this becomes:

$$(q - l)^4 = 256\,(1 - q^2)(1 - l^2) \text{ into}$$

	1	q	q^2	q^3
1			+ 405	
l		+ 486		− 9
l^2	+ 405		− 270	
l^3		− 9		+ 16

which is equivalent to the form given *Fund. Nova*, p. 67. The equation may also be written

$$(q - l)^4 = 256\,(1 - q^2)(1 - l^2)\,[16ql\,(9 - ql)^2 + 9\,(45 - ql)\,(q - l)^2].$$

Properties of the Modular Equation for n an odd prime.
Art. Nos. 274 to 277.

274. The cubic, quintic and septic transformations supply illustrations of certain properties of the modular equation for any odd prime value of n. It may be convenient to mention here that the equation has been further calculated for the odd prime values 11, 13, 17 and 19, by Sohnke, in the Memoir, Equationes modulares pro transformatione functionum ellipticarum, *Crelle*, t. XVI (1836), pp. 97—130; the results are given in a tabular form in my Memoir on the transformation of elliptic functions, *Phil. Trans.* t. 164 (1874), pp. 397—456.

The degree in u, v respectively is $= n + 1$.

275. The equation remains unaltered if for u, v we write therein $-u$, $-v$ respectively.

Connected herewith we have an important property not *explicitly* noticed by Jacobi. In general an equation $F(u, v) = 0$ of the order ν in u and v respectively can be transformed into an equation of the order 2ν, in u^2, v^2 respectively: viz. the transformed equation is

$$F(u, v)\, F(-u, v)\, F(u, -v)\, F(-u, -v) = 0,$$

where the left-hand side is a rational and integral function of u^2, v^2 of the order 2ν in these quantities respectively. But as regards the modular equation, since $F(-u, -v) = F(u, v)$, and therefore also $F(-u, v) = F(u, -v)$, the transformed equation may be written $F(u, v)\, F(u, -v) = 0$, and it is thus an equation in u^2, v^2 of the order ν, $= n + 1$, only. It has just been seen how in the cases $n = 3$ and $n = 5$, we obtain equations not only in (u^2, v^2), but also in (u^4, v^4) and in (u^8, v^8), of the same order, 4, 6, in these quantities respectively: and the same thing might easily be shown in the case $n = 7$.

276. The modular equation remains unaltered when for u, v we write therein v, $(-)^{\frac{1}{8}(n^2-1)}\, u$; viz. $n = 3$ or 5, $(v, -u)$, but $n = 7$, (v, u) in place of (u, v). Taking the equation in

(u^2, v^2), (u^4, v^4) or (u^8, v^8) this merely means that the equation is symmetrical as regards the two variables, but as regards the original form as an equation between (u, v), we have, as just stated, $n \equiv 3$ or $5 \pmod 8$ a skew symmetry, but $n \equiv 1$ or $7 \pmod 8$ a complete symmetry.

The above change u, v into $[v, (-)^{\frac{1}{8}(n^2-1)} u]$ changes the multiplier M into $\dfrac{(-)^{\frac{1}{2}(n-1)}1}{nM}$, and it thus appears that, given the expression of the multiplier in terms of (u, v), we can deduce the modular equation: thus, $n = 3$,

$$M = \frac{v}{v + 2u^3}, \quad \frac{-1}{3M} = \frac{-u}{-u + 2v^3},$$

whence $$(2u^3 + v)(2v^3 - u) - 3uv = 0,$$

the modular equation. And so also, $n = 5$,

$$M = \frac{v(1 - uv^3)}{v - u^5}, \quad \frac{1}{5M} = \frac{-u(1 + u^3 v)}{-u - v^5},$$

whence $$5uv(1 - uv^3)(1 + u^3 v) - (v - u^5)(v^5 + u) = 0,$$

the modular equation.

277. The modular equation remains unaltered on changing therein u, v into $\dfrac{1}{u}$, $\dfrac{1}{v}$ respectively.

The modular equation also remains unaltered on changing therein k, λ into k', λ' respectively, that is u^8, v^8 into $1 - u^8$, $1 - v^8$; this appears from the equations expressed in terms of $q = 1 - 2k^2$ and $l = 1 - 2\lambda^2$; viz. by the change in question q, l are changed into $-q$, $-l$; and the equation remains unaltered.

Two Transformations leading to Multiplication. Art. No. 278.

278. It appears from the property stated in No. 276 that we can by a twice-repeated transformation obtain a multiplication, thus, $n = 3$,

$$y = \frac{v(v+2u^3)x + u^6x^3}{v^2 + vu^2(v+2u^3)x^2}$$

gives

$$\frac{dy}{\sqrt{1-y^2 \, . \, 1-v^8y^2}} = \frac{v+2u^3}{v} \frac{dx}{\sqrt{1-x^2 \, . \, 1-u^8x^2}};$$

and writing $(v, -u)$ for (u, v), and (z, y) for (y, x),

$$z = \frac{u(u-2v^3)y + v^6y^3}{u^2 + uv^2(u-2v^3)y^2}$$

gives

$$\frac{dz}{\sqrt{1-z^2 \, . \, 1-u^8z^2}} = \frac{u-2v^3}{u} \frac{dy}{\sqrt{1-y^2 \, . \, 1-v^8y^2}},$$

$$= -3 \frac{dx}{\sqrt{1-x^2 \, . \, 1-u^8x^2}}.$$

279. Similarly, $n = 5$,

$$y = \frac{v(v-u^5)x + u^5(u^2+v^2)(v-u^5)x^3 + u^{10}(1-uv^3)x^5}{v^2(1-uv^3) + uv^2(u^2+v^2)(v-u^5)x^2 + u^6v^2(v-u^5)x^4}$$

gives

$$\frac{dy}{\sqrt{1-y^2 \, . \, 1-v^8y^2}} = \frac{v-u^5}{v(1-uv^3)} \frac{dx}{\sqrt{1-x^2 \, . \, 1-u^8x^2}},$$

and

$$z = \frac{u(u+v^5)y - v^5(u^2+v^2)(u+v^5)y^3 + v^{10}(1+u^3v)y^5}{u^2(1+u^3v) - u^2v(u^2+v^2)(u+v^5)y^2 + u^2v^6(u+v^5)y^4}$$

gives

$$\frac{dz}{\sqrt{1-z^2 \, . \, 1-u^8z^2}} = \frac{u+v^5}{u(1+u^3v)} \frac{dy}{\sqrt{1-y^2 \, . \, 1-v^8y^2}},$$

whence

$$\frac{dz}{\sqrt{1-z^2 \, . \, 1-u^8z^2}} = 5 \frac{dx}{\sqrt{1-x^2 \, . \, 1-u^8x^2}}.$$

The Multiplier M. Art. Nos. 280 to 284.

280. The above-mentioned values of M, $\frac{1}{M}$ lead to convenient expressions of nM^2; thus

$$n=3,\qquad 3M^2 = \frac{v\,(2v^3-u)}{u\,(2u^3+v)},$$

$$n=5,\qquad 5M^2 = \frac{v}{u}\,\frac{u+v^5}{v-u^5}\,\frac{1-uv^3}{1+u^3v},$$

$$n=7,\qquad 7M^2 = -\frac{v\,(u-v^7)}{u\,(v-u^7)}.$$

It will be shown that we have in general

$$nM^2 = \frac{\lambda\lambda'^2\,dk}{kk'^2\,d\lambda},\ = \frac{v\,(1-v^8)\,du}{u\,(1-u^8)\,dv},$$

or, what is the same thing, if $\phi=0$ be the modular equation, then

$$-nM^2 = v(1-v^8)\frac{d\phi}{dv} \div u(1-u^8)\frac{d\phi}{du},$$

a formula which is here to be verified in the three cases $n=3$, $n=5$ and $n=7$.

281. In the case $n=3$, we have

$$M = \frac{v}{v+2u^3} = \frac{2v^3-u}{3u},$$

also

$$\frac{du}{dv} = \frac{2v^3-u+3u^3v^2}{2u^3+v-3u^2v^3},$$

and the equation becomes

$$\frac{2v^3-u}{2u^3+v} = \frac{1-v^8}{1-u^8}\cdot\frac{2v^3-u+3u^3v^2}{2u^3+v-3u^2v^3}.$$

But writing $3 = \frac{(2v^3-u)(2u^3+v)}{uv}$, then in the last fraction the numerator becomes $=(2v^3-u)(1+u^2v^2+2u^5v)$, and the denominator $= (2u^3 + v)(1 + u^2v^2 - 2uv^5)$: and the equation thus is

$$1 = \frac{1 - v^8}{1 - u^8} \cdot \frac{1 + u^2v^2 + 2u^5v}{1 + u^2v^2 - 2uv^5}.$$

But we have

$$\begin{aligned} 1 - u^8 &= (1 + u^4)\,[1 - v^4 + 2uv\,(1 - u^2v^2)], \\ &= 1 - u^4v^4 + u^4 - v^4 + 2uv\,(1 + u^4)(1 - u^2v^2), \\ &= 1 - u^4v^4 + 2u^5v\,(1 - u^2v^2), \\ &= (1 - u^2v^2)(1 + u^2v^2 + 2u^5v), \end{aligned}$$

and similarly

$$1 - v^8 = (1 - u^2v^2)(1 + u^2v^2 - 2uv^5),$$

which prove the theorem.

282. In the case $n = 5$ we have

$$M = \frac{v\,(1 - uv^3)}{v - u^5} = \frac{u + v^5}{5u\,(1 + u^3v)},$$

and the equation becomes

$$\frac{(1 - uv^3)(u + v^5)}{(v - u^5)(1 + u^3v)} = \frac{1 - v^8}{1 - u^8}\,\frac{du}{dv}.$$

The modular equation may be written (by No. 273)

$$(u^2 - v^2)^6 = 16\,u^2v^2\,(1 - u^8)(1 - v^8),$$

whence differentiating and multiplying by $u^2 - v^2$, and reducing, we have

$$\begin{aligned} &6uv\,(1 - u^8)(1 - v^8)(udu - vdv) \\ &\quad = u\,(u^2 - v^2)(1 - u^8)(1 - 5v^8)\,dv + v\,(u^2 - v^2)(1 - v^8)(1 - 5u^8)\,du, \end{aligned}$$

or, as this may be written,

$$\begin{aligned} &v\,(1 - v^8)(5u^2 - u^{10} + v^2 - 5u^8v^2)\,du \\ &\quad = u\,(1 - u^8)(5v^2 - v^{10} + u^2 - 5u^2v^8)\,dv, \end{aligned}$$

that is

$$\frac{v}{u}\,\frac{du}{dv}\,\frac{1 - v^8}{1 - u^8} = \frac{5v^2 - v^{10} + u^2 - 5u^2v^8}{5u^2 - u^{10} + v^2 - 5u^8v^2};$$

or, observing that from the modular equation, we obtain

$$5u^2 - u^{10} + v^2 - 5u^8v^2 = (1 - u^4v^4)(v^2 + 5u^2 + 4u^5v),$$

$$5v^2 - v^{10} + u^2 - 5u^2v^8 = (1 - u^4v^4)(u^2 + 5v^2 - 4uv^5),$$

this is

$$\frac{v}{u}\frac{du}{dv}\frac{1 - v^8}{1 - u^8} = \frac{u^2 + 5v^2 - 4uv^5}{v^2 + 5u^2 + 4u^5v},$$

and the equation to be verified is

$$\frac{v(1 - uv^3)(u + v^5)}{u(v - u^5)(1 + u^3v)} = \frac{u^2 + 5v^2 - 4uv^5}{v^2 + 5u^2 + 4u^5v}.$$

283. Write

$$A = u + v^5,\ B = u(1 + u^3v),\ C = v - u^5,\ D = v(1 - uv^3),$$

then we have

$$u^2 + 5v^2 - 4uv^5 = uA + 5vD,$$

$$v^2 + 5u^2 + 4u^5v = vC + 5uB,$$

and the equation becomes

$$\frac{AD}{BC} = \frac{uA + 5vD}{vC + 5uB},$$

or, what is the same thing,

$$vACD + 5uABD = uABC + 5vCBD:$$

but from the modular equation $5BD = AC$, and substituting this value and throwing out the factor AC, the equation becomes $vD + uA = uB + vC$, which is true since each side is $= u^2 + v^2$.

284. In the case $n = 7$, we have

$$M = \frac{v(1 - uv)(1 - uv + u^2v^2)}{v - u^7} = -\frac{u - v^7}{7u(1 - uv)(1 - uv + u^2v^2)},$$

and the formula is

$$-\frac{u - v^7}{v - u^7} = \frac{1 - v^8}{1 - u^8}\frac{du}{dv}.$$

Starting from the modular equation

$$\phi, = (1-u^8)(1-v^8)-(1-uv)^8, = 0,$$

we have

$$\tfrac{1}{8}\frac{d\phi}{dv} = -v^7(1-u^8)+u(1-uv)^7;$$

and thence

$$\begin{aligned}\tfrac{1}{8}(1-v^8)\frac{d\phi}{dv} &= -v^7(1-u^8)(1-v^8)+(1-v^8)\,u\,(1-uv)^7,\\ &= -v^7(1-uv)^8+(1-v^8)\,u\,(1-uv)^7,\\ &= \quad (1-uv)^7(u-v^7).\end{aligned}$$

And similarly

$$\tfrac{1}{8}(1-u^8)\frac{d\phi}{du} = \quad (1-uv)^7\,(v-u^7);$$

whence

$$\frac{1-v^8}{1-u^8}\frac{du}{dv},\ = -(1-v^8)\frac{d\phi}{dv} \div (1-u^8)\frac{d\phi}{du},\ = -\frac{u-v^7}{v-u^7},$$

the formula in question.

Further theory of the Cubic Transformation.
Art. Nos. 285 to 294.

285. The cubic transformation may be considered from a converse point of view. Writing $x = \text{sn}\,(u, k)$, $s = \text{sn}\,(3u, k)$, we have

$$s = \frac{3x\left(1-\frac{x^2}{\alpha^2}\right)\left(1-\frac{x^2}{\beta^2}\right)\left(1-\frac{x^2}{\gamma^2}\right)\left(1-\frac{x^2}{\delta^2}\right)}{(1-k^2\alpha^2x^2)(1-k^2\beta^2x^2)(1-k^2\gamma^2x^2)(1-k^2\delta^2x^2)},$$

where

$$\alpha = \text{sn}\,\frac{4K}{3}, \qquad \beta = \text{sn}\,\frac{4iK'}{3},$$

$$\gamma = \text{sn}\,\frac{4K+4iK'}{3},\ \delta = \text{sn}\,\frac{-4K+4iK'}{3},$$

these being the roots of

$$3 - 4(1+k^2)x^2 + 6k^2x^4 - k^4x^8 = 0;$$

and it is to be shown that this relation between z, x may be decomposed into two transformation equations between (y, x) and (z, y) respectively.

286. We take these to be

$$y = \frac{\frac{x}{M}\left(1 - \frac{x^2}{\alpha^2}\right)}{1 - k^2\alpha^2x^2},\quad z = \frac{3My\left(1 - \frac{y^2}{\theta^2}\right)}{1 - \lambda^2\theta^2y^2},$$

giving respectively

$$\frac{M\,dy}{\sqrt{1-y^2 \,.\, 1-\lambda^2y^2}} = \frac{dx}{\sqrt{1-x^2 \,.\, 1-k^2x^2}},$$

and

$$\frac{dz}{\sqrt{1-z^2 \,.\, 1-k^2z^2}} = \frac{3M\,dy}{\sqrt{1-y^2 \,.\, 1-\lambda^2y^2}},$$

where observe that α, which enters into the relation between y, x, being as above the real root $\operatorname{sn}\frac{4K}{3}$, the equation between y, x is a first transformation, and consequently that the relation between z, y ought to come out a second transformation.

287. Writing

$$\sqrt{1-\alpha^2} = \operatorname{cn}\frac{4K}{3},\quad \sqrt{1-k^2\alpha^2} = \operatorname{dn}\frac{4K}{3},$$

we have

$$\operatorname{sn}\frac{8K}{3} = \operatorname{sn}\left(4K - \frac{4K}{3}\right) = -\operatorname{sn}\frac{4K}{3},$$

that is

$$2\sqrt{1-\alpha^2}\,\sqrt{1-k^2\alpha^2} = -(1-k^2\alpha^4),$$

and similarly

$$2\sqrt{1-\beta^2}\,\sqrt{1-k^2\beta^2} = -(1-k^2\beta^4).$$

Also

$$\gamma\delta = \operatorname{sn}\frac{4iK'+4K}{3}\operatorname{sn}\frac{4iK'-4K}{3} = \frac{\beta^2-\alpha^2}{1-k^2\alpha^2\beta^2},$$

$$\gamma+\delta = \frac{2\beta\sqrt{1-\alpha^2}\sqrt{1-k^2\alpha^2}}{1-k^2\alpha^2\beta^2} = \frac{-\beta\,(1-k^2\alpha^4)}{1-k^2\alpha^2\beta^2},$$

and thence

$$\beta+(\gamma+\delta) = \frac{k^2\alpha^2\beta\,(\alpha^2-\beta^2)}{1-k^2\alpha^2\beta^2},$$

$$\beta\,(\gamma+\delta)+\gamma\delta = \frac{-\alpha^2\,(1-k^2\alpha^2\beta^2)}{1-k^2\alpha^2\beta^2};$$

that is

$$\beta+\gamma+\delta = -k^2\alpha^2\beta\gamma\delta,$$

$$\gamma\delta+\delta\beta+\beta\gamma = -\alpha^2,$$

or, what is the same thing,

$$\frac{1}{\gamma\delta}+\frac{1}{\delta\beta}+\frac{1}{\beta\gamma} = -k^2\alpha^2,$$

$$\frac{1}{\beta}+\frac{1}{\gamma}+\frac{1}{\delta} = -\frac{\alpha^2}{\beta\gamma\delta};$$

and, moreover, since

$$3\left(1-\frac{x^2}{\alpha^2}\right)\left(1-\frac{x^2}{\beta^2}\right)\left(1-\frac{x^2}{\gamma^2}\right)\left(1-\frac{x^2}{\delta^2}\right)$$
$$= 3-4\,(1+k^2)\,x^2+6k^2x^4-k^4x^8,$$

we have

$$\frac{3}{\alpha^2\beta^2\gamma^2\delta^2} = -k^4, \text{ or } \alpha^2\beta^2\gamma^2\delta^2 = -\frac{3}{k^4}.$$

288. *Determination of* $y = \dfrac{\dfrac{x}{M}\left(1-\dfrac{x^2}{\alpha^2}\right)}{1-k^2\alpha^2x^2}$, *leading to*

$$\frac{M\,dy}{\sqrt{1-y^2\,.\,1-\lambda^2y^2}} = \frac{dx}{\sqrt{1-x^2\,.\,1-k^2x^2}}.$$

We determine M so that $x=1$, $y=1$ shall be corresponding values, viz. we have

$$1 = \frac{\frac{1}{M}\left(1-\frac{1}{\alpha^2}\right)}{1-k^2\alpha^2}, \text{ or } M = -\frac{1-\alpha^2}{\alpha^2(1-k^2\alpha^2)},$$

and then (denom. $= 1-k^2\alpha^2x^2$), writing

$$1-y = (1-k^2\alpha^2x^2) - \frac{x}{M}\left(1-\frac{x^2}{\alpha^2}\right) \qquad (\div),$$

$$= (1-x)\left\{1 - x\left(\frac{1}{M}-1\right) - \frac{x^2}{M\alpha^2}\right\} \qquad (\div);$$

the term in { } is taken to be a perfect square, $=\left(1-\frac{x}{f}\right)^2$ suppose, viz. this being so we have

$$\frac{2}{f} = -\frac{1-k^2\alpha^4}{1-\alpha^2}\left(\text{or } \frac{1}{f} = \frac{\sqrt{1-k^2\alpha^2}}{\sqrt{1-\alpha^2}}\right),$$

$$\frac{1}{f^2} = -\frac{1}{M\alpha^2}, = \frac{1-k^2\alpha^2}{1-\alpha^2},$$

which agree; and then

$$1-y = (1-x)\left(1-\frac{x}{f}\right)^2 \qquad (\div),$$

whence also

$$1+y = (1+x)\left(1+\frac{x}{f}\right)^2 \qquad (\div).$$

We next determine λ, so that x being changed into $\frac{1}{kx}$ y shall be changed into $\frac{1}{\lambda y}$: we thus have

$$\frac{1}{\lambda y} = \frac{1}{Mk^3\alpha^4x} \cdot \frac{1-k^2\alpha^2x^2}{1-\frac{x^2}{\alpha^2}},$$

or, multiplying by y,

$$\frac{1}{\lambda} = \frac{1}{M^2k^3\alpha^4},$$

that is

$$\lambda = M^2k^3\alpha^4, = \frac{k^3(1-\alpha^2)^2}{(1k-{}^2\alpha^2)^2}.$$

Observe that, α being real, we have $1-\alpha^2 < 1-k^2\alpha^2$, and hence $\lambda < k^3$, viz. we pass from a modulus k to a smaller modulus λ.

And then the expressions for $1-y$ and $1+y$ lead to

$$1-\lambda y = (1-kx)(1-k\alpha x)^2 \qquad (+),$$

$$1+\lambda y = (1+kx)(1+k\alpha x)^2 \qquad (+),$$

so that we have the required equation

$$\frac{M\,dy}{\sqrt{1-y^2\,.\,1-\lambda^2y^2}} = \frac{dx}{\sqrt{1-x^2\,.\,1-k^2x^2}}.$$

289. *Modular equation.*

Next, for finding the modular equation, we have

$$\lambda = \frac{k^3(1-\alpha^2)^2}{(1-k^2\alpha^2)^2}, \text{ or } \sqrt{\lambda k} = \frac{k^2(1-\alpha^2)}{1-k^2\alpha^2},$$

$$\lambda'^2 = \frac{1}{(1-k^2\alpha^2)^4}\left[(1-k^2\alpha^2)^4 - k^6(1-\alpha^2)^4\right],$$

where the term in [ ] is

$$1 - 4k^2\alpha^2 + 6k^4\alpha^4 - 4k^6\alpha^6 + k^8\alpha^8$$

$$- k^6 + 4k^6\alpha^2 - 6k^6\alpha^4 + 4k^6\alpha^6 - k^6\alpha^8,$$

$$= (1-k^2)\left[1 + k^2 + k^4 - 4(k^2+k^4)\alpha^2 + 6k^4\alpha^4 - k^6\alpha^8\right],$$

$$= (1-k^2)\left[(1-k^2)^2 + k^2\{3 - 4(1+k^2)\alpha^2 + 6k^2\alpha^4 - k^4\alpha^8\}\right],$$

$$= (1-k^2)^3;$$

that is

$$\lambda' = \frac{k'^3}{(1-k^2\alpha^2)^2}, \text{ or } \sqrt{\lambda' k'} = \frac{k'^2}{1-k^2\alpha^2};$$

and hence

$$\sqrt{\lambda k} + \sqrt{\lambda' k'} = \frac{k'^2 + k^2(1-\alpha^2)}{1-k^2\alpha^4},$$

that is

$$\sqrt{\lambda k} + \sqrt{\lambda' k'} = 1,$$

the required equation.

290. We have next (θ being arbitrary)

$$1 - \frac{y}{\theta} = 1 - k^2\alpha^2 x^2 - \frac{x}{M\theta}\left(1 - \frac{x^2}{\alpha^2}\right) \qquad (\div),$$

(denom. as before $= 1 - k^2\alpha^2 x^2$).

And taking

$$\theta = -\frac{1}{M}\frac{\beta\gamma\delta}{\alpha^2}, \text{ that is } -\frac{1}{M\theta} = \frac{\alpha^2}{\beta\gamma\delta},$$

then $$1 - \frac{y}{\theta} = 1 - k^2\alpha^2 x^2 + \frac{\alpha^2 x}{\beta\gamma\delta}\left(1 - \frac{x^2}{\alpha^2}\right) \qquad (\div),$$

$$= 1 + \frac{\alpha^2 x}{\beta\gamma\delta} - k^2\alpha^2 x^2 - \frac{x^3}{\beta\gamma\delta} \qquad (\div),$$

$$= \left(1 - \frac{x}{\beta}\right)\left(1 - \frac{x}{\gamma}\right)\left(1 - \frac{x}{\delta}\right) \qquad (\div),$$

and similarly

$$1 + \frac{y}{\theta} = \left(1 + \frac{x}{\beta}\right)\left(1 + \frac{x}{\gamma}\right)\left(1 + \frac{x}{\delta}\right) \qquad (\div);$$

also, changing x, y into $\frac{1}{kx}, \frac{1}{\lambda y}$,

$$1 - \lambda\theta y = (1 - k\beta x)(1 - k\gamma x)(1 - k\delta x) \qquad (\div),$$

$$1 + \lambda\theta y = (1 + k\beta x)(1 + k\gamma x)(1 + k\delta x) \qquad (\div);$$

consequently

$$\frac{y\left(1 - \frac{y^2}{\theta^2}\right)}{1 - \lambda^2\theta^2 y^2} = \frac{\frac{x}{M}\left(1 - \frac{x^2}{\alpha^2}\right)\left(1 - \frac{x^2}{\beta^2}\right)\left(1 - \frac{x^2}{\gamma^2}\right)\left(1 - \frac{x^2}{\delta^2}\right)}{(1 - k^2\alpha^2 x^2)(1 - k^2\beta^2 y^2)(1 - k^2\gamma^2 x^2)(1 - k^2\delta^2 x^2)}.$$

We have

$$\theta = -\frac{1}{M}\frac{\beta\gamma\delta}{\alpha^2}, \text{ hence } \theta^2 = \frac{1}{M^2}\frac{\alpha^4\beta^2\gamma^2\delta^2}{\alpha^8}, = \frac{-3}{k^4\alpha^3 M^2};$$

but $\lambda = M^2k^3\alpha^4$, hence $\lambda\theta^2 = -\dfrac{3}{k\alpha^2}$, and $\lambda^2\theta^2 = -3M^2k^2\alpha^2$; whence, putting for shortness $M, = -\dfrac{1-\alpha^2}{\alpha^2(1-k^2\alpha^2)}, = -\dfrac{A}{\alpha^2 B}$, we have

$$\theta^2 = \frac{-3B^2}{k^4\alpha^2A^2},\quad \lambda\theta^2 = -\frac{3}{k\alpha^2},\quad \lambda^2\theta^2 = \frac{-3k^2A^2}{\alpha^2B^2}.$$

291. It is to be shown that θ is connected with λ as α is with k; viz. that we have

$$3 - 4(1+\lambda^2)\theta^2 + 6\lambda^2\theta^4 - \lambda^4\theta^8 = 0.$$

Substituting for θ^2, $\lambda^2\theta^2$ and $\lambda\theta^2$ their values, the expression on the left-hand side is

$$= -\frac{3}{k^4\alpha^8A^2B^2}\left[(27 - 18k^2\alpha^4 - k^4\alpha^8)A^2B^2 - 4\alpha^6(B^4 + k^6A^4)\right],$$

$(A = 1-\alpha^2,\ B = 1-k^2\alpha^2)$, viz. the term in [ ] is a function $(1, \alpha^2)^8$ the coefficients of which are

$$\begin{array}{l}
27,\\
-54-54k^2,\\
27+90k^2+27k^4,\\
-4-18k^2-18k^4-4k^6,\\
-2k^2-46k^4-2k^6,\\
14k^4+14k^6,\\
-k^4+10k^6-k^8,\\
-2k^6-2k^8,\\
-k^8,
\end{array}$$

and this is equal to the product of $3-4(1+k^2)\alpha^2+6k^2\alpha^4-k^4\alpha^8$ by a function $(1, \alpha^2)^4$, the coefficients of which are

$$\begin{array}{l}
9,\\
-6-6k^2,\\
1-4k^2+k^4,\\
2k^2+2k^4,\\
k^4,
\end{array}$$

viz. this other factor is $=[3-(1+k^2)\alpha^2-k^2\alpha^4]^2$. The first factor vanishes and we have thus the required relation

$$3-4(1+\lambda^2)\theta^2+6\lambda^2\theta^4-\lambda^4\theta^8=0.$$

We have $\theta=-\dfrac{1}{M}\dfrac{\beta\gamma\delta}{\alpha^3}$, where M, α, $\gamma\delta$ are all real but β is a pure imaginary, hence also θ is a pure imaginary. Now the equation in z, y corresponds with the differential relation

$$\frac{dz}{\sqrt{1-z^2\,.\,1-k^2z^2}}=\frac{3Mdy}{\sqrt{1-y^2\,.\,1-\lambda^2y^2}},$$

and we thence see that θ must denote one of the quantities

$$\operatorname{sn}\frac{4\Lambda}{3},\ \operatorname{sn}\frac{4i\Lambda'}{3},\ \operatorname{sn}\frac{4\Lambda+4i\Lambda'}{3},\ \operatorname{sn}\frac{-4\Lambda+4i\Lambda'}{3};$$

and, being as just shown, a pure imaginary, it clearly denotes $\operatorname{sn}\dfrac{4i\Lambda'}{3}$, viz. the transformation from z to y is a second transformation.

Writing now

$$z=\frac{\dfrac{y}{N}\left(1-\dfrac{y^2}{\theta^2}\right)}{1-\lambda^2\theta^2y^2},$$

we may determine N so that corresponding values shall be $z=1$, $y=-1$ (or $z=-1$, $y=1$), viz. this will be the case if

$$1=\frac{-\dfrac{1}{N}\left(1-\dfrac{1}{\theta^2}\right)}{1-\lambda^2\theta^2},\ \text{or say } N=\frac{1-\theta^2}{\theta^2(1-\lambda^2\theta^2)},$$

and the value of N thus determined will be $=\dfrac{1}{3M}$. To verify this we have to prove the equation

$$\theta^2(1-\lambda^2\theta^2)=3M(1-\theta^2).$$

Substituting for θ^2, $\lambda\theta^2$ and M their values, the equation is

$$\frac{3}{A^2B^2\alpha^2k^4}\left[-\alpha^4(B^3-k^4A^3)+3AB(B-k^2A)\right]=0,$$

$(A=1-\alpha^2,\ B=1-k^2\alpha^2$, as before).

We have $B-k^2A=1-k^2$, and the term $B^2-k^4A^2$ contains this same factor. Omitting the factor in question, $1-k^2$, the term in { } is

$$=-\alpha^2(1+k^2-3k^2\alpha^2+k^4\alpha^6)+3\{1-(1+k^2)\alpha^2+k^2\alpha^4\},$$

viz. this is

$$=3-4(1+k^2)\alpha^2+6k^2\alpha^4-k^4\alpha^8,$$

which is $=0$, and the theorem is thus proved.

292. Starting from the equation

$$z=\frac{3My\left(1-\dfrac{y^2}{\theta^2}\right)}{1-\lambda^2\theta^2y^2},$$

where $1, -1$ are corresponding values of z, y, and

$$3-4(1+\lambda^2)\theta^2+6\lambda^2\theta^4-\lambda^4\theta^8=0,$$

we have

$$1+z=(1-y)\left(1-\frac{y}{g}\right)^2 \qquad (\div),$$

$$1-z=(1+y)\left(1+\frac{y}{g}\right)^2 \qquad (+),$$

$$1+kz=(1-\lambda y)(1-\lambda gy)^2 \qquad (\div),$$

$$1-kz=(1+\lambda y)(1+\lambda gy)^2 \qquad (\div),$$

where $\quad$ denom. $=1-\lambda^2\theta^2y^2$.

viz. in obtaining the above we have

$$1+z=1-\lambda^2\theta^2y^2+3My\left(1-\frac{y^2}{\theta^2}\right) \qquad (+),$$

$$=(1-y)\left\{1+(3M+1)y+\frac{3M}{\theta^2}y^2\right\} \qquad (+),$$

$$=(1-y)\left(1-\frac{y}{g}\right)^2 \qquad (\div),$$

that is

$$-\frac{2}{g}=3M+1, \quad =\frac{1-\lambda^2\theta^4}{1-\theta^2},$$

$$\frac{1}{g^2}=\frac{3M}{\theta^2}, \quad =\frac{1-\lambda^2\theta^4}{1-\theta^2},$$

which agree. We have therefore

$$\frac{dz}{\sqrt{1-z^2 . 1-k^2z^2}} = \frac{3M dy}{\sqrt{1-y^2 . 1-\lambda^2 y^2}},$$

and the proof is thus completed.

293. The investigation would have been very similar if, in the formula

$$y = \frac{\frac{x}{M}\left(1-\frac{x^2}{a^2}\right)}{1-k^2a^2x^2},$$

a had denoted any other root of the modular equation, or, what is the same thing, if α were replaced by any other root β, γ or δ: there would have been in each case a corresponding equation in (z, y) giving by its combination with the assumed equation the triplication. In particular if the root had been β, then the equation in x, y would have been a second transformation and the corresponding equation in (z, y) a first transformation. But if the root had been γ or δ, then in either case the equation in (x, y) and the corresponding equation in (y, z) would have been each an imaginary transformation.

294. Returning to the quantities α, β, γ, δ, which denote

$$\text{sn}\,\frac{4K}{3},\ \text{sn}\,\frac{4iK'}{3},\ \text{sn}\,\frac{4K+4iK'}{3},\ \text{sn}\,\frac{-4K+4iK'}{3},$$

respectively the two equations obtained in No. 287 belong to a system which may be written

$$\begin{array}{ll|l}
\alpha^2 = & . \quad . \quad . -\beta\gamma-\beta\delta-\gamma\delta, & k^2\alpha^2\beta\gamma\delta = \ . -\beta-\gamma-\delta, \\
\beta^2 = & . -\alpha\gamma+\alpha\delta \ . \quad . +\gamma\delta, & k^2\alpha\beta^2\gamma\delta = \alpha \ . +\gamma-\delta, \\
\gamma^2 = & \alpha\beta \ . -\alpha\delta \ . +\beta\delta \ . \ , & k^2\alpha\beta\gamma^2\delta = \alpha-\beta \ . +\delta, \\
\delta^2 = & -\alpha\beta+\alpha\gamma \ . +\beta\gamma \ . \quad . \ , & k^2\alpha\beta\gamma\delta^2 = \alpha+\beta-\gamma \ . \ .
\end{array}$$

But $\alpha^2\beta^2\gamma^2\delta^2 = -\frac{3}{k^4}$, or if for shortness $s = i\sqrt{3}$, then we may write $\alpha\beta\gamma\delta = -\frac{s}{k^2}$ or $k^2\alpha\beta\gamma\delta = -s$, and the last set of equations becomes

$$\begin{aligned} s\alpha - \beta - \gamma - \delta &= 0, \\ \alpha + s\beta + \gamma - \delta &= 0, \\ \alpha - \beta + s\gamma + \delta &= 0, \\ \alpha + \beta - \gamma + s\delta &= 0, \end{aligned}$$

which must be equivalent to two equations only: in fact the equations may also be written

$$\begin{array}{lllll} 2\alpha & & +(s-1)\gamma & +(s+1)\delta & =0, \\ . & 2\beta & -(s+1)\gamma & +(s-1)\delta & =0, \\ -(s-1)\alpha & +(s+1)\beta + & 2\gamma & . & =0, \\ -(s+1)\alpha & -(s-1)\beta & . & +2\delta & =0, \end{array}$$

which linearly determine any two of the quantities in terms of the remaining two, for instance α and β in terms of γ and δ: but then, substituting for α and β their values, the third and fourth equations are satisfied identically.

A General Form of the Cubic Transformation.
Art. Nos. 295, 296.

295. Consider the two quartic functions

$$X = (a, b, c, d, e)(x, 1)^4,\ X' = (a', b', c', d', e')(x', 1)^4,$$

we may imagine the variables x, x' connected by a cubic transformation so as to give rise to a differential relation

$$\frac{M dx'}{\sqrt{X'}} = \frac{dx}{\sqrt{X}},$$

and this being so the modular equation will be given as a relation between the absolute invariants of these two quartic functions, viz. writing as usual I, J

$$(= ae - 4bd + 3c^2,\ ace - ad^2 - b^2e + 2bcd - c^3, \text{ respectively})$$

for the invariants of X, and similarly I', J' for those of X', then the absolute invariants are $\Omega = 1 - 27\frac{J^2}{I^3}$, and

$$\Omega' = 1 - 27\frac{J'^2}{I'^3}.$$

Supposing the function U linearly transformed into $1-x^2 . 1-k^2x^2$, and similarly U' linearly transformed into $1-y^2 . 1-\lambda^2y^2$: then it has been seen that the relation between k^2, λ^2 can be obtained by the elimination of α from the equations

$$k^2=\frac{\alpha^3(2+\alpha)}{1+2\alpha},\quad \lambda^2=\frac{\alpha(2+\alpha)^3}{(1+2\alpha)^3},$$

or, what is the same thing, writing $-\beta=\frac{2+\alpha}{1+2\alpha}$, we have α, β connected by the equation

$$2\alpha\beta+\alpha+\beta+2=0,$$

and then $$k^2=-\alpha^3\beta,\ \lambda^2=-\alpha\beta^3.$$

The theory of linear transformations gives

$$\Omega=\frac{108k^2(1-k^2)^4}{(k^4+14k^2+1)^3},\quad \Omega'=\frac{108\lambda^2(1-\lambda^2)^4}{(\lambda^4+14\lambda^2+1)^3},$$

the question therefore is between these equations to eliminate $\alpha, \beta, k^2, \lambda^2$ so as to obtain a relation between Ω, Ω'.

290. By considerations which I cannot now recall I was led to assume

$$\alpha'=\tfrac{1}{2}\frac{(1+2\alpha)(2+\alpha)(1-\alpha)^4}{(1+4\alpha+\alpha^2)^3},\quad \beta'=\tfrac{1}{2}\frac{(1+2\beta)(2+\beta)(1-\beta)^4}{(1+4\beta+\beta^2)^3}.$$

The equation between α, β gives

$$\begin{aligned}1+2\beta&=-3\div(1+2\alpha),\\ 2+\beta&=3\alpha\div(1+2\alpha),\\ 1-\beta&=3(1+\alpha)\div(1+2\alpha),\\ 1+4\beta+\beta^2&=-3(1+4\alpha+\alpha^2)\div(1+2\alpha)^2;\end{aligned}$$

and we thence have

$$\beta'=\frac{27\alpha(1+\alpha)^4}{2(1+4\alpha+\alpha^2)^3},$$

viz. in virtue of the identity

$$(1+2\alpha)(2+\alpha)(1-\alpha)^4+27\alpha(1+\alpha)^4=2(1+4\alpha+\alpha^2)^3,$$

we find $$\alpha'+\beta'=1.$$

We then have

$$k^2 = \alpha^3 (2+\alpha) \div (1+2\alpha),$$

$$k'^2 - 1 = (\alpha - 1)(\alpha+1)^3 \div (1+2\alpha),$$

$$\left.\begin{aligned} k^4 + 14k^2 + 1 &= \alpha^6(2+\alpha)^2 + 14\alpha^3(2+\alpha)(2\alpha+1) + (2\alpha+1)^2, \\ &= (\alpha^2+4\alpha+1)(\alpha^6+3\alpha^4+16\alpha^3+3\alpha^2+1) \end{aligned}\right\} \div (1+2\alpha)^2,$$

and consequently

$$\Omega = \frac{108\,\alpha^3(1+2\alpha)(2+\alpha)(\alpha-1)^4(\alpha+1)^{12}}{(\alpha^2+4\alpha+1)^3(\alpha^6+3\alpha^4+16\alpha^3+3\alpha^2+1)^3}.$$

But

$$\alpha' = \tfrac{1}{2}(1+2\alpha)(2+\alpha)(1-\alpha)^4 \qquad \div (\alpha^2+4\alpha+1)^3,$$

and thence

$$\left.\begin{aligned} 1-\alpha' &= (1+4\alpha+\alpha^2)^3 - \tfrac{1}{2}(1+2\alpha)(2+\alpha)(1-\alpha)^4 \\ &= \tfrac{27}{2}\,\alpha(1+\alpha)^4 \end{aligned}\right\} \div \quad ,,$$

$$\left.\begin{aligned} 1+8\alpha' &= (1+4\alpha+\alpha^2)^3 + 4(1+2\alpha)(2+\alpha)(1-\alpha)^4 \\ &= 9(\alpha^6+3\alpha^4+16\alpha^3+3\alpha^2+1) \end{aligned}\right\} \div \quad ,,$$

whence

$$\Omega = \frac{64\alpha'(1-\alpha')^3}{(1+8\alpha')^3},$$

and similarly

$$\Omega' = \frac{64\beta'(1-\beta')^3}{(1+8\beta')^3},$$

where $\alpha' + \beta' = 1$. Writing $\alpha' = \frac{1}{2} + \theta$, and therefore $\beta' = \frac{1}{2} - \theta$, we have

$$(5+8\theta)^3\,\Omega = 4(1+2\theta)(1-2\theta)^3,$$

$$(5-8\theta)^3\,\Omega' = 4(1+2\theta)^3(1-2\theta),$$

and the elimination of θ from these equations gives the required relation between Ω, Ω'.

Proof of the Equation $nM^2 = \dfrac{\lambda\lambda'^2}{kk'^2}\dfrac{dk}{d\lambda}$. Art. Nos. 297 to 299.

297. The proof depends on the formulæ for the differentiation of the complete functions referred to at the end of Chap. IV.

We deduce

$$\frac{d}{dk}\frac{K'}{K} = -\frac{\tfrac{1}{2}\pi}{K^2 kk'^2},$$

and similarly, if Λ, Λ' are the complete functions in the first transformation, we have

$$\frac{d}{d\lambda}\frac{\Lambda'}{\Lambda} = -\frac{\tfrac{1}{2}\pi}{\Lambda^2\lambda\lambda'^2}.$$

But
$$\frac{\Lambda'}{\Lambda} = n\frac{K'}{K},$$

and we thus obtain

$$\frac{d\lambda}{\lambda\lambda'^2\Lambda^2} = \frac{ndk}{kk'^2K^2}, \text{ or say } \frac{d\lambda}{\lambda\lambda'^2}\cdot\frac{K^2}{n\Lambda^2} = \frac{dk}{kk'^2}.$$

But we have also $\Lambda = \dfrac{K}{nM}$, that is $\dfrac{K^2}{n\Lambda^2} = nM^2$, and consequently

$$nM^2\cdot\frac{d\lambda}{\lambda\lambda'^2} = \frac{dk}{kk'^2}, \text{ or say } nM^2 = \frac{\lambda\lambda'^2}{kk'^2}\frac{dk}{d\lambda},$$

or writing $k = u^4$, $\lambda = v^4$, this is

$$nM^2 = \frac{v(1-v^8)}{u(1-u^8)}\frac{du}{dv}.$$

298. M is given as a rational function of (u, v), the same function in the first and in every other transformation; and if we imagine $\dfrac{du}{dv}$ expressed from the modular equation as a rational function of (u, v), and substitute these values of M and $\dfrac{du}{dv}$, the resulting equation must be true in virtue of the modular equation, viz. it must contain as a factor the modular equation. And this being so, it follows conversely that the expression of M^2, viz.

$$nM^2 = \frac{\lambda\lambda'^2}{kk'^2}\frac{dk}{d\lambda},$$

holds good, not for the first transformation only, but for every transformation of the order n.

299. Jacobi, *Fund. Nova*, p. 74, effects the generalisation from different considerations. Writing in the first instance $Q = aK + bK'$, $Q' = a'K + b'K'$, where a, b, a', b' are constants, he finds

$$\frac{d}{dk}\frac{Q'}{Q} = -\frac{\frac{1}{2}\pi(ab' - a'b)}{Q^2 kk'^2},$$

and similarly if $L = \alpha\Lambda + \beta\Lambda'$, $L' = \alpha'\Lambda + \beta'\Lambda'$, where α, β, α', β' are constants, then

$$\frac{d}{d\lambda}\frac{L'}{L} = -\frac{\frac{1}{2}\pi(\alpha\beta' - \alpha'\beta)}{L^2\lambda\lambda'^2},$$

viz. these correspond to the formulæ of the last No. with only Q, Q', L, L' in place of K, K', Λ, Λ' respectively. But then using the equations

$$\alpha\Lambda + i\beta\Lambda' = \frac{aK + ibK'}{nM},$$

$$\alpha'\Lambda' + i\beta'\Lambda = \frac{a'K' + ib'K}{nM},$$

where $aa' + bb' = 1$, $\alpha\alpha' + \beta\beta' = 1$, (see end of Chap. VII.), the equations become

$$d\frac{Q'}{Q} = -\frac{\frac{1}{2}n\pi dk}{kk'^2 Q^2}, \quad d\frac{L'}{L} = -\frac{\frac{1}{2}\pi d\lambda}{\lambda\lambda'^2 L^2},$$

or since $\dfrac{Q'}{Q} = \dfrac{L'}{L}$, $\dfrac{Q}{L} = nM$, we have as before $nM^2 = \dfrac{\lambda\lambda'^2 dk}{kk'^2 d\lambda}$.

Differential Equation satisfied by the multiplier M. Art. No. 300.

300. We have, No. 76, writing K instead of F,

$$kk'^2\frac{d^2K}{dk^2} + (1 - 3k^2)\frac{dK}{dk} - kK = 0,$$

and similarly if λ, Λ belong to the first transformation,

$$\lambda\lambda'^2\frac{d^2\Lambda}{d\lambda^2} + (1 - 3\lambda^2)\frac{d\Lambda}{d\lambda} - \lambda\Lambda = 0.$$

These equations may be written

$$\frac{d}{dk}\left(\frac{kk'^2 dK}{dk}\right) - kK = 0, \quad \frac{d}{d\lambda}\left(\frac{\lambda\lambda'^2 d\Lambda}{d\lambda}\right) - \lambda\Lambda = 0.$$

But $M = \dfrac{K}{\Lambda}$, that is $K = M\Lambda$, or substituting in the first equation

$$\Lambda\left\{kk'^2\frac{d^2M}{dk^2} + (1-3k^2)\frac{dM}{dk} - kM\right\}$$
$$+\frac{d\Lambda}{dk}\left\{2kk'^2\frac{dM}{dk} + (1-3k^2)\,M\right\} + kk'^2M\frac{d^2\Lambda}{dk^2} = 0,$$

which multiplied by M may be written

$$M\Lambda\left\{kk'^2\frac{d^2M}{dk^2} + (1-3k^2)\frac{dM}{dk} - kM\right\} + \frac{d}{dk}\left(\frac{M^2kk'^2d\Lambda}{dk}\right) = 0.$$

But $M^2 = \dfrac{\lambda\lambda'^2dk}{nkk'^2d\lambda}$, whence the second term is

$$\frac{1}{n}\frac{d}{dk}\left(\frac{\lambda\lambda'^2d\Lambda}{d\lambda}\right), = \frac{1}{n}\frac{d\lambda}{dk}\frac{d}{d\lambda}\left(\frac{\lambda\lambda'^2d\Lambda}{d\lambda}\right),$$

viz. this is $= \dfrac{1}{n}\dfrac{d\lambda}{dk}\lambda\Lambda$: the whole equation thus divides by Λ, and it becomes

$$M\left\{kk'^2\frac{d^2M}{dk^2} + (1-3k^2)\frac{dM}{dk} - kM\right\} + \frac{1}{n}\frac{\lambda d\lambda}{dk} = 0.$$

We have $M^2 = \dfrac{\lambda\lambda'^2dk}{nkk'^2d\lambda}$, and if we use this equation to eliminate $\dfrac{d\lambda}{dk}$, we obtain

$$\left\{kk'^2\frac{d^2M}{dk^2} + (1-3k^2)\frac{dM}{dk} - kM\right\} + \frac{\lambda^2\lambda'^2}{kk'^2}\frac{1}{n^2M^3} = 0,$$

a differential equation of the second order satisfied by the multiplier M considered as a function of k. (*Fund. Nova*, p. 77.) Observe that this equation contains n, viz. it depends on the order of the transformation.

It is in the proof assumed that λ belongs to the first transformation: but it may be seen as in No. 298, (or we may as in No. 299 by using Q, L in place of K, Λ respectively show) that the theorem is true for any root whatever of the modular equation.

Differential Equation of the third order satisfied by the modulus λ.
Art. Nos. 301 to 305.

301. If we use the same equation $M^2 = \frac{\lambda\lambda'^2 dk}{nkk'^2 d\lambda}$ to eliminate M from the foregoing differential equation, then since the terms of

$$M\left\{kk'^2\frac{d^2M}{dk^2} + (1-3k^2)\frac{dM}{dk} - kM\right\}$$

all contain the factor $\frac{1}{n}$, which occurs also in the remaining term $\frac{1}{n}\frac{\lambda d\lambda}{dk}$ of the equation, this factor divides out, and we obtain an equation involving k, λ, but independent of n: viz. observing that the equation in M may be written

$$M\left\{\frac{d}{dk}\left(\frac{kk'^2dM}{dk}\right) - kM\right\} + \frac{1}{n}\frac{\lambda d\lambda}{dk} = 0,$$

and putting for convenience $M^2 = \frac{1}{n}\Omega^2$, that is

$$\Omega = \sqrt{\frac{\lambda\lambda'^2dk}{kk'^2d\lambda}},$$

the equation in question is

$$\Omega\left\{\frac{d}{dk}\left(kk'^2\frac{d\Omega}{dk}\right) - k\Omega\right\} + \frac{\lambda d\lambda}{dk} = 0,$$

or, what is the same thing,

$$\Omega\frac{d}{dk}\left(kk'^2\frac{d\Omega}{dk}\right) - k\Omega^2 + \frac{\lambda^2\lambda'^2}{kk'^2}\cdot\frac{1}{\Omega^2} = 0,$$

where Ω is a given function of k, λ, and λ is any root whatever of the modular equation.

302. In this form dk is taken to be constant (that is, k to be the independent variable), but taking dk, $d\lambda$ to be each variable (in effect k, λ to be functions of a new variable), the equation may be written

$$\frac{\Omega}{(dk)^2}\left[dk\,d(kk'^2d\Omega) - d^2k\,.\,kk'^2d\Omega\right] - k\Omega^2 + \frac{\lambda^2(1-\lambda^2)}{k(1-k^2)}\frac{1}{\Omega^2} = 0,$$

and after all reductions we arrive at Jacobi's form

$$3\{(dk)^2(d^2\lambda)^2-(d\lambda)^2(d^2k)^2\}-2dk\,d\lambda\,(dk\,d^3\lambda-d\lambda\,d^3k)$$
$$+(dk)^2(d\lambda)^2\left\{\left(\frac{1+k^2}{k-k^3}\right)^2(dk)^2-\left(\frac{1+\lambda^2}{\lambda-\lambda^3}\right)^2(d\lambda)^2\right\}=0.$$

303. It may be remarked that if

$$\frac{dk}{kk'^2}=dp,\ \text{that is } p=\log\frac{k}{k'},$$

and therefore $k^2=\dfrac{e^{2p}}{1+e^{2p}},\quad k'^2=\dfrac{1}{1+e^{2p}},\quad kk'=\dfrac{e^p}{1+e^{2p}},$

$$\frac{d\lambda}{\lambda\lambda'^2}=dq,\ \text{that is } q=\log\frac{\lambda}{\lambda'},$$

and therefore $\lambda^2=\dfrac{e^{2q}}{1+e^{2q}},\quad \lambda'^2=\dfrac{1}{1+e^{2q}},\quad \lambda\lambda'=\dfrac{e^q}{1+e^{2q}},$

we have

$$\Omega=\sqrt{\frac{dp}{dq}},$$

and the equation then is

$$\Omega\frac{d^2\Omega}{dp^2}=k^2k'^2\Omega^4-\frac{\lambda^2\lambda'^2}{\Omega^2},$$

which is readily converted into

$$\frac{2p'q'(q'p'''-p'q''')-3(q'^2p''^2-p'^2q''^2)}{4(p'q')^2}$$
$$=\left(\frac{e^p}{1+e^{2p}}\right)^2\cdot\frac{p'}{q'}-\left(\frac{e^q}{1+e^{2q}}\right)^2\cdot\frac{q'}{p'},$$

where p', p'', p''' and q', q'', q''', are the derived functions of p, q with respect to the independent variable.

304. The equation in No. 301 is easily verified in the case of the quadric transformation: we have here $\lambda=\dfrac{1-k'}{1+k'}$; and we thence find $\Omega=\dfrac{\sqrt{2}}{1+k'},\ \dfrac{d\lambda}{dk}=\dfrac{2(1-k')^{\frac{1}{2}}}{k'(1+k')^{\frac{3}{2}}}$, and the equation takes the form

$$\frac{\sqrt{2}}{1+k'}\left\{\frac{k}{k'}\frac{d}{dk'}\left[k^2k'\frac{d}{dk'}\left(\frac{\sqrt{2}}{1+k'}\right)\right]-\frac{\sqrt{2}k}{1+k'}\right\}+\frac{2(1-k')^{\frac{1}{2}}}{k'(1+k')^{\frac{3}{2}}}=0,$$

viz. dividing by 2, and reducing, this is

$$\frac{1}{1+k'}\left\{\frac{k}{k'}\frac{d}{dk'}\left(\frac{-k'+k'^2}{1+k'}\right)-\frac{k}{1+k'}\right\}+\frac{(1-k')^{\frac{3}{2}}}{k'(1+k')^{\frac{5}{2}}}=0.$$

But the first term is

$$\frac{1}{1+k'}\left\{\frac{k}{k'}\cdot\frac{-1+2k'+k'^2}{(1+k')^2}-\frac{k}{1+k'}\right\},$$

$$=\frac{k}{k'(1+k')^3}(-1+k'),\ =\frac{-(1-k')^{\frac{3}{2}}}{k'(1+k')^{\frac{5}{2}}},$$

and the equation is verified.

In the case of the cubic transformation, the equation in Jacobi's form No. 302, might be verified (although not without some difficulty) by means of the expressions, No. 261,

$$k^2=\frac{\alpha^3(2+\alpha)}{1+2\alpha},\ \lambda^2=\alpha\left(\frac{2+\alpha}{1+2\alpha}\right)^3,$$

of the moduli k, λ in terms of a parameter α: but the verification in the next following case of the quintic equation would apparently be very difficult. Jacobi remarks that if a method existed for finding the algebraical solutions of a differential equation, then, by means of the foregoing differential equation alone, it would be possible to obtain the modular equation in the transformation of any order n whatever: but, the mere verifications being so difficult, it does not appear that anything can be done in this manner in regard to the modular equations.

A relation involving M, K, Λ, E, G. Art. No. 305.

305. Immediately connected with what precedes we have a result which will be useful in the sequel: we have

$$\frac{dK}{dk}=\frac{1}{kk'^2}(E-k'^2K),$$

that is

$$\frac{dk}{k}-\frac{E}{kk'^2K}dk+\frac{dK}{K}=0,$$

and similarly if Λ, G are the complete functions to the modulus λ ($\Lambda = F_1\lambda$, as before, $G = E_1\lambda$), then

$$\frac{d\lambda}{\lambda} - \frac{G}{\lambda\lambda'^2\Lambda}d\lambda + \frac{d\Lambda}{\Lambda} = 0.$$

Hence establishing the equation

$$\frac{d\lambda}{\lambda} - \frac{Gd\lambda}{\lambda\lambda'^2\Lambda} + \frac{d\Lambda}{\Lambda} = \frac{dk}{k} - \frac{Edk}{kk'^2K} + \frac{dK}{K},$$

and observing that $M = \frac{1}{n}\frac{K}{\Lambda}$ and therefore $\frac{dM}{M} = \frac{dK}{K} - \frac{d\Lambda}{\Lambda}$, we obtain

$$\frac{d\lambda}{\lambda} - \frac{Gd\lambda}{\lambda\lambda'^2\Lambda} - \frac{dM}{M} = \frac{dk}{k} - \frac{Edk}{kk'^2K},$$

viz. this is

$$\frac{d\lambda}{\lambda\lambda'^2}\left\{\lambda'^2 - \frac{G}{\Lambda} - \frac{\lambda\lambda'^2}{M}\frac{dM}{d\lambda}\right\} = \frac{dk}{kk'^2}\left(k'^2 - \frac{E}{K}\right),$$

or, eliminating dk, $d\lambda$ by the relation $\frac{d\lambda}{\lambda\lambda'^2} = \frac{1}{nM^2}\frac{dk}{kk'^2}$, this is

$$\frac{1}{nM^2}\left\{\lambda'^2 - \frac{G}{\Lambda} - \frac{\lambda\lambda'^2}{M}\frac{dM}{d\lambda}\right\} = k'^2 - \frac{E}{K},$$

which is the result in question. Observe that $\frac{dM}{d\lambda}$ is the total differential coefficient, viz. if M is taken to be a function of k, λ, then in the differentiation, k must be treated as a function of λ. The equation, as involving not only K, Λ but also E, G, is in its actual form only true for the first transformation, and it does not readily appear how it should be modified in the case where λ is any root whatever.

CHAPTER IX.

JACOBI'S PARTIAL DIFFERENTIAL EQUATIONS FOR THE FUNCTIONS H, Θ, AND FOR THE NUMERATORS AND DENOMINATORS IN THE MULTIPLICATION AND TRANSFORMATION OF THE ELLIPTIC FUNCTIONS sn u, cn u, dn u.

Outline of the Results. Art. Nos. 306 to 309.

306. THE functions Θu, Hu have an important application to the theory of multiplication, and theoretically a like one to the theory of transformation. To explain this, recalling the formulæ

$$\sqrt{k}\,\text{sn}\,u = Hu \qquad (+),$$

$$\sqrt{\frac{k}{k'}}\,\text{cn}\,u = H(u+K) \qquad (+),$$

$$\frac{1}{\sqrt{k'}}\,\text{dn}\,u = \Theta(u+K) \qquad (+);$$

where
$$\text{denom.} = \Theta u,$$

and considering first the case of multiplication, it has already been seen that considering the expressions of

$$\sqrt{k}\,\text{sn}\,nu, \quad \sqrt{\frac{k}{k'}}\,\text{cn}\,nu, \quad \frac{1}{\sqrt{k'}}\,\text{dn}\,nu,$$

in terms of sn u, the three numerators and the denominator of these functions are respectively

$= Hnu\,\Theta^{n^2-1}0,\ H(nu+K)\,\Theta^{n^2-1}0,\ \Theta(nu+K)\,\Theta^{n^2-1}0,\ \Theta nu\,\Theta^{n^2-1}0,$
each divided by $\Theta^{n^2}u$: where for shortness $\Theta 0$ is written instead of its value $=\sqrt{\dfrac{2k'K}{\pi}}$.

307. The corresponding formulæ in the transformation of the order n are that considering the expressions of

$$\sqrt{\lambda}\,\text{sn}\left(\frac{u}{M},\lambda\right),\quad \sqrt{\frac{\lambda}{\lambda'}}\,\text{cn}\left(\frac{u}{M},\lambda\right),\quad \frac{1}{\sqrt{\lambda'}}\,\text{dn}\left(\frac{u}{M},\lambda\right),$$

in terms of sn u, the three numerators and the denominator are respectively $=$

$$H\left(\frac{u}{M},\lambda\right)\Theta_1^{\,n-1}0,\ H\left(\frac{u}{M}+\Lambda,\lambda\right)\Theta_1^{\,n-1}0,\ \Theta\left(\frac{u}{M}+\Lambda,\lambda\right)\Theta_1^{\,n-1}0,$$

$$\text{and } \Theta\left(\frac{u}{M},\lambda\right)\Theta_1^{\,n-1}0,$$

each divided by $\Theta^n u$; where for shortness $\Theta_1 0$ is written instead of its value $=\sqrt{\frac{2\lambda'\Lambda}{\pi}}$: the proof need not be at present considered. Observe that for $u=0$ the denominator is

$$\Theta_1^n 0 + \Theta^n 0, = \left\{\frac{\lambda'\Lambda}{k'K}\right\}^{\frac{1}{2}n}.$$

Now the functions Θu, Hu, $\Theta(u+K)$, $H(u+K)$, each satisfy as will be shown a certain partial differential equation which in its most simple form is $\frac{d^2\sigma}{dv^2}-4\frac{d\sigma}{d\omega}=0$, where the variables are $\omega, =\frac{\pi K'}{K}$, and $v, =\frac{\pi u}{2K}$, Jacobi, *Crelle*, t. III. (1828) p. 306. And we hence deduce a partial differential equation satisfied by the foregoing numerator- and denominator-functions, as well in the case of transformation as in that of multiplication: viz. if, in the case of multiplication by n, we write $\nu=n^2$, but in the case of the transformation of the nth order $\nu=n$, then (in one of several forms) this equation is (Jacobi, *Crelle*, t. IV (1829) p. 185)

$$(1-ax^2+x^4)\frac{d^2s}{dx^2}+(\nu-1)(ax-2x^3)\frac{ds}{dx}$$

$$+\nu(\nu-1)x^2s-2\nu(a^2-4)\frac{ds}{da}=0,$$

in which equation the variables are $x, =\sqrt{k}\,\text{sn}\,u$, and $a, =k+\frac{1}{k}$.

308. The form is specially applicable to the denominator of the three functions of nu, for this is a rational and integral function of k and $\text{sn}^2 u$, which when we introduce therein x, $=\sqrt{k}\,\text{sn}\,u$, becomes a function of x and k, which is unaltered when k is changed into $\frac{1}{k}$, and is therefore a rational and integral function of x and a: and it is for the like reason specially applicable to the numerator of $\sqrt{k}\,\text{sn}\,nu$ when n is an odd number. But the form is not in other cases the most convenient one; for instance as regards the numerators of $\sqrt{\frac{k}{k'}}\,\text{cn}\,u$, $\frac{1}{\sqrt{k'}}\,\text{dn}\,u$, these do not thus become rational in regard to a, and it would be better to have k as a variable in place of a; and in the case where the numerator contains as a factor an irrational function $\text{cn}\,u$, $\text{dn}\,u$ or $\text{cn}\,u\,\text{dn}\,u$ of $\text{sn}\,u$, it is proper instead of z to consider z divided by such irrational factor, that is the other factor, rational in regard to $\text{sn}\,u$. But making the suitable modifications the formula is for multiplication a very convenient one: viz. we can by means of it actually determine the numerator- and denominator-functions.

309. But for transformation the formula is practically useless; for observe that λ is therein regarded as a function of k, that is of a; viz. the modular equation must be taken to be known. Supposing that it is known, we cannot even then determine by means of it the numerator- and denominator-functions; for in seeking a solution by the method of indeterminate coefficients the coefficients of the several powers of x would be functions of (u, v) not only unknown, but in form indeterminate (as admitting of modification by means of the modular equation):—and even when the actual expression of z as a function of (x, u, v) is known, as of course it is for the cubic, quintic, &c. transformations, it is, from the complexity of the modular equations, by no means easy to verify the formula: the process is in fact one of difficulty even in the case of the cubic transformation $n=3$. This of course in no wise diminishes the interest of the result;

and the investigation of it being substantially identical in the two cases of transformation and multiplication, it is proper not to separate them.

Partial Differential Equation satisfied by Θu.

Art. Nos. 310 to 312.

310. It is to be shown that the function

$$\sigma, = \Theta u, = \sqrt{\frac{2k'K}{\pi}}\, e^{-\frac{1}{2}\frac{E}{K}u^2 + \int_0 du \int_0 du\, \mathrm{dn}^2 u}$$

satisfies the differential equation

$$\frac{d^2\sigma}{du^2} - 2u\left(k'^2 - \frac{E}{K}\right)\frac{d\sigma}{du} + 2kk'^2\frac{d\sigma}{dk} = 0.$$

We have

$$\frac{d\sigma}{du} = \left(\int_0 du\, \mathrm{dn}^2 u - \frac{E}{K}u\right)\sigma,$$

$$= \left\{u\left(k'^2 - \frac{E}{K}\right) + k^2\int_0 du\, \mathrm{cn}^2 u\right\}\sigma,$$

$$\frac{d^2\sigma}{du^2} = \left[\mathrm{dn}^2 u - \frac{E}{K} + \left\{u\left(k'^2 - \frac{E}{K}\right) + k^2\int_0 du\, \mathrm{cn}^2 u\right\}^2\right]\sigma,$$

$$\frac{d\sigma}{dk} = \left[\frac{1}{2KK'}\frac{d.KK'}{dk} - \tfrac{1}{2}u^2\frac{d}{dk}\frac{E}{K} + \int_0 du\int_0 du\frac{d}{dk}\mathrm{dn}^2 u\right]\sigma.$$

311. The success of the process depends on a transformation of the double integral

$$\int_0 du \int_0 du \frac{d}{dk}\mathrm{dn}^2 u.$$

We have, see No. 128,

$$\frac{d}{dk}\mathrm{dn}\, u = \frac{k^2}{k'^2}\mathrm{sn}\, u\, \mathrm{cn}\, u \int_0 \mathrm{cn}^2 u\, du - \frac{k}{k'^2}\mathrm{sn}^2 u\, \mathrm{dn}\, u,$$

whence

$$\frac{d}{dk}\mathrm{dn}^2 u = -\frac{2k}{k'^2}\left\{\mathrm{sn}^2 u\, \mathrm{dn}^2 u - k^2\mathrm{sn}\, u\, \mathrm{cn}\, u\, \mathrm{dn}\, u \int_0 du\, \mathrm{cn}^2 u\right\},$$

$$= -\frac{2k}{k'^2}\left\{\mathrm{sn}^2 u\, \mathrm{dn}^2 u + \tfrac{1}{2}k^2\left(\frac{d}{du}\mathrm{cn}^2 u\right)\int_0 du\, \mathrm{cn}^2 u\right\},$$

and thence

$$\int_0 du \int_0 du \,\frac{d}{dk}\,\mathrm{dn}^2 u = -\frac{2k}{k'^2}\Big\{\int_0 du \int_0 du\, \mathrm{sn}^2 u\, \mathrm{dn}^2 u$$

$$+ \tfrac{1}{2} k^2 \int_0 du \left(\mathrm{cn}^2 u \int_0 du\, \mathrm{cn}^2 u - \int_0 du\, \mathrm{cn}^4 u\right)\Big\},$$

$$= -\frac{k}{k'^2}\Big\{\int_0 du \int_0 du\, (2\,\mathrm{sn}^2 u\, \mathrm{dn}^2 u - k^2\, \mathrm{cn}^4 u)$$

$$+ \tfrac{1}{2} k^2 \left(\int_0 du\, \mathrm{cn}^2 u\right)^2\Big\}.$$

But we have

$$\frac{d^2}{du^2}\,\mathrm{sn}^2 u = 2\,(\mathrm{cn}^2 u\, \mathrm{dn}^2 u - \mathrm{sn}^2 u\, \mathrm{dn}^2 u - k^2\, \mathrm{sn}^2 u\, \mathrm{cn}^2 u),$$

$$= 2\,(k'^2 - 2\,\mathrm{sn}^2 u\, \mathrm{dn}^2 u + k^2\, \mathrm{cn}^4 u),$$

or multiplying by du^2 and integrating twice

$$\mathrm{sn}^2 u = k'^2 u^2 - 2\int_0 du \int_0 du\, (2\,\mathrm{sn}^2 u\, \mathrm{dn}^2 u - k^2\, \mathrm{cn}^4 u),$$

whence at length

$$\int_0 du \int_0 du\, \frac{d}{dk}\,\mathrm{dn}^2 u = -\tfrac{1}{2} k u^2 + \tfrac{1}{2}\frac{k}{k'^2}\,\mathrm{sn}^2 u - \frac{k^3}{2k'^2}\left(\int_0 du\, \mathrm{cn}^2 u\right)^2,$$

the required value of the integral.

312. Resuming the investigation, we have

$$\frac{d}{dk}\,Kk' = \frac{E-K}{kk'}, \quad \frac{d}{dk}\frac{E}{K} = \frac{1}{kk'^2}\left\{k'^2\left(\frac{2E}{K} - 1\right) - \frac{E^2}{K^2}\right\},$$

and hence

$$\frac{d\sigma}{dk} = \frac{1}{2kk'^2}\left\{\frac{E}{K} - \mathrm{dn}^2 u + \left(k'^2 - \frac{E}{K}\right)^2 u^2 - k^4 \left(\int_0 du\, \mathrm{cn}^2 u\right)^2\right\}\sigma.$$

Substituting the foregoing values of $\frac{d^2\sigma}{du^2}$, $\frac{d\sigma}{du}$, $\frac{d\sigma}{dk}$ in the differential equation, the several terms destroy each other, and we thus have the equation in question.

Same Equation satisfied by Hu, $\Theta(u+K)$, $H(u+K)$.
Art. Nos. 313, 314.

313. The equation

$$\frac{d^2\sigma}{du^2} - 2u\left(k'^2 - \frac{E}{K}\right)\frac{d\sigma}{du} + 2kk'^2\frac{d\sigma}{dk} = 0$$

is satisfied by $\sigma = \Theta u$; write for a moment $u + iK' = v$, then the equation

$$\frac{d^2\sigma_1}{dv^2} - 2v\left(k'^2 - \frac{E}{K}\right)\frac{d\sigma_1}{dv} + 2kk'^2\frac{d\sigma_1}{dk} = 0$$

is satisfied by $\sigma_1 = \Theta v$, $= \Theta(u + iK')$; and transforming to the variable u, we have

$$\frac{d\sigma_1}{du} = \frac{d\sigma_1}{dv}, \quad \frac{d^2\sigma_1}{du^2} = \frac{d^2\sigma}{dv^2}, \quad \frac{d\sigma_1}{dk} = \frac{d\sigma_1}{dk} + \frac{d\sigma_1}{dv}\frac{dv}{dk},$$

that is

$$\frac{d\sigma_1}{dv} = \frac{d\sigma_1}{du}, \quad \frac{d^2\sigma_1}{dv^2} = \frac{d^2\sigma_1}{du^2}, \quad \frac{d\sigma_1}{dk} = \frac{d\sigma_1}{dk} - \frac{d\sigma_1}{dv}\frac{idK'}{dk};$$

whence the equation is

$$\frac{d^2\sigma_1}{du^2} - \left[2(u + iK')\left(k'^2 - \frac{E}{K}\right) + 2kk'^2\frac{idK'}{dk}\right]\frac{d\sigma_1}{du} + 2kk'^2\frac{d\sigma_1}{dk} = 0,$$

which is at once reduced to

$$\frac{d^2\sigma_1}{du^2} - \left[2u\left(k'^2 - \frac{E}{K}\right) - \frac{\pi i}{K}\right]\frac{d\sigma_1}{du} + 2kk'^2\frac{d\sigma_1}{dk} = 0.$$

It is easy to show that this equation is satisfied by $\sigma_1 = e^{\frac{\pi(2iu - K')}{4K}}$, $= Q$ suppose.

Hence, assuming $\sigma_1 = Q\sigma$, we find

$$\frac{d^2\sigma}{du^2} + \left[\frac{2}{Q}\frac{dQ}{du} - 2u\left(k'^2 - \frac{E}{K}\right) + \frac{\pi i}{K}\right]\frac{d\sigma}{du} + 2kk'^2\frac{d\sigma}{dk} = 0;$$

or observing that $\frac{2}{Q}\frac{dQ}{du} = -\frac{i\pi}{K}$, this is the original equation in σ: hence this equation is satisfied by

$$\sigma = -ie^{\frac{\pi(2iu - K')}{4K}}\Theta(u + iK'),$$

that is by $\sigma = Hu$.

314. Write for a moment $u + K = v$, the equation

$$\frac{d^2\sigma_1}{dv^2} - 2v\left(k'^2 - \frac{E}{K}\right)\frac{d\sigma_1}{dv} + 2kk'^2\frac{d\sigma_1}{dk} = 0$$

is satisfied by $\sigma_1 = \Theta v$ or Hv, that is $= \Theta(u+K)$ or $H(u+K)$. But transforming to the new variable u, we have

$$\frac{d\sigma_1}{du} = \frac{d\sigma_1}{dv}, \quad \frac{d^2\sigma_1}{du^2} = \frac{d^2\sigma_1}{dv^2}, \quad \frac{d\sigma_1}{dk} = \frac{d\sigma_1}{dk} + \frac{d\sigma_1}{dv}\frac{dv}{dk},$$

that is

$$\frac{d\sigma_1}{dv} = \frac{d\sigma_1}{du}, \quad \frac{d^2\sigma_1}{dv^2} = \frac{d^2\sigma}{du^2}, \quad \frac{d\sigma_1}{dk} = \frac{d\sigma_1}{dk} - \frac{d\sigma_1}{du}\frac{dK}{dk}$$

$$= \frac{d\sigma_1}{dk} + \frac{K}{kk'^2}\left(k'^2 - \frac{E}{K}\right)\frac{d\sigma_1}{du}$$

as the values to be substituted in the differential equation: viz. this becomes

$$\frac{d^2\sigma_1}{du^2} - \left\{2(u+K)\left(k'^2 - \frac{E}{K}\right) - 2K\left(k'^2 - \frac{E}{K}\right)\right\}\frac{d\sigma_1}{du} + 2kk'^2\frac{d\sigma_1}{dk} = 0,$$

the original equation with σ_1 for σ. We thus see that the equation in σ is satisfied not only by the values Θu, Hu, but also by the values $\Theta(u+K)$, $H(u+K)$, or, what is the same thing, by the denominator (Θu) and the numerators of $\sqrt{k}\,\text{sn}\,u$, $\sqrt{\frac{k}{k'}}\,\text{cn}\,u$ and $\frac{1}{\sqrt{k'}}\,\text{dn}\,u$.

Differential Equation satisfied by $\Theta\left(\frac{u}{M}, \lambda\right)$, &c.

Art. Nos. 315, 316.

315. Considering now the new modulus λ and the multiplier M in the first transformation (of order n) write $v = \frac{u}{M}$, and consider the equation

$$\frac{d^2\sigma_1}{dv^2} - 2v\left(\lambda'^2 - \frac{G}{\Lambda}\right)\frac{d\sigma_1}{dv} - 2\lambda\lambda'^2\frac{d\sigma_1}{d\lambda} = 0,$$

$(G = E_1(\lambda)$, the same function of λ that E is of $k)$ satisfied of course by

$$\sigma_1 = \Theta(v, \lambda), \quad H(v, \lambda), \quad \Theta(v+\Lambda, \lambda), \quad H(v+\Lambda, \lambda).$$

Transforming to the new variables u, k, we have

$$\frac{d\sigma_1}{du}=\frac{1}{M}\frac{d\sigma_1}{dv},\quad \frac{d^2\sigma_1}{du^2}=\frac{1}{M^2}\frac{d^2\sigma_1}{dv^2},\quad \frac{d\sigma_1}{dk}=-\frac{u}{M^2}\frac{dM}{dk}\frac{d\sigma_1}{dv}+\frac{d\sigma_1}{d\lambda}\frac{d\lambda}{dk},$$

and thence

$$\frac{d\sigma_1}{dv}=M\frac{d\sigma_1}{du},\quad \frac{d^2\sigma_1}{dv^2}=M^2\frac{d^2\sigma_1}{du^2},\quad \frac{d\sigma_1}{d\lambda}=\left(\frac{d\sigma_1}{dk}+\frac{u}{M}\frac{dM}{dk}\frac{d\sigma_1}{du}\right)\frac{dk}{d\lambda},$$

$$=\frac{d\sigma_1}{dk}\frac{dk}{d\lambda}+\frac{u}{M}\frac{dM}{d\lambda}\frac{d\sigma_1}{du},$$

and the equation thus becomes

$$\frac{d^2\sigma_1}{du^2}-\frac{2u}{M^2}\left\{\left(\lambda'^2-\frac{G}{\Lambda}\right)-\frac{\lambda\lambda'^2}{M}\frac{dM}{d\lambda}\right\}\frac{d\sigma_1}{du}+\frac{2\lambda\lambda'^2}{M^2}\frac{dk}{d\lambda}\frac{d\sigma_1}{dk}=0.$$

316. We have

$$\frac{1}{M^2}\left\{\left(\lambda'^2-\frac{G}{\Lambda}\right)-\frac{\lambda\lambda'^2}{M}\frac{dM}{d\lambda}\right\}=n\left(k'^2-\frac{E}{K}\right),$$

$$\frac{\lambda\lambda'^2}{M^2}\frac{dk}{d\lambda}\qquad\qquad =nkk'^2,$$

and the equation thus is

$$\frac{d^2\sigma_1}{du^2}-2nu\left(k'^2-\frac{E}{K}\right)\frac{d\sigma_1}{du}+2nkk'^2\frac{d\sigma_1}{dk}=0.$$

Hence, writing σ for σ_1, the equation

$$\frac{d^2\sigma}{du^2}-2nu\left(k'^2-\frac{E}{K}\right)\frac{d\sigma}{du}+2nkk'^2\frac{d\sigma}{dk}=0$$

is satisfied by

$$\sigma=\Theta\left(\frac{u}{M},\lambda\right),\ H\left(\frac{u}{M},\lambda\right),\ \Theta\left(\frac{u}{M}+\Lambda,\lambda\right),\ H\left(\frac{u}{M}+\Lambda,\lambda\right).$$

New form of the two Differential Equations.
Art. Nos. 317, 318.

317. The connexion of the two equations

$$\frac{d^2\sigma}{du^2}-2u\left(k'^2-\frac{E}{K}\right)\frac{d\sigma}{du}+2kk'^2\frac{d\sigma}{dk}=0,$$

and

$$\frac{d^2\sigma}{du^2}-2nu\left(k'^2-\frac{E}{K}\right)\frac{d\sigma}{du}+2nkk'^2\frac{d\sigma}{dk}=0,$$

may be established in a different manner thus: writing in the first equation

$$\omega = \frac{\pi K'}{K}, \quad v = \frac{\pi u}{2K},$$

then observing that

$$\frac{d}{dk}\frac{K'}{K} = \frac{1}{K^2 kk'^2}(KK' - KE' - K'E), = -\frac{\pi}{2K^2 kk'^2},$$

we have

$$\frac{d\sigma}{du} = \frac{\pi}{2K}\frac{d\sigma}{dv}, \quad \frac{d^2\sigma}{du^2} = \frac{\pi^2}{4K^2}\frac{d^2\sigma}{dv^2},$$

$$\frac{d\sigma}{dk} = \frac{v}{kk'^2}\left(k^2 - \frac{E}{K}\right)\frac{d\sigma}{dv} - \frac{\pi^2}{2K^2 kk'^2}\frac{d\sigma}{d\omega},$$

and the equation becomes

$$\frac{d^2\sigma}{dv^2} - 4\frac{d\sigma}{d\omega} = 0,$$

satisfied by $\sigma = \Theta u$, &c.

318. Writing in the second equation

$$\omega = \frac{n\pi K'}{K}, \quad v = \frac{n\pi u}{2K},$$

this is in like manner transformed into the same equation $\frac{d^2\sigma}{dv^2} - 4\frac{d\sigma}{d\omega} = 0$. Hence whatever function of $\frac{u}{K}$ and $\frac{K'}{K}$ satisfies the first equation, the same function of $\frac{nu}{K}$ and $\frac{nK'}{K}$ satisfies the second equation. Let λ be the modulus in the first transformation of the nth order, and Λ, Λ' the complete functions, $\frac{\Lambda'}{\Lambda} = \frac{nK'}{K}$ and $\Lambda = \frac{K}{nM}$, that is $\frac{nK'}{K} = \frac{\Lambda'}{\Lambda}$ and $\frac{nu}{K} = \frac{u}{M}$; or the second equation is satisfied by the same function of $\frac{u}{M}$, $\frac{\Lambda'}{\Lambda}$. Hence the first equation being satisfied by $\sigma = \Theta u$, &c., the second equation is satisfied by

$$\sigma = \Theta\left(\frac{u}{M},\ \lambda\right), \text{ \&c.}$$

Partial Differential Equations satisfied by the Numerators and Denominator. Art. Nos. 319 to 327.

319. Start now with the equation

$$\frac{d^2\Sigma}{du^2} - 2nu\left(k'^2 - \frac{E}{K}\right)\frac{d\Sigma}{du} + 2nkk'^2\frac{d\Sigma}{dk} = 0,$$

satisfied by $\Sigma = \Theta\left(\frac{u}{M}, \lambda\right)$, &c.

And assume

$$\Sigma = (\tfrac{1}{2}\pi)^{\frac{1}{2}(n-1)}(Kk')^{-\frac{1}{2}(n-1)}\,\Theta^n u \,.\, z,$$

say for shortness this is

$$= C\Omega\sigma^n \,.\, z,$$

(where σ denotes Θu and consequently satisfies the equation

$$\frac{d^2\sigma}{du^2} - 2u\left(k'^2 - \frac{E}{K}\right)\frac{d\sigma}{du} + 2kk'^2\frac{d\sigma}{dk} = 0).$$

We find

$$\begin{aligned} z\Bigg[\Omega n\left\{\sigma^{n-1}\frac{d^2\sigma}{du^2} + (n-1)\,\sigma^{n-2}\left(\frac{d\sigma}{du}\right)^2\right\} \\ - 2nu\left(k'^2 - \frac{E}{K}\right)\Omega n\sigma^{n-1}\frac{d\sigma}{du} + 2nkk'^2\frac{d}{dk}(\Omega\sigma^n)\Bigg] \\ + \frac{dz}{du}\left[\Omega \,.\, 2n\sigma^{n-1}\frac{d\sigma}{du} - 2nu\left(k'^2 - \frac{E}{K}\right)\Omega\sigma^n\right] \\ + \frac{d^2z}{du^2}\Omega\sigma^n \\ + \frac{dz}{dk}2nkk'^2\,\Omega\sigma^n = 0, \end{aligned}$$

where in the coefficient of z we write for $\frac{d^2\sigma}{du^2}$ its value thereby changing this coefficient into

$$\Omega\sigma^n\left[n(n-1)\left\{\frac{1}{\sigma^2}\left(\frac{d\sigma}{du}\right)^2-2u\left(k'^2-\frac{E}{K}\right)\frac{1}{\sigma}\frac{d\sigma}{du}+2kk'^2\cdot\frac{1}{\sigma}\frac{d\sigma}{dk}\right\}\right.$$
$$\left.+2n\,kk'^2\cdot\frac{1}{\Omega}\frac{d\Omega}{dk}\right],$$

or in the term $\frac{1}{\Omega}\frac{d\Omega}{dk}$ substituting for Ω its value $=(KK')^{-\frac{1}{2}(n-1)}$, this is

$$=\Omega\sigma^n\,.\,n(n-1)\left\{\frac{1}{\sigma^2}\left(\frac{d\sigma}{du}\right)^2-2u\left(k'^2-\frac{E}{K}\right)\frac{1}{\sigma}\frac{d\sigma}{du}\right.$$
$$\left.+2kk'^2\frac{1}{\sigma}\frac{d\sigma}{dk}-\frac{kk'}{K}\frac{d}{dk}(Kk')\right\}.$$

Hence dividing the whole equation by $\Omega\sigma^n$ it becomes

$$\frac{d^2z}{du^2}$$
$$+\frac{dz}{du}\cdot 2n\left\{\frac{1}{\sigma}\frac{d\sigma}{du}-u\left(k'^2-\frac{E}{K}\right)\right\}$$
$$+\quad z\,.\,n(n-1)\left\{\frac{1}{\sigma^2}\left(\frac{d\sigma}{du}\right)^2-2u\left(k'^2-\frac{E}{K}\right)\frac{1}{\sigma}\frac{d\sigma}{du}\right.$$
$$\left.+2kk'^2\frac{1}{\sigma}\frac{d\sigma}{dk}-\frac{kk'}{K}\frac{d}{dk}Kk'\right\}$$
$$+\frac{dz}{dk}\cdot 2nkk'^2=0.$$

320. Recurring to the investigation in regard to the function σ, $=\Theta u$, we have

$$\frac{1}{\sigma}\frac{d\sigma}{du}-u\left(k'^2-\frac{E}{K}\right)=k^2\textstyle\int_0 du\,\mathrm{cn}^2u;$$

whence

$$\frac{1}{\sigma^2}\left(\frac{d\sigma}{du}\right)^2-2u\left(k'^2-\frac{E}{K}\right)\frac{1}{\sigma}\frac{d\sigma}{du}=-u^2\left(k'^2-\frac{E}{K}\right)^2+k^4\left(\textstyle\int_0 du\,\mathrm{cn}^2u\right)^2.$$

Also

$$2kk'^2\frac{1}{\sigma}\frac{d\sigma}{dk}=\frac{E}{K}-\mathrm{dn}^2u+u^2\left(k'^2-\frac{E}{K}\right)^2-k^4\left(\textstyle\int_0 du\,\mathrm{cn}^2u\right)^2,$$

$$-\frac{kk'}{K}\frac{d}{dk}Kk'=1-\frac{E}{K}.$$

Adding these several quantities, the coefficient of $n(n-1)z$ is $1-\text{dn}^2u, =k^2\,\text{sn}^2u$; the coefficient of $\frac{dz}{du}$ has also been found; and the equation thus becomes

$$\frac{d^2z}{du^2}+2nk'^2\left(\int_0 \text{cn}^2u\,du\right)\frac{dz}{du}+n(n-1)k^2\,\text{sn}^2u\,.\,z+2nkk'^2\frac{dz}{dk}=0,$$

which equation is consequently satisfied by

$$z=\left(\frac{2k'K}{\pi}\right)^{\frac{1}{2}(n-1)}\frac{1}{\Theta^n(u)}\Theta\left(\frac{u}{M},\lambda\right), \text{ \&c.}$$

321. It is to be further remarked that if we had started with

$$\frac{d^2\Sigma}{du^2}-2n^2u\left(k'^2-\frac{E}{K}\right)\frac{d\Sigma}{du}+2n^2kk'^2\frac{d\Sigma}{dk}=0,$$

which equation is obviously satisfied by $\Sigma=\Theta(nu)$, &c., and had then assumed

$$\Sigma=(\tfrac{1}{2}\pi)^{\frac{1}{2}(n^2-1)}(Kk')^{-\frac{1}{2}(n^2-1)}\Theta^{n^2}(u)\,.\,z,$$

we should at every step of the investigation have had n^2 in place of n, and should finally have arrived at the equation

$$\frac{d^2z}{du^2}+2n^2k^2\left(\int_0 \text{cn}^2u\,du\right)\frac{dz}{du}+n^2(n^2-1)k^2\,\text{sn}^2u\,.\,z$$

$$+2n^2kk'^2\frac{dz}{dk}=0,$$

which equation is consequently satisfied by

$$z=\left(\frac{2k'K}{\pi}\right)^{\frac{1}{2}(n^2-1)}\frac{1}{\Theta^n(u)}\Theta(nu), \text{ \&c.}$$

322. It will be convenient to include the two equations in the common form

$$\frac{d^2z}{du^2}+2\nu k^2\left(\int_0 \text{cn}^2u\,du\right)\frac{dz}{du}+\nu(\nu-1)k^2\,\text{sn}^2u\,.\,z+2\nu kk'^2\frac{dz}{dk}=0,$$

where for the transformation equation $\nu = n$, and for the multiplication equation $\nu = n^2$.

323. Write in this equation $x = \sqrt{k}\,\text{sn}\,u$.

We have

$$dx = \sqrt{k}\,\text{cn}\,u\,\text{dn}\,u\,du + \Omega\,dk,$$

if for a moment

$$\begin{aligned}
\Omega &= \frac{d}{dk}(\sqrt{k}\,\text{sn}\,u),\\
&= \frac{1}{2\sqrt{k}}\,\text{sn}\,u + \sqrt{k}\,\frac{1}{k'^2}\{k\,\text{sn}\,u\,\text{cn}^2 u - k\,\text{cn}\,u\,\text{dn}\,u \textstyle\int_0 \text{cn}^2 u\,du\},\\
&= \frac{1}{2\sqrt{k}}\,\text{sn}\,u\left(1 + \frac{2k^2}{k'^2}\,\text{cn}^2 u\right) - \frac{k\sqrt{k}}{k'^2}\,\text{cn}\,u\,\text{dn}\,u \textstyle\int_0 \text{cn}^2 u\,du,\\
&= \frac{1}{2k'^2\sqrt{k}}\,\text{sn}\,u\,(1 + k^2 - 2k^2\,\text{sn}^2 u) - \frac{k\sqrt{k}}{k'^2}\,\text{cn}\,u\,\text{dn}\,u \textstyle\int_0 \text{cn}^2 u\,du,
\end{aligned}$$

and hence

$$\frac{dz}{du} = \sqrt{k}\,\text{cn}\,u\,\text{dn}\,u\,\frac{dz}{dx},$$

$$\frac{dz}{dk} = \frac{dz}{dk} + \Omega\,\frac{dz}{dx},$$

where on the right-hand side $\frac{dz}{dk}$ is the new value of this differential coefficient, viz. that belonging to the assumption $z =$ a function of x, k, or (as we may express this) $z = z(x, k)$. And thence also

$$\begin{aligned}
\frac{d^2 z}{du^2} &= k\,\text{cn}^2 u\,\text{dn}^2 u\,\frac{d^2 z}{dx^2} + \sqrt{k}\,\frac{dz}{dx}\,\frac{d}{du}(\text{cn}\,u\,\text{dn}\,u)\\
&= k\,\text{cn}^2 u\,\text{dn}^2 u\,\frac{d^2 z}{dx^2}\\
&\quad - \sqrt{k}\,\text{sn}\,u\,(1 + k^2 - 2k^2\,\text{sn}^2 u)\,\frac{dz}{dx}.
\end{aligned}$$

Substituting, the equation becomes

$$\frac{d^2z}{dx^2} . k \, \text{cn}^2 u \, \text{dn}^2 u$$

$$+ \frac{dz}{dx} . - \sqrt{k} \, \text{sn} \, u \, (1 + k^2 - 2k^2 \, \text{sn}^2 u)$$

$$+ \frac{dz}{dx} . \, 2\nu k'^2 \sqrt{k} \, \text{cn} \, u \, \text{dn} \, u \int_0 \text{cn}^2 u \, du$$

$$+ \, z . \nu (\nu - 1) k^2 \, \text{sn}^2 u$$

$$+ \frac{dz}{dx} \{ \nu \sqrt{k} \, \text{sn} \, u \, (1 + k^2 - 2k^2 \, \text{sn}^2 u) - 2\nu k'^2 \sqrt{k} \, \text{cn} \, u \, \text{dn} \, u \int_0 \text{cn}^2 u \, du \}$$

$$+ \frac{dz}{dk} . 2\nu k k'^2 = 0,$$

where the term involving the integral disappears, and two other terms combine together; viz. the result is

$$\frac{d^2z}{dx^2} . k \, \text{cn}^2 u \, \text{dn}^2 u$$

$$+ \frac{dz}{dx} (\nu - 1) \sqrt{k} \, \text{sn} \, u \, (1 + k^2 - 2k^2 \, \text{sn}^2 u)$$

$$+ \, z . \nu (\nu - 1) k^2 \, \text{sn}^2 u$$

$$+ \frac{dz}{dk} . 2\nu k k'^2 = 0,$$

in which equation sn u should be replaced by its value $\frac{x}{\sqrt{k}}$. Introducing at the same time in place of k the quantity $a, = k + \frac{1}{k}$, the equation becomes

$$(1 - ax^2 + x^4) \frac{d^2z}{dx^2} + (\nu - 1)(ax - 2x^3) \frac{dz}{dx}$$
$$+ \nu(\nu - 1) x^2 z - 2\nu (a^2 - 4) \frac{dz}{da} = 0,$$

where I recall that the variables are $x = \sqrt{k} \, \text{sn} \, u$ and $a = k + \frac{1}{k}$.

The equation is satisfied by the numerators and the denominator of $\sqrt{\lambda}\,\mathrm{sn}\left(\frac{u}{M}, \lambda\right)$, $\sqrt{\frac{\lambda}{\lambda'}}\,\mathrm{cn}\left(\frac{u}{M}, \lambda\right)$, $\frac{1}{\sqrt{\lambda'}}\,\mathrm{dn}\left(\frac{u}{M}, \lambda\right)$ in the transformation of the order n, or ($\nu = n^2$) by the numerators and denominator of $\sqrt{k}\,\mathrm{sn}\,nu$, $\sqrt{\frac{k}{k'}}\,\mathrm{cn}\,nu$, $\frac{1}{\sqrt{k'}}\,\mathrm{dn}\,nu$.

324. As already remarked, the formula is not practically useful in the transformation-theory, but it is so for multiplication. As regards this last theory it has been observed that although with respect to the denominator-function, and the numerator of sn nu when n is an odd number, there is great elegance in taking as above the variable to be $\sqrt{k}\,\mathrm{sn}\,u$, and in introducing a in place of k, yet that for the other functions, this is not the case, and it seems better to have as the variables ξ, $=\mathrm{sn}\,u$, and k. The transformation is of course easily effected, viz. writing $\xi = \frac{x}{\sqrt{k}}$, we find

$$\frac{dz}{dx} = \frac{1}{\sqrt{k}}\,\frac{dz}{d\xi},$$

$$\frac{dz}{dk} = \frac{dz}{dk} - \frac{\xi}{2k}\,\frac{dz}{d\xi},$$

where on the right-hand side $\frac{dz}{dk}$ is the value belonging to the assumption $z = z\,(\xi, k)$. Hence also $\frac{d^2z}{dx^2} = \frac{1}{k}\,\frac{d^2z}{d\xi^2}$, and the equation, finally restoring therein x in place of ξ, becomes

$$\frac{d^2z}{dx^2}(1 - x^2 \,.\, 1 - k^2x^2) + \frac{dz}{dx}[(2\nu k^2 - 1 - k^2)\,x - 2\,(\nu - 1)\,k^2x^3]$$

$$+ z \,.\, \nu(\nu - 1)\,k^2x^2 + 2\nu k\,(1 - k^2)\,\frac{dz}{dk} = 0\,;$$

viz. x is here $=\mathrm{sn}\,u$, and the equation is satisfied by the numerators and denominator of $\sqrt{k}\,\mathrm{sn}\,nu$, $\sqrt{\frac{k}{k'}}\,\mathrm{cn}\,nu$, $\frac{1}{\sqrt{k'}}\,\mathrm{dn}\,nu$.

We may of course get rid of the exterior factors, and thus obtain a system of four equations, viz.

$$\frac{d^2z}{dx^2}(1-x^2\,.\,1-k^2x^2)+\frac{dz}{dx}[(2\nu k^2-1-k^2)\,x-2\,(\nu-1)\,k^2x^3]$$

$$+z\,[(\nu^2-\nu)\,k^2x^2+A]+\frac{dz}{dk}\,.\,2\nu k\,(1-k^2)=0;$$

where $A=\nu(1-k^2)$, equation satisfied by numerator of sn nu,

$A=\nu$	,	,,	,,	,,	cn nu,
$A=\nu k^2$	,	,,	,,	,,	dn nu,
$A=0$	,	,,	,,	denom. of each function.	

323. For instance $n=2$, $\nu=4$, the equations are

$$\frac{d^2z}{dx^2}(1-x^2\,.\,1-k^2x^2)+\frac{dz}{dx}[(7k^2-1)\,x-6k^2x^3]$$

$$+z\left\{12k^2x^2+\begin{matrix}4\,(1-k^2),\\ 4\quad ,\\ 4k^2\quad ,\\ 0\quad ,\end{matrix}\right\}+\frac{dz}{dk}\,.\,8k\,(1-k^2)=0,$$

satisfied by

$$\begin{aligned} z &= x\sqrt{1-x^2\,.\,1-k^2x^2},\\ &= 1-\ \ 2x^2+k^2x^4,\\ &= 1-2k^2x^2+k^2x^4,\\ &= 1\qquad\quad -k^2x^4, \text{ respectively.}\end{aligned}$$

As to the first equation, observe that writing for the moment $X=(1-x^2\,.\,1-k^2x^2)$, we have $z=x\sqrt{X}$, and thence

$$\frac{dz}{dx}=\frac{1-2\,(1+k^2)\,x^2+3k^2x^4}{\sqrt{X}},$$

$$\frac{d^2z}{dx^2}=\frac{1}{X\sqrt{X}}\{(-3-3k^2)x+(2+14k^2+2k^4)x^3+(-9k^2-9k^4)x^5+6k^4x^7\},$$

$$\frac{dz}{dk}=\frac{1}{\sqrt{X}}(-kx^3+kx^5).$$

Hence the first term $X\dfrac{d^2z}{dx^2}$ contains in the denominator, like the other terms, $\sqrt{X}$ only, and omitting this denominator the equation is

			x^1	x^3	x^5	x^7
first term = ...			$-3-3k^2$,	$2+14k^2+2k^4$,	$-9k^2-9k^4$,	$+6k^4$
$(-1+7k^2)x-6k^2x^3$	.	$1-2(1+k^2)x^2+3k^2x^4$	$-1+7k^2$,	$2-18k^2-14k^4$,	$9k^2+33k^4$,	$-18k^4$
$4-4k^2+12k^2x^2$	.	$x-(1+k^2)x^3+k^2x^5$	$4-4k^2$,	$-4+12k^2+4k^4$,	$-8k^2-16k^4$,	$+12k^4$
$8(k-k^3)$	.	$-kx^3+kx^5$		$-8k^2+8k^4$,	$8k^2-8k^4$,	
			0	0	0	0

326. For the other equations the verification is easier.

Second equation.

			x^0	x^2	x^4	x^6
$1-(1+k^2)x^2+k^2x^4$	.	$-4+12k^2x^2$	-4,	$4+16k^2$,	$-16k^2-12k^4$,	$+12k^4$
$(-1+7k^2)x-6k^2x^3$	.	$-4x+4k^2x^3$		$4-28k^2$,	$20k^2+28k^4$,	$-24k^4$
$4+12k^2x^2$	.	$1-2x^2+k^2x^4$	4,	$-8+12k^2$,	$-20k^2$,	$+12k^4$
$8(k-k^3)$		$2kx^4$			$16k^2-16k^4$,	
			0	0	0	0

Third equation.

	x^0	x^2	x^4	x^6
$1-(1+k^2)x^2+k^2x^4. \quad -4k^2+12k^2x^2$	$-4k^2$,	$16k^2+4k^4$,	$-12k^2-16k^4$,	$12k^4$
$(-1+7k^2)x-6k^2x^3. \quad -4k^2x+4k^2x^3$		$4k^2-28k^4$,	$-4k^2+52k^4$,	$-24k^4$
$4k^2+12k^2x^2. \quad 1-2k^2x^2+k^2x^4$	$4k^2$,	$12k^2-8k^4$,	$-20k^4$,	$+12k^4$
$8(k-k^3). \quad -4kx^2+2kx^4$		$-32k^2+32k^4$,	$16k^2-16k^4$	
	0	0	0	0

Fourth equation.

	x^2	x^4	x^6
$1-(1+k^2)x^2+k^2x^4. \quad -12k^2x^2$	$-12k^2$,	$+12k^2+12k^4$,	$-12k^4$
$+(-1+7k^2)x-6k^2x^3. \quad -4k^2x^3$		$+4k^2-28k^4$,	$+24k^4$
$+ \quad 12k^2x^3. \quad 1-k^2x^4$	$+12k^2$,		$-12k^4$
$+ \quad 8(k-k^3). \quad -2kx^4$		$-16k^2+16k^4$	
	0	0	0

327. But we may further develope the system of formulæ. When n is even the numerator of $\operatorname{sn} nu$ contains the factor $\sqrt{1-x^2 . 1-k^2x^2}$; and when n is odd the numerator of $\operatorname{cn} nu$ contains the factor $\sqrt{1-x^2}$, and that of $\operatorname{dn} nu$ the factor $\sqrt{1-k^2x^2}$: and we may in the several cases find the differential equation satisfied by the other, or rational, factor. There is no difficulty in the investigation: the results are, n even,

$$\frac{d^2z}{dx^2}(1-x^2 . 1-k^2x^2)+\frac{dz}{dx}\left[(-3+(2\nu-3)k^2)x+(-2\nu+6)k^2x^3\right]$$
$$+z\left[(\nu-1)-(\nu+1)k^2+(\nu-2)(\nu-3)k^2x^2\right]+2\nu(k-k^3)\frac{dz}{dk}=0,$$

satisfied by numerator of $\operatorname{sn} nu$ omitting the factor $\sqrt{1-x^2 . 1-k^2x^2}$: for example, $n=2$ $(\nu=4)$ the equation is

$$\frac{d^2z}{dx^2}(1-x^2 . 1-k^2x^2)+\frac{dz}{dx}\left[(5k^2-3)x-2k^2x^3\right]$$
$$+z\left[(3-5k^2)+2k^2x^2\right]+8(k-k^3)\frac{dz}{dk}=0,$$

satisfied by $z=x$:

and, n odd,

$$\frac{d^2z}{dx^2}(1-x^2 . 1-k^2x^2)+\frac{dz}{dx}\left[(-3+(2\nu-1)k^2)x+(-2\nu+4)k^2x^3\right]$$
$$+z(\nu-1)\left[1+(\nu-2)k^2x^2\right]+2\nu(k-k^3)\frac{dz}{dk}=0,$$

satisfied by numerator of $\operatorname{cn} nu$ omitting the factor $\sqrt{1-x^2}$; for example, $n=1$ $(\nu=1)$ the equation is satisfied by $z=1$;

$$\frac{d^2z}{dx^2}(1-x^2 . 1-k^2x^2)+\frac{dz}{dx}\left[(-1+(2\nu-3)k^2)x+(-2\nu+4)k^2x^3\right]$$
$$+z(\nu-1)k^2\left[1+(\nu-2)x^2\right]+2\nu(k-k^3)\frac{dz}{dk}=0,$$

satisfied by numerator of $\operatorname{dn} nu$ omitting the factor $\sqrt{1-k^2x^2}$; for example, $n=1$ $(\nu=1)$ the equation is satisfied by $z=1$.

Verification for the Cubic Transformation.
Art. Nos. 328 to 335.

328. To show how the formulæ apply to the case of transformation, suppose $n=3$; then writing

$$x=\sqrt{k}\,\mathrm{sn}(u, k),\quad y=\sqrt{\lambda}\,\mathrm{sn}\left(\frac{u}{M}, \lambda\right),$$

we have
$$y=\frac{x\left\{\frac{v}{u^3}(v+2u^3)+x^2\right\}}{1+\frac{v}{u^3}(v+2u^3)\,x^2}.$$

Hence multiplying numerator and denominator by a factor A, the denominator is

$$=A\left\{1+\frac{v}{u^3}(v+2u^3)\,x^2\right\};$$

writing $x=0$, and observing that in this case the denominator should be $=\sqrt{\frac{\lambda'\Lambda}{k'K}}$, or what is the same thing $=\sqrt{\frac{\lambda'}{k'.3M}}$; we find $A=\sqrt{\frac{\lambda'}{3k'M}}$, or say $A=\sqrt{\frac{\lambda'}{k'M}}$.

329. We have
$$\frac{\lambda'^2}{k'^2M^2}=\frac{1-v^8}{1-u^8}\,\frac{(v+2u^3)^2}{v^2},$$

or observing that the modular equation may be written

$$(v^3-u)(v+2u^3)=u(v-u^3),$$

this is

$$\frac{\lambda'^2}{k'^2M^2}=\frac{1-v^8}{1-u^8}\cdot\frac{u^2(v-u^3)^2}{v^2(v^3-u)^2},\ =\frac{(1+u^2v^2-2uv^3)\,u^2\,(v-u^3)^2}{(1+u^2v^2+2u^3v)\,v^2\,(v^3-u)^2};$$

but from the same equation we have

$$(1+u^2v^2-2uv^3)\,u^2=(v^3-u)^2,$$
$$(1+u^2v^2+2u^3v)\,v^2=(v+u^3)^2,$$

whence the fraction is $=\left(\frac{v-u^3}{v+u^3}\right)^2$, or we have

$$\frac{\lambda'}{k'M} = \frac{v - u^3}{v + u^3}, \text{ and therefore } A = \sqrt{\frac{v - u^3}{v + u^3}}.$$

It thus appears that we have the function

$$z, = \sqrt{\frac{v - u^3}{v + u^3}} \left\{1 + \frac{v}{u^3}(v + 2u^3) x^2\right\},$$

satisfying the equation

$$6x^2 z + 2(cx - 2x^3)\frac{dz}{dx} + (1 - cx^2 + x^4)\frac{d^2 z}{dx^2} - 6(c^2 - 4)\frac{dz}{dc} = 0,$$

or, what is the same thing, the equation

$$6x^2 z + 2\left\{\left(k + \frac{1}{k}\right)x - 2x^3\right\}\frac{dz}{dx} + \left\{1 - \left(k + \frac{1}{k}\right)x^2 + x^4\right\}\frac{d^2 z}{dx^2} + 6k'^2\frac{dz}{dk} = 0.$$

Writing the foregoing value of z in the form $A + Bx^2$, the equations to be satisfied by the coefficients A, B are

$$B + 3k'^2\frac{dA}{dk} = 0,$$

$$3A + \left(k + \frac{1}{k}\right)B + 3k'^2\frac{dB}{dk} = 0.$$

330. We have $k = u^4$, $k'^2 = 1 - u^8$, and in general, for any function Ω of (u, v)

$$k'^2\frac{d\Omega}{dk} = \frac{1 - u^8}{4u^3}\left\{\frac{d\Omega}{du} + \frac{d\Omega}{dv}\,\frac{1 - v^8}{1 - u^8}\,\frac{2u^3 + v}{2v^3 - u}\right\},$$

$$= \frac{1}{4u^3(2v^3 - u)}\left\{(1 - u^8)(2v^3 - u)\frac{d\Omega}{du} + (1 - v^8)(2u^3 + v)\frac{d\Omega}{dv}\right\},$$

$$= \frac{1 - u^2 v^2}{4u^3(2v^3 - u)}\left\{(1 + u^2 v^2 + 2u^3 v)(2v^3 - u)\frac{d\Omega}{du}\right.$$

$$\left. + (1 + u^2 v^2 - 2uv^3)(2u^3 + v)\frac{d\Omega}{dv}\right\},$$

or if, as will be convenient, we write

$$u^4 = \alpha,\ \ v^4 = \beta,\ \ uv = \theta,\ \text{then}$$

$$k'^2 \frac{d\Omega}{dk} = \frac{1-\theta^2}{4\alpha(2\beta-\theta)}\left\{(1+2\alpha\theta+\theta^2)(2\beta-\theta)\,u\frac{d\Omega}{du}\right.$$
$$\left. + (1-2\beta\theta+\theta^2)(2\alpha+\theta)\,v\frac{d\Omega}{dv}\right\}.$$

331. In particular if $\Omega, = A, = \sqrt{\dfrac{v-u^3}{v+u^3}}$,

then $\log A = \tfrac{1}{2}\log(v-u^3) - \tfrac{1}{2}\log(v+u^3)$,

and thence

$$\frac{1}{A}\frac{dA}{du} = \frac{-\tfrac{3}{2}u^2}{v-u^3} - \frac{\tfrac{3}{2}u^2}{v+u^3},\ = \frac{-3u^2v}{v^2-u^6},$$

$$\frac{1}{A}\frac{dA}{dv} = \frac{\tfrac{1}{2}}{v-u^3} - \frac{\tfrac{1}{2}}{v+u^3},\ = \frac{u^3}{v^2-u^6}.$$

Hence

$$k'^2\frac{dA}{dk} = \frac{1-\theta^2}{4\alpha(2\beta-\theta)}\,\frac{u^3v}{v^2-u^6}\{-3(1+2\alpha\theta+\theta^2)(2\beta-\theta)$$
$$+ (1-2\beta\theta+\theta^2)(2\alpha+\theta)\}\,A.$$

But $\dfrac{u^3v}{v^2-u^6} = \dfrac{\theta^3}{v^4-u^6v^2} = \dfrac{\theta^3}{\beta-\alpha\theta^2}$, and the modular equation is $\alpha-\beta+2\theta-2\theta^3=0$; whence $\beta-\alpha\theta^2 = (\alpha+2\theta)(1-\theta^2)$: hence

$$\frac{1-\theta^2}{4\alpha(2\beta-\theta)}\,\frac{u^3v}{v^2-u^6} = \frac{\theta^3}{4\alpha(2\beta-\theta)(\alpha+2\theta)},$$

and consequently

$$k'^2\frac{dA}{dk} = \frac{\theta^3}{4\alpha(2\beta-\theta)(\alpha+2\theta)}\{-3(1+2\alpha\theta+\theta^2)(2\beta-\theta)$$
$$+ (1-2\beta\theta+\theta^2)(2\alpha+\theta)\}\,A.$$

Also

$$B = \frac{v}{u^2}(v+2u^3)\,A,\ = \frac{\theta}{\alpha}(2\alpha+\theta)\,A.$$

332. Hence the first equation to be verified is

$$4(2\alpha+\theta) + \frac{3\theta^3}{(2\beta-\theta)(\alpha+2\theta)}\{-3(1+2\alpha\theta+\theta^2)(2\beta-\theta)$$
$$+ (1-2\beta\theta+\theta^2)(2\alpha+\theta)\} = 0.$$

We have

$$(2\alpha+\theta)(2\beta-\theta)=4\theta^4-(2\alpha-2\beta)\theta-\theta^2, =3\theta^2,$$

by the modular equation; hence the equation is

$$4(\alpha+2\theta)-3(1+2\alpha\theta+\theta^2)(2\beta-\theta)+(1-2\beta\theta+\theta^2)(2\alpha+\theta),$$

viz. this is

$$6\alpha-6\beta+(12-16\alpha\beta)\theta+(8\alpha-8\beta)\theta^2+4\theta^3=0.$$

But from the modular equation $\beta=\alpha+2\theta-2\theta^3$, on substituting for β this value the equation becomes

$$-16\alpha^2\theta-32\alpha\theta^2+32\alpha\theta^4+16\theta^5=0,$$

viz. this is $$-\alpha^2-2\alpha\theta+2\alpha\theta^3+\theta^4=0,$$

which is in fact the equation $\theta^4=\alpha\beta, =\alpha(\alpha+2\theta-2\theta^3)$.

333. For the second equation, writing for convenience $B=QA$, this is

$$3A+\left(k+\frac{1}{k}\right)QA+3k'^2.QA\left\{\frac{1}{Q}\frac{dQ}{dk}+\frac{1}{A}\frac{dA}{dk}\right\}=0,$$

or if for the term $3k'^2.Q\dfrac{dA}{dk}$ we substitute its value from the first equation, $=-QB$, that is $=-Q^2A$, then throwing out the factor A, the equation becomes

$$3+\left(k+\frac{1}{k}\right)Q-Q^2+3k'^2\frac{dQ}{dk}=0,$$

which should therefore be satisfied by $Q=\dfrac{v^2}{u^2}+2uv$: viz. this is

$$3+\left(u^4+\frac{1}{u^4}\right)Q-Q^2+\frac{3(1-\theta^8)}{4\alpha(2\beta-\theta)}\left\{(1+2\alpha\theta+\theta^2)(2\beta-\theta)\,u\frac{dQ}{du}\right.$$
$$\left.+(1-2\beta\theta+\theta^2)(2\alpha+\theta)\,v\frac{dQ}{dv}\right\}=0,$$

or, what is the same thing, it is

$$3+\left(u^4+\frac{1}{u^4}\right)Q-Q^2+\frac{3(1-\theta^8)}{2\alpha(2\beta-\theta)}\left\{(1+2\alpha\theta+\theta^2)(2\beta-\theta)\left(-\frac{\beta}{\theta^2}+\theta\right)\right.$$
$$\left.+(1-2\beta\theta+\theta^2)(2\alpha+\theta)\left(\frac{\beta}{\theta^2}+\theta\right)\right\}=0.$$

We have $Q = \frac{\beta}{\theta^2} + 2\theta$, and then

$$3 + \left(\alpha + \frac{1}{\alpha}\right) Q - Q^2 = \frac{1}{\alpha^2}(1-\theta^2)^2(2\alpha+\theta)^2,$$

viz. this will be the case if

$$3\alpha^2 + (\alpha^3+\alpha)\left(\frac{\beta}{\theta^2}+2\theta\right) - \alpha^2\left(\frac{\beta}{\theta^2}+2\theta\right)^2 = (1-\theta^2)^2(2\alpha+\theta)^2,$$

or since $\alpha\left(\frac{\beta}{\theta^2}+2\theta\right) = \theta^2 + 2\alpha\theta$ it is

$$3\alpha^2 + (\alpha^2+1)(\theta^2+2\alpha\theta) - (\theta^2+2\alpha\theta)^2 = (1-\theta^2)^2(2\alpha+\theta)^2,$$

which is to be verified.

334. The equation $\theta^4 + 2\alpha\theta^3 - 2\alpha\theta - \alpha^2 = 0$ gives

$$3\alpha^2 = (-\theta^3 + 2\alpha)(\theta + 2\alpha),$$

thereby reducing the identity to

$$-\theta^3 + 2\alpha + (\alpha^2+1)\theta - \theta^2(2\alpha+\theta) = (1-\theta^2)^2(2\alpha+\theta),$$

that is

$$\theta(\alpha^2-\theta^2) + (1-\theta^2)(2\alpha+\theta) = (1-\theta^2)^2(2\alpha+\theta),$$

or

$$\alpha^2 - \theta^2 = (-\theta+\theta^3)(2\alpha+\theta),$$

viz. this is $\alpha^2 = -2\alpha\theta + 2\alpha\theta^3 + \theta^4$, the equation in question.

The equation is thus

$$(1-\theta^2)(2\alpha+\theta)^2 + \frac{3\alpha}{2(2\beta-\theta)}\left\{(1+2\alpha\theta+\theta^2)(2\beta-\theta)\left(-\frac{\beta}{\theta^2}+\theta\right) + (1-2\beta\theta+\theta^2)(2\alpha+\theta)\left(\frac{\beta}{\theta^2}+\theta\right)\right\} = 0,$$

or multiplying by $2(2\beta-\theta)$ and observing as before that $(2\alpha+\theta)(2\beta-\theta) = 3\theta^2$, this is

$$2\theta^2(1-\theta^2)(2\alpha+\theta) + \alpha\left\{(1+2\alpha\theta+\theta^2)(2\beta-\theta)\left(-\frac{\beta}{\theta^2}+\theta\right) + (1-2\beta\theta+\theta^2)(2\alpha+\theta)\left(\frac{\beta}{\theta^2}+\theta\right)\right\} = 0,$$

or, what is the same thing, it is

$$2\theta(1-\theta^2)(2\alpha+\theta) + [(1+2\alpha\theta+\theta^2)(2\beta-\theta)(\alpha-\theta) + (1-2\beta\theta+\theta^2)(2\alpha+\theta)(\alpha+\theta)] = 0.$$

Multiplying out, this is

$$(2\alpha^2 + 2\alpha\beta) + \theta(6\alpha - 2\beta) + \theta^2(4 - 8\alpha\beta) - 4\beta\theta^3 = 0,$$

or, what is the same thing,

$$\alpha^2 + 3\alpha\theta + 2\theta^2 + \beta(\alpha - \theta - 4\alpha\theta^2 - 2\theta^3) = 0,$$

viz. substituting for β its value, this is

$$\alpha^2 + 3\alpha\theta + 2\theta^2 + (\alpha - \theta - 4\alpha\theta^2 - 2\theta^3)(\alpha + 2\theta - 2\theta^3) = 0,$$

or working out it is

$$2\alpha^2 + 4\alpha\theta - 4\alpha^2\theta^2 - 12\alpha\theta^3 - 2\theta^4 + 8\alpha\theta^5 + 4\theta^6 = 0,$$

viz. this is

$$(\alpha^2 + 2\alpha\theta - 2\alpha\theta^3 - \theta^4)(2 - 4\theta^2) = 0,$$

which is right.

335. The foregoing differential equation, written in the form

$$3 + \left(k + \frac{1}{k}\right) Q - Q^2 + 3(1 - k^2)\frac{dQ}{dk} = 0,$$

is further considered in my two papers "On a differential equation in the theory of Elliptic Functions," *Messenger of Mathematics*, vol. IV. (1874) pp. 69 and 110, and in the last of them it is shown that the equation can be integrated generally: the process is, by the assumption

$$Q = -3(1 - k^2)\frac{1}{z}\frac{dz}{dk},$$

to transform the equation into a linear equation of the second order

$$3(1 - k^2)\frac{d^2z}{dk^2} + \frac{1 - 5k^2}{k}\frac{dz}{dk} - \frac{1}{1 - k^2}z = 0;$$

we have a particular solution of the original equation in Q, and therefore a particular solution of this equation in z; whence by a known method, the general solution can be obtained.

The result is expressed in terms of a variable

$$\gamma_1 = \frac{1}{\sqrt{\alpha}}\sqrt{2 + \alpha \,.\, 1 + 2\alpha},$$

where α is given in terms of k by the equation, *ante* No. 260,

$$k^2 = \frac{\alpha^2(2 + \alpha)}{1 + 2\alpha}.$$

CHAPTER X.

TRANSFORMATION FOR AN ODD AND IN PARTICULAR AN ODD-PRIME ORDER: DEVELOPMENT OF THE THEORY BY MEANS OF THE n-DIVISION OF THE COMPLETE FUNCTIONS.

THE algebraical theory of the transformation has been explained; in the present Chapter it is shown how, by means of formulæ depending on the n-division of the complete functions, the prescribed algebraical conditions are satisfied; and that we thus obtain the actual expressions of the transformed functions $\operatorname{sn}\left(\frac{u}{M}, \lambda\right)$, &c.

The general Theory. Art. Nos. 336 to 341.

336. We have n an odd number; m, m' any positive integers having no common divisor which also divides n;

$$\omega = \frac{mK + m'iK'}{n};$$

s a positive integer extending from 1 to $\frac{1}{2}(n-1)$; and when any expression depending on s is enclosed within [ ], this signifies that the product of the $\frac{1}{2}(n-1)$ terms is to be taken. The formulæ for the new modulus λ and multiplier M are assumed to be

$$\lambda = \quad k^n \quad [\operatorname{sn}(K - 4s\omega)]^4,$$

$$M = (-)^{\frac{1}{2}(n-1)} [\operatorname{sn}(K - 4s\omega)]^2 \div [\operatorname{sn} 4s\omega]^2,$$

and we then assume between y and x a relation expressed in the several forms:

$$y = \frac{x}{M}\left[1 - \frac{x^2}{\text{sn}^2 4s\omega}\right] \qquad (+),$$

$$1 - y = (1 - x)\left[1 - \frac{x}{\text{sn}(K - 4s\omega)}\right]^2 \qquad (+),$$

$$1 + y = (1 + x)\left[1 + \frac{x}{\text{sn}(K - 4s\omega)}\right]^2 \qquad (+),$$

$$1 - \lambda y = (1 - kx)\left[1 - kx \,\text{sn}(K - 4s\omega)\right]^2 \qquad (+),$$

$$1 + \lambda y = (1 + kx)\left[1 + kx \,\text{sn}(K - 4s\omega)\right]^2 \qquad (+),$$

where $\qquad \text{denom.} = [1 - k^2 \text{sn}^2 4s\omega \,.\, x^2].$

It has of course to be shown that the different expressions of y as a function of x are consistent with each other: but assuming that this is so, it at once follows that

$$\frac{dy}{\sqrt{1 - y^2 \,.\, 1 - \lambda^2 y^2}} = \frac{1}{M}\frac{dx}{\sqrt{1 - x^2 \,.\, 1 - k^2 x^2}},$$

and consequently that, writing $x = \text{sn}(u, k)$, we have

$$y = \text{sn}\left(\frac{u}{M}, \lambda\right).$$

337. We start from the equation

$$(1 - y) + (1 + y) = (1 - x)\left[1 - \frac{x}{\text{sn}(K - 4s\omega)}\right]^2$$
$$+ (1 + x)\left[1 + \frac{x}{\text{sn}(K - 4s\omega)}\right]^2,$$

and show that, λ and M being assumed as above, this value of y leads to the other equations of the system.

In the first place, it is clear that the assumed expression of $\frac{1 - y}{1 + y}$ gives for y a value of the form

$$y = \frac{x(1, x^2)^{\frac{1}{2}(n-1)}}{(1, x^2)^{\frac{1}{2}(n-1)}};$$

and if we show that y is $=0$ for $x = \pm \operatorname{sn} 4t\omega$, and $=\infty$ for $x = \dfrac{\pm 1}{k \operatorname{sn} 4t\omega}$, ($t$ any integer from 1 to $\tfrac{1}{2}(n-1)$,) then writing s in place of t, clearly the actual value will be

$$y = \frac{x}{C}\left[1 - \frac{x^2}{\operatorname{sn}^2 4s\omega}\right] \div \left[1 - k^2 \operatorname{sn}^2 4s\omega \, . \, x^2\right].$$

Moreover, if in the assumed expression of $\dfrac{1-y}{1+y}$ we write $x = 1$, we find $y = 1$: hence the last-mentioned value of y must for $x = 1$ reduce itself to $y = 1$; and we thus find

$$1 = \frac{1}{C}\left[1 - \frac{1}{\operatorname{sn}^2 4s\omega}\right] \div \left[1 - k^2 \operatorname{sn}^2 4s\omega\right];$$

viz. $$C = (-)^{\frac{1}{2}(n-1)}\left[\operatorname{cn} . 4s\omega\right]^2 \div \left[\operatorname{sn} 4s\omega \, . \operatorname{dn} 4s\omega\right]^2;$$

or, what is the same thing,

$$C = (-)^{\frac{1}{2}(n-1)}\left[\operatorname{sn}(K - 4s\omega)\right]^2 \div \left[\operatorname{sn} 4s\omega\right]^2;$$

viz. $C = M$; and the required expression of y is thus shown to be true. Combining it with the assumed expression of $\dfrac{1-y}{1+y}$, we at once obtain the required expressions of $1-y$ and $1+y$.

338. It then appears that the change of x into $\dfrac{1}{kx}$ changes y into $\dfrac{1}{\lambda y}$: viz. writing $\dfrac{1}{kx}$ for x the expression for y becomes

$$\frac{1}{Mkx}\left[1 - \frac{1}{k^2 x^2 \operatorname{sn}^2 4s\omega}\right] \div \left[1 - \frac{\operatorname{sn}^2 4s\omega}{x^2}\right];$$

viz. this is

$$= \frac{1}{Mkx}\left[\frac{1 - k^2 \operatorname{sn}^2 4s\omega \, . \, x^2}{k^2 x^2 \operatorname{sn}^2 4s\omega}\right]\left[\frac{x^2}{\operatorname{sn}^2 4s\omega - x^2}\right];$$

or, what is the same thing,

$$= \frac{1}{Mkx}\,\frac{\left[1 - k^2 \operatorname{sn}^2 4s\omega \, . \, x^2\right]}{\left[k^2 \operatorname{sn}^2 4s\omega\right]\left[\operatorname{sn}^2 4s\omega\right]}\;\frac{1}{\left[1 - \dfrac{x^2}{\operatorname{sn}^2 4s\omega}\right]};$$

or finally it is

$$= \frac{1}{M^2 k^n [\text{sn}\, 4s\omega]^4} \cdot [1 - k^2 \text{sn}^2\, 4s\omega \,.\, x^2] + \frac{x}{M}\left[1 - \frac{x^2}{\text{sn}^2\, 4s\omega}\right];$$

viz. observing that $\lambda = M^2 k^n [\text{sn}\, 4s\omega]^4$,

this is $= \dfrac{1}{\lambda y}$.

Lastly, in the expressions for $1 - y$ and $1 + y$, making the above changes x into $\dfrac{1}{kx}$ and y into $\dfrac{1}{\lambda y}$, and combining with the value of y, we obtain the required expressions for $1 - \lambda y$, $1 + \lambda y$; and the system of formulæ is thus completed.

339. We have to prove the subsidiary theorem, viz. that, starting with the assumed value of $\dfrac{1-y}{1+y}$, the values of x for which y becomes $= 0$ and $= x$ respectively are as stated above. And for this purpose it is to be shown that, x being taken $= \text{sn}\, u$, the formula may be written

$$\frac{1-y}{1+y} = \frac{[1 - \text{sn}\,(u + 4s'\omega)]}{[1 + \text{sn}\,(u + 4s\,\omega)]};$$

s' being any positive integer from 0 to $n - 1$, and the [ ]'s denoting the product of the $n - 1$ terms accordingly.

For suppose this proved, then changing u into $u + 4\omega$, each factor is changed into that which immediately follows it; except only the last factor $1 \mp \text{sn}\,(u + 4\,(n-1)\,\omega)$, which is changed into $1 \mp \text{sn}(u + 4n\omega)$; but, ω being as above, we have $\text{sn}(u + 4n\omega) = \text{sn}\, u$; or the last factor becomes $1 \mp \text{sn}\, u$, viz. this is the first factor: hence the value of the product is unaltered.

340. Now for $u = 0$ we have $x = 0$, and therefore (from the original assumed value of $\dfrac{1-y}{1+y}$), $y = 0$: hence also $y = 0$ for $u = 4\omega$, $8\omega \ldots 4\,(n-1)\,\omega$, that is for $x = \text{sn}\, 4\omega$, $\text{sn}\, 8\omega, \ldots$ $\text{sn}\, 4\,(n-1)\,\omega$: or since in general $\text{sn}\, 4\,(n-t)\,\omega = -\,\text{sn}\, 4t\omega$, we have $y = 0$ for $x = \pm\, \text{sn}\, 4\omega$, $\pm\, \text{sn}\, 8\omega, \ldots \pm\, \text{sn}\, 2\,(n-1)\,\omega$.

Similarly for $u = iK'$ we have $x = \infty$, and therefore $y = \infty$: hence also $y = \infty$ for $u = iK' + 4\omega,\ iK' + 8\omega \dots iK' + 4(n-1)\omega$, that is $x = \text{sn}\,(iK' + 4\omega), \dots \text{sn}\,(iK' + 4\,(n-1)\,\omega)$; or, what is the same thing, $x = \text{sn}\,(iK' \pm 4\omega) \dots \text{sn}\,(iK' \pm 2\,(n-1)\,\omega)$ say for $x = \text{sn}\,(iK' \pm 4s\omega)$, which is $= \dfrac{\pm 1}{k\,\text{sn}\,4s\omega}$: hence $y = 0$, and $y = \infty$, respectively for the required series of values.

341. To prove the formula

$$\frac{1-y}{1+y} = \frac{[1-\text{sn}\,(u+4s'\omega)]}{[1+\text{sn}\,(u+4s'\omega)]},$$

we have in general

$$[1+\text{sn}\,(u+a)]\,[1+\text{sn}\,(u-a)] \div \text{cn}^2 a = \left\{1 + \frac{\text{sn}\,u}{\text{sn}\,(K-a)}\right\}^2 \quad (\div),$$

$$[1-\text{sn}\,(u+a)]\,[1-\text{sn}\,(u-a)] \div \text{cn}^2 a = \left\{1 - \frac{\text{sn}\,u}{\text{sn}\,(K-a)}\right\}^2 \quad (\div),$$

where $\qquad \text{denom.} = 1 - k^2\,\text{sn}^2 u\,\text{sn}^2 a.$

Hence

$$\frac{[1-\text{sn}\,(u+a)]\,[1-\text{sn}\,(u-a)]}{[1+\text{sn}\,(u+a)]\,[1+\text{sn}\,(u+a)]} =$$

$$\left\{1 - \frac{\text{sn}\,u}{\text{sn}\,(K-a)}\right\}^2 \div \left\{1 + \frac{\text{sn}\,u}{\text{sn}\,(K-a)}\right\}^2.$$

Write herein successively $a = 4\omega, 8\omega, \dots 2\,(n-1)\,\omega$: take on each side the product of all the terms, and multiply each side of the resulting equation by $\dfrac{1-\text{sn}\,u}{1+\text{sn}\,u}$: then observing that

$$\text{sn}\,(u - 4s\omega) = \text{sn}\,(u + 4\,(n-s)\,\omega),$$

and supposing as before that s' has every integer value from 0 to $n-1$, the equation becomes

$$[1-\text{sn}\,(u+4s'\omega)] \div [1+\text{sn}\,(u+4s'\omega)]$$

$$= (1-\text{sn}\,u)\left[1 - \frac{\text{sn}\,u}{\text{sn}\,(K-4s\omega)}\right]^2 \div (1+\text{sn}\,u)\left[1 + \frac{\text{sn}\,u}{\text{sn}\,(K-4s\omega)}\right]^2,$$

viz. writing $\text{sn}\,u = x$, the right-hand side is $(1-y) \div (1+y)$: and the equation in question is thus proved.

Additional Formulæ. Art. Nos. 342 to 347.

342. We may in addition to the foregoing formulæ write

$$\sqrt{1-y^2} = \sqrt{1-x^2}\left[1-\frac{x^2}{\operatorname{sn}^2(K-4s\omega)}\right] \quad (\div),$$

$$\sqrt{1-\lambda^2y^2} = \sqrt{1-k^2x^2}\,[1-k^2x^2\operatorname{sn}^2(K-4s\omega)]\ (\div),$$

where as before

$$\text{denom.} = [1-k^2x^2\operatorname{sn}^2 4s\omega].$$

And of course writing $x = \operatorname{sn} u$, the values of y, $\sqrt{1-y^2}$, $\sqrt{1-\lambda^2y^2}$ are $\operatorname{sn}\left(\frac{u}{M}, \lambda\right)$, $\operatorname{cn}\left(\frac{u}{M}, \lambda\right)$, $\operatorname{dn}\left(\frac{u}{M}, \lambda\right)$.

343. The expressions for y, $\sqrt{1-y^2}$, $\sqrt{1-k^2y^2}$, writing therein $x = \operatorname{sn} u$, may be transformed in the same manner as the expression for $(1-y) \div (1+y)$. We have for instance

$$\operatorname{sn}(u+a)\operatorname{sn}(u-a) = -\operatorname{sn}^2 a\left(1-\frac{\operatorname{sn}^2 u}{\operatorname{sn}^2 a}\right) \div (1-k^2\operatorname{sn}^2 a\operatorname{sn}^2 u),$$

and hence writing successively $a = 4\omega, 8\omega, \ldots 2(n-1)\omega$, and proceeding as before we find ($s' = 0$ to $n-1$ as before, or, what is the same thing but is rather more convenient, $s' = -\frac{1}{2}(n-1)$ to $+\frac{1}{2}(n-1)$,)

$$y = \frac{(-)^{\frac{1}{2}(n-1)}}{M}[\operatorname{sn}(u+4s'\omega)] \div [\operatorname{sn} 4s\omega]^2,$$

and similarly

$$\sqrt{1-y^2} = \qquad [\operatorname{cn}(u+4s'\omega)] \div [\operatorname{cn} 4s\omega]^2,$$

$$\sqrt{1-\lambda^2y^2} = \qquad [\operatorname{dn}(u+4s'\omega)] \div [\operatorname{dn} 4s\omega]^2.$$

344. From the former expression of $\sqrt{1-\lambda^2y^2}$ putting therein $y = 1$, we deduce a value of λ', which (observing that $\frac{\operatorname{dn}(K-4s\omega)}{\operatorname{dn} 4s\omega} = \frac{k'}{\operatorname{dn}^2 4s\omega}$) may be written

$$\lambda' = k'^n \div [\operatorname{dn} 4s\omega]^4,$$

and combining herewith the values of λ, M we obtain various formulæ in regard to the new modulus and the multiplier:

$$\frac{(-)^{\frac{1}{2}(n-1)}}{M}\sqrt{\frac{\lambda}{k^n}} = [\text{sn } 4s\omega]^2,$$

$$\sqrt{\frac{\lambda k'^n}{\lambda' k^n}} = [\text{cn } 4s\omega]^2,$$

$$\sqrt{\frac{k'^n}{\lambda'}} = [\text{dn } 4s\omega]^2,$$

$$\sqrt{\frac{\lambda}{k^n}} = [\text{sn } (K - 4s\omega)]^2,$$

$$\frac{(-)^{\frac{1}{2}(n-1)}}{M}\sqrt{\frac{\lambda\lambda' k'^{n-2}}{k^n}} = [\text{cn } (K - 4s\omega)]^2,$$

$$\sqrt{\lambda' k'^{n-2}} = [\text{dn } (K - 4s\omega)]^2.$$

345. We may now write down the system of formulæ

$$\lambda = k^n [\text{sn } (K - 4s\omega)]^4,$$

$$\lambda' = k'^n \div [\text{dn } 4s\omega]^4,$$

$$M = (-)^{\frac{1}{2}(n-1)} [\text{sn } (K - 4s\omega)]^2 \div [\text{sn } 4s\omega]^2,$$

$$\text{sn}\left(\frac{u}{M}, \lambda\right) = \frac{\text{sn } u}{M}\left[1 - \frac{\text{sn}^2 u}{\text{sn}^2 4s\omega}\right](\div),$$

$$= \sqrt{\frac{k^n}{\lambda}}\,[\text{sn } (u + 4s'\omega)],$$

$$\text{cn}\left(\frac{u}{M}, \lambda\right) = \text{cn } u\left[1 - \frac{\text{sn}^2 u}{\text{sn}^2 (K - 4s\omega)}\right](\div),$$

$$= \sqrt{\frac{\lambda' k^n}{\lambda k'^n}}\,[\text{cn } (u + 4s'\omega)],$$

$$\text{dn}\left(\frac{u}{M}, \lambda\right) = \text{dn } u\left[1 - k^2 \text{sn}^2 (K - 4s\omega)\, \text{sn}^2 u\right](\div),$$

$$= \sqrt{\frac{\lambda'}{k'^n}}\,[\text{dn } (u + 4s'\omega)],$$

$$1-\operatorname{sn}\left(\frac{u}{M},\lambda\right)=(1-\operatorname{sn}u)\left[1-\frac{\operatorname{sn}u}{\operatorname{sn}(K-4s\omega)}\right]^2(+),$$

$$1+\operatorname{sn}\left(\frac{u}{M},\lambda\right)=(1+\operatorname{sn}u)\left[1+\frac{\operatorname{sn}u}{\operatorname{sn}(K-4s\omega)}\right]^2(+),$$

$$1-\lambda\operatorname{sn}\left(\frac{u}{M},\lambda\right)=(1-k\operatorname{sn}u)\left[1-k\operatorname{sn}(K-4s\omega)\operatorname{sn}u\right]^2(+),$$

$$1+\lambda\operatorname{sn}\left(\frac{u}{M},\lambda\right)=(1+k\operatorname{sn}u)\left[1+k\operatorname{sn}(K-4s\omega)\operatorname{sn}u\right]^2(+),$$

$$\text{Denom.}=\left[1-k^2\operatorname{sn}^2 4s\omega\operatorname{sn}^2 u\right].$$

346. To obtain a different group of formulæ, observe that the equation between y, x may be written

$$x\left[x^2-\operatorname{sn}^2 4s\omega\right]-\frac{\lambda}{kM}y\left[x^2-\frac{1}{k^2\operatorname{sn}^2 4s\omega}\right]=0,$$

which is of the form $x\,(x^2,1)^{\frac{1}{2}(n-1)}-(x^2,1)^{\frac{1}{2}(n-1)}=0$, where the coefficient of the highest power x^n is $=1$;

and that the roots of this equation are

$$x=\operatorname{sn}u,\ \operatorname{sn}(u+4\omega)\ldots,\ \operatorname{sn}(u+4(n-1)\omega);$$

whence we have the identity

$$x\left[x^2-\operatorname{sn}^2 4s\omega\right]-\frac{\lambda}{kM}\operatorname{sn}\left(\frac{u}{M},\lambda\right)\left[x^2-\frac{1}{k^2\operatorname{sn}^2 4s\omega}\right]$$
$$=\left[x-\operatorname{sn}(u+4s'\omega)\right];$$

and comparing the terms in x^{n-1} we have

$$\Sigma\operatorname{sn}(u+4s'\omega)=\frac{\lambda}{kM}\operatorname{sn}\left(\frac{u}{M},\lambda\right);$$

and similarly

$$\Sigma\operatorname{cn}(u+4s'\omega)=\frac{(-)^{\frac{1}{2}(n-1)}\lambda}{kM}\operatorname{cn}\left(\frac{u}{M},\lambda\right),$$

$$\Sigma\operatorname{dn}(u+4s'\omega)=\frac{(-)^{\frac{1}{2}(n-1)}}{M}\operatorname{dn}\left(\frac{u}{M},\lambda\right),$$

$$\Sigma\operatorname{tn}(u+4s'\omega)=\frac{\lambda'}{k'M}\operatorname{tn}\left(\frac{u}{M},\lambda\right).$$

in all which formulæ s' extends from 0 to $n-1$, or, what is the same thing, from $-\frac{1}{2}(n-1)$ to $+\frac{1}{2}(n-1)$.

In the first equation the left-hand side may be written

$$= \operatorname{sn} u + \Sigma \{\operatorname{sn}(u+4s\omega) + \operatorname{sn}(u-4s\omega)\}; \qquad s = 1 \text{ to } \tfrac{1}{2}(n-1),$$

viz. this is

$$= \operatorname{sn} u + \Sigma \frac{2 \operatorname{cn} 4s\omega \operatorname{dn} 4s\omega \,.\, \operatorname{sn} u}{1 - k^2 \operatorname{sn}^2 4s\omega \,.\, \operatorname{sn}^2 u},$$

and making the like changes in the other equations we find

$$\frac{\lambda}{kM} \operatorname{sn}\left(\frac{u}{M}, \lambda\right) = \operatorname{sn} u \left\{1 + 2\Sigma \frac{\operatorname{cn} 4s\omega \operatorname{dn} 4s\omega}{1 - k^2 \operatorname{sn}^2 4s\omega \operatorname{sn}^2 u}\right\},$$

$$\frac{(-)^{\frac{1}{2}(n-1)}\lambda}{kM} \operatorname{cn}\left(\frac{u}{M}, \lambda\right) = \operatorname{cn} u \left\{1 + 2\Sigma \frac{\operatorname{cn} 4s\omega}{1 - k^2 \operatorname{sn}^2 4s\omega \operatorname{sn}^2 u}\right\},$$

$$\frac{(-)^{\frac{1}{2}(n-1)}}{M} \operatorname{dn}\left(\frac{u}{M}, \lambda\right) = \operatorname{dn} u \left\{1 + 2\Sigma \frac{\operatorname{dn} 4s\omega}{1 - k^2 \operatorname{sn}^2 4s\omega \operatorname{sn}^2 u}\right\},$$

$$\frac{\lambda'}{k'M} \operatorname{tn}\left(\frac{u}{M}, \lambda\right) = \operatorname{tn} u \left\{1 + 2\Sigma \frac{\operatorname{dn} 4s\omega \operatorname{cn}^2 u}{\operatorname{cn}^2 4s\omega - \operatorname{dn}^2 4s\omega \operatorname{sn}^2 u}\right\}.$$

347. The last formula, which is of a different form from the others, depends on

$$\operatorname{tn}(u+a) + \operatorname{tn}(u-a), = \frac{\operatorname{sn}(u+a)\operatorname{cn}(u-a) + \operatorname{sn}(u-a)\operatorname{cn}(u+a)}{\operatorname{cn}(u+a)\operatorname{cn}(u-a)},$$

where the numerator, $= \sin\{\operatorname{am}(u+a) + \operatorname{am}(u-a)\}$, is

$$= 2 \operatorname{sn} u \operatorname{cn} u \operatorname{dn} a, \qquad (\div)$$

and the denominator is

$$= \operatorname{cn}^2 a - \operatorname{dn}^2 a \operatorname{sn}^2 u, \qquad (\div)$$

the common denominator,

$$= 1 - k^2 \operatorname{sn}^2 a \operatorname{sn}^2 u, \text{ disappearing.}$$

The 2ω-formulæ. Art. Nos. 348 to 351.

348. The above may be called 4ω-formulæ: we may change them into 2ω-formulæ. For this purpose observe that the series of values

$$\operatorname{sn}(u+4\omega),\quad \operatorname{sn}(u+8\omega)\ \ldots\quad \operatorname{sn}(u+2\overline{n-1})\,\omega$$

is in a different order

$$=(-)^m\operatorname{sn}(u-2\omega),\ \operatorname{sn}(u+4\omega),\ (-)^m\operatorname{sn}(u-6\omega)\ldots\ \pm\operatorname{sn}(u\pm\overline{n-1}\omega),$$

where the last term is $\operatorname{sn}(u+\overline{n-1}\omega)$ or $(-)^m\operatorname{sn}(u-\overline{n-1}\omega)$, according as $n-1$ is evenly even or oddly even.

To prove this, write $4t+2t'=2n$, then

$$u+4t\omega-(u-2t'\omega)=2n\omega,\ =2mK+2m'iK',$$

whence $\operatorname{sn}(u+4t\omega)=(-)^m\operatorname{sn}(u-2t'\omega)$. If $n-1$ be evenly even, $=4\nu$, then giving t every value from 1 to $\frac{1}{4}(n-1)$, $4t$ is less than n, and the term is retained in its original form; but giving t the remaining values from $\frac{1}{4}(n+3)$ to $\frac{1}{2}(n-1)$, the corresponding values of t' are from 1 to $\frac{1}{2}(n-3)$, and the term $\operatorname{sn}(u+4t\omega)$ is changed into $(-)^m\operatorname{sn}(u-2t'\omega)$. So if $n-1$ be oddly even, $=4\nu-2$, then giving t every value from 1 to $\frac{1}{4}(n-3)$, $4t$ is less than n, and the term is retained in its original form; but giving t the remaining values from $\frac{1}{4}(n+1)$ to $\frac{1}{2}(n-1)$ the corresponding values of t' are from 1 to $\frac{1}{2}(n-1)$, and the term $\operatorname{sn}(u+4t\omega)$ is changed into $(-)^m\operatorname{sn}(u-2t'\omega)$. We have thus the theorem.

349. Repeating the result, and writing down the analogous results for cn and dn,

$$\left.\begin{array}{l}\text{series } \operatorname{sn}(u+4\omega),\ \operatorname{sn}(u+8\omega)\ldots\quad \operatorname{sn}(u+2\overline{n-1}\omega)\\ \text{is in a different order}\\ \quad=(-)^m\operatorname{sn}(u-2\omega),\ \operatorname{sn}(u+4\omega),\ (-)^m\operatorname{sn}(u-6\omega)\ldots\\ \qquad\qquad\qquad\pm\operatorname{sn}(u\pm\overline{n-1}\omega);\end{array}\right\}$$

series $\text{cn}\,(u+4\omega),\ \text{cn}\,(u+8\omega)\ldots\quad \text{cn}\,(u+2\overline{n-1}\omega)$
is in a different order
$$=(-)^{m+m'}\,\text{cn}\,(u-2\omega),\ \ \text{cn}\,(u+4\omega),\ (-)^{m-m'}\,\text{cn}\,(u-6\omega)\ldots \pm\text{cn}\,(u\pm\overline{n-1}\omega);$$

series $\text{dn}\,(u+4\omega),\ \text{dn}\,(u+8\omega)\ldots\quad \text{dn}\,(u+\overline{2n-1}\omega)$
is in a different order
$$(-)^{m'}\,\text{dn}\,(u-2\omega),\ \ \text{dn}\,(u+4\omega),\ (-)^{m'}\,\text{dn}\,(u-6\omega)\ldots \pm\text{dn}(u\pm\overline{n-1}\omega).$$

350. It will be at once seen that these formulæ, on writing therein $u=0$, give for the series of sn, cn, dn of 4ω, 8ω, &c. the several values

$$(-)^{m+1}\ \text{sn}\,2\omega,\ \ \text{sn}\,4\omega,\ (-)^{m+1}\ \text{sn}\,6\omega\ldots\quad \mp\text{sn}\,(n-1)\omega,$$
$$(-)^{m+m'}\,\text{cn}\,2\omega,\ \ \text{cn}\,4\omega,\ (-)^{m+m'}\,\text{cn}\,6\omega\ldots\quad \pm\text{cn}\,(n-1)\omega,$$
$$(-)^{m'}\ \text{dn}\,2\omega,\ \text{dn}\,4\omega,\ (-)^{m'}\ \text{dn}\,6\omega\ldots\quad \pm\text{dn}\,(n-1)\omega.$$

The results are also required for $u=K$: as to this, observe that in general

$$\text{sn}\,(K+a)=-\text{sn}\,(-K+a)=\ \ \text{sn}\,(K-a)\,;$$
$$\text{cn}\,(K+a)=-\text{cn}\,(-K+a)=-\text{cn}\,(K-a)\,;$$
$$\text{dn}\,(K+a)=\ \ \text{dn}\,(-K+a)=\ \ \text{dn}\,(K-a).$$

Hence we see that

series $\text{sn}\,(K+4\omega),\ \text{sn}\,(K+8\omega)\ldots\quad \text{sn}\,(K+\overline{2n-1}\omega)$
is in a different order
$$(-)^{m}\,\text{sn}\,(K+2\omega),\ \ \text{sn}\,(K+4\omega),\ (-)^{m}\,\text{sn}\,(K+6\omega)\ldots \pm\text{sn}\,(K+\overline{n-1}\omega)\,;$$

series $\text{cn}\,(K+4\omega),\ \text{cn}\,(K+8\omega)\ldots\quad \text{cn}\,(K+2\overline{n-1}\omega)$
is in a different order
$$(-)^{m+m'+1}\,\text{cn}\,(K+2\omega),\ \ \text{cn}\,(K+4\omega),\ (-)^{m+m'+1}\,\text{cn}\,(K+6\omega)\ldots \pm\text{cn}(K+\overline{n-1}\omega)\,;$$

$$\left.\begin{array}{l}\text{series dn}\,(K+4\omega),\ \text{dn}\,(K+8\omega)\ldots\qquad \text{dn}\,(K+2n-\overline{1}\omega)\\ \text{is in a different order}\\ \quad(-)^{m'}\,\text{dn}\,(K+2\omega),\ \text{dn}\,(K+4\omega),\ (-)^{m'}\,\text{dn}\,(K+6\omega)\ldots\\ \qquad\qquad\qquad\qquad\qquad \pm\,\text{dn}\,(K+\overline{n-1}\omega)\,;\end{array}\right\}$$

in each of which formulæ we may for ω write $-\omega$.

351. It will be observed that in the formulæ which contain only $\text{sn}\,u$, $\text{cn}\,u$, $\text{dn}\,u$ (i.e. which do not contain $\text{sn}\,(u+4s'\omega)$ &c.) and squared functions such as $\text{sn}^2 4s\omega$, &c., the change of form is effected simply by writing 2ω instead of 4ω: in the other formulæ there are signs to be changed, and it is safer to retain the 4ω-formulæ, making the change of form only if and when it is required.

We have thus:

$$\lambda = k^n\,[\text{sn}\,(K-2s\omega)]^4,$$

$$\lambda' = k'^n \div [\text{dn}\,2s\omega]^4,$$

$$M = [\text{sn}\,(K-2s\omega)]^2 \div [\text{sn}\,2s\omega]^2,$$

$$\text{sn}\left(\frac{u}{M},\ \lambda\right) = \frac{\text{sn}\,u}{M}\left[1-\frac{\text{sn}^2 u}{\text{sn}^2 2s\omega}\right]\ (\div),$$

$$\text{cn}\left(\frac{u}{M},\ \lambda\right) = \text{cn}\,u\left[1-\frac{\text{sn}^2 u}{\text{sn}^2 (K-2s\omega)}\right]\ (\div),$$

$$\text{dn}\left(\frac{u}{M},\ \lambda\right) = \text{dn}\,u\,[1-k^2\,\text{sn}^2\,(K-2s\omega)\,\text{sn}^2 u]\ (\div),$$

$$\text{denom.} = \qquad [1-k^2\,\text{sn}^2\,2s\omega\,\text{sn}^2 u]\,;$$

but I do not write down the other formulæ in their 2ω-form.

The change from the 4ω- to the 2ω-formulæ is, as will appear, a very essential one, and it is important to take notice of it.

n an odd-prime; the Real Transformations, First and Second. Art. No. 352 to 359.

352. We have $\omega = \dfrac{mK + m'iK'}{n}$, where m and m' are positive and negative integers having no common divisor which also divides n. It is convenient to take n an odd-prime: there are here $n+1$ distinct transformations corresponding to $n+1$ values of ω which may be taken to be

$$\frac{K}{n},\ \frac{iK'}{n},\ \frac{K+iK'}{n},\ \frac{K+2iK'}{n},\ \ldots\ \frac{K+(n-1)\,iK'}{n};$$

or to be

$$\frac{K}{n},\ \frac{iK'}{n},\ \frac{K+iK'}{n},\ \frac{2K+iK'}{n},\ \ldots\ \frac{(n-1)\,K+iK'}{n};$$

or again to be

$$\frac{K}{n},\ \frac{iK'}{n},\ \frac{K \pm iK'}{n},\ \ldots\qquad \frac{K \pm \frac{1}{2}(n-1)\,iK'}{n}.$$

Two of these transformations are real: the former of them corresponding to the value $\omega = \dfrac{K}{n}$, and called the first transformation, is a transformation to a modulus λ which is less than k; the latter of them corresponding to the value $\omega = \dfrac{iK'}{n}$, and called the second transformation, is a transformation to a modulus λ_1 which is greater than k.

First Transformation, $\omega = \dfrac{K}{n}$ *(to a smaller modulus* λ*).*

353. The general formulæ apply at once to this case, but it is convenient to slightly alter them by omitting the factor $(-)^{\frac{1}{2}(n-1)}$ which presents itself in M. This comes to writing $(-)^{\frac{1}{2}(n-1)}y$ in place of y: so that in the new formulæ $x=1$, in place of giving $y=1$, gives $y=(-)^{\frac{1}{2}(n-1)}1$, or say $y=\pm 1$, the upper sign answering to an evenly even value of $n-1$ and the

lower sign to an oddly even value of $n-1$. It will be convenient to give the formulæ as well in the 4ω- as in the 2ω- form.

In the formulæ which contain $\pm$ or $\mp$ the upper sign is to be taken when $n-1$ is evenly even, the lower sign when it is oddly even.

354. For the conversion of 4ω- into 2ω-formulæ, observe that the series

$$\operatorname{sn}\left(u+\frac{4K}{n}\right),\ \operatorname{sn}\left(u+\frac{8K}{n}\right)\ldots\ \operatorname{sn}\left(u+\frac{2(n-1)K}{n}\right)$$

is in a different order

$$=-\operatorname{sn}\left(u-\frac{2K}{n}\right),\ \operatorname{sn}\left(u+\frac{4K}{n}\right),\ -\operatorname{sn}\left(u-\frac{6K}{n}\right)\ldots$$
$$\pm\operatorname{sn}\left(u\pm\frac{(n-1)K}{n}\right);$$

the series

$$\operatorname{cn}\left(u+\frac{4K}{n}\right),\ \operatorname{cn}\left(u+\frac{8K}{n}\right)\ldots\ \operatorname{cn}\left(u+\frac{2(n-1)K}{n}\right)$$

is in a different order

$$=-\operatorname{cn}\left(u-\frac{2K}{n}\right),\ \operatorname{cn}\left(u+\frac{4K}{n}\right),\ -\operatorname{cn}\left(u-\frac{6K}{n}\right)\ldots$$
$$\pm\operatorname{cn}\left(u\pm\frac{(n-1)K}{n}\right);$$

and the series

$$\operatorname{dn}\left(u+\frac{4K}{n}\right),\ \operatorname{dn}\left(u+\frac{8K}{n}\right)\ldots\ \operatorname{dn}\left(u+\frac{2(n-1)K}{n}\right)$$

is in a different order

$$=\operatorname{dn}\left(u-\frac{2K}{n}\right),\ \operatorname{dn}\left(u+\frac{4K}{n}\right),\ \operatorname{dn}\left(u-\frac{6K}{n}\right)\ldots$$
$$\operatorname{dn}\left(u\pm\frac{(n-1)K}{n}\right).$$

In all the formulæ s has the different integer values from 1 to $\frac{1}{2}(n-1)$, and s' the different integer values from $-\frac{1}{2}(n-1)$ to $+\frac{1}{2}(n-1)$; or as regards the 4ω-formulæ, we may consider s' as having the different integer values from 0 to $(n-1)$.

353. *First Transformation, to a smaller modulus* λ.

$\lambda = k^n \left[\mathrm{sn}\left(K - \frac{4sK}{n}\right)\right]^4,$	$= k^n \left[\mathrm{sn}\left(K - \frac{2sK}{n}\right)\right]^4,$
$\lambda' = k'^n \div \left[\mathrm{dn}\,\frac{4sK}{n}\right]^4,$	$= k'^n \div \left[\mathrm{dn}\,\frac{2sK}{n}\right]^4,$
$M = \left[\mathrm{sn}\left(K - \frac{4sK}{n}\right)\right]^2 \div \left[\mathrm{sn}\,\frac{4sK}{n}\right]^2,$	$= \left[\mathrm{sn}\left(K - \frac{2sK}{n}\right)\right]^2 \div \left[\mathrm{sn}\,\frac{2sK}{n}\right]^2,$
$\mathrm{sn}\left(\frac{u}{M}, \lambda\right) = \frac{\mathrm{sn}\,u}{M}\left[1 - \frac{\mathrm{sn}^2 u}{\mathrm{sn}^2 \frac{4sK}{n}}\right](\div),$	$= \frac{\mathrm{sn}\,u}{M}\left[1 - \frac{\mathrm{sn}^2 u}{\mathrm{sn}^2 \frac{2sK}{n}}\right](\div),$
$\mathrm{cn}\left(\frac{u}{M}, \lambda\right) = \mathrm{cn}\,u\left[1 - \frac{\mathrm{sn}^2 u}{\mathrm{sn}^2\left(K - \frac{4sK}{n}\right)}\right](\div),$	$= \mathrm{cn}\,u\left[1 - \frac{\mathrm{sn}^2 u}{\mathrm{sn}^2\left(K - \frac{2sK}{n}\right)}\right](\div),$
$\mathrm{dn}\left(\frac{u}{M}, \lambda\right) = \mathrm{dn}\,u\left[1 - k^2\,\mathrm{sn}^2\left(K - \frac{4sK}{n}\right)\mathrm{sn}^2 u\right](\div),$	$= \mathrm{dn}\,u\left[1 - k^2\,\mathrm{sn}^2\left(K - \frac{2sK}{n}\right)\mathrm{sn}^2 u\right](\div),$
denom. $= \left[1 - k^2\,\mathrm{sn}^2\,\frac{4sK}{n}\,\mathrm{sn}^2 u\right].$	$= \left[1 - k^2\,\mathrm{sn}^2\,\frac{2sK}{n}\,\mathrm{sn}^2 u\right].$

$$\operatorname{sn}\left(\frac{u}{M},\lambda\right)=(-)^{\frac{1}{2}(n-1)}\sqrt{\frac{k^n}{\lambda}}\left[\operatorname{sn}\left(u+\frac{4s'K'}{n}\right)\right], \qquad =(-)^{\frac{1}{2}(n-1)}\sqrt{\frac{k^n}{\lambda}}\left[\operatorname{sn}\left(u+\frac{2s'K'}{n}\right)\right],$$

$$\operatorname{cn}\left(\frac{u}{M},\lambda\right)=\sqrt{\frac{\lambda' k^n}{\lambda k'^n}}\left[\operatorname{cn}\left(u+\frac{4s'K'}{n}\right)\right], \qquad =\sqrt{\frac{\lambda' k^n}{\lambda k'^n}}\left[\operatorname{cn}\left(u+\frac{2s'K'}{n}\right)\right],$$

$$\operatorname{dn}\left(\frac{u}{M},\lambda\right)=\sqrt{\frac{\lambda'}{k'^n}}\left[\operatorname{dn}\left(u+\frac{4s'K'}{n}\right)\right], \qquad =\sqrt{\frac{\lambda'}{k'^n}}\left[\operatorname{dn}\left(u+\frac{2s'K'}{n}\right)\right],$$

$$1\mp\operatorname{sn}\left(\frac{u}{M},\lambda\right)=(1-\operatorname{sn}u)\left[1-\frac{\operatorname{sn}u}{\operatorname{sn}\left(K-\frac{4sK}{n}\right)}\right]^s(\div), \qquad =(1-\operatorname{sn}u)\left[1-\frac{(-)^s\operatorname{sn}u}{\operatorname{sn}\left(K-\frac{2sK}{n}\right)}\right]^s(\div),$$

$$1\pm\operatorname{sn}\left(\frac{u}{M},\lambda\right)=(1+\operatorname{sn}u)\left[1+\frac{\operatorname{sn}u}{\operatorname{sn}\left(K-\frac{4sK}{n}\right)}\right]^s(\div), \qquad =(1+\operatorname{sn}u)\left[1+\frac{(-)^s\operatorname{sn}u}{\operatorname{sn}\left(K-\frac{2sK}{n}\right)}\right]^s(\div),$$

$$1\mp\lambda\operatorname{sn}\left(\frac{u}{M},\lambda\right)=(1-k\operatorname{sn}u)\left[1-k\operatorname{sn}\left(K-\frac{4sK}{n}\right)\operatorname{sn}u\right]^s(\div), \qquad =(1-k\operatorname{sn}u)\left[1-(-)^s k\operatorname{sn}\left(K-\frac{2sK}{n}\right)\operatorname{sn}u\right]^s(\div),$$

$$1\pm\lambda\operatorname{sn}\left(\frac{u}{M},\lambda\right)=(1+k\operatorname{sn}u)\left[1+k\operatorname{sn}\left(K-\frac{4sK}{n}\right)\operatorname{sn}u\right]^s(\div). \qquad =(1+k\operatorname{sn}u)\left[1+(-)^s k\operatorname{sn}\left(K-\frac{2sK}{n}\right)\operatorname{sn}u\right]^s(\div),$$

denom. same as on preceding page.

$$\frac{\lambda}{k M}\operatorname{sn}\left(\frac{u}{M}, \lambda\right) = \operatorname{sn} u \left\{1 + 2\Sigma \frac{\operatorname{cn}\frac{4sK}{n}\operatorname{dn}\frac{4sK}{n}}{1 - k^2 \operatorname{sn}^2\frac{4sK}{n}\operatorname{sn}^2 u}\right\} = \operatorname{sn} u \left\{1 + 2\Sigma \frac{(-)^s \operatorname{cn}\frac{2sK}{n}\operatorname{dn}\frac{2sK}{n}}{1 - k^2 \operatorname{sn}^2\frac{2sK}{n}\operatorname{sn}^2 u}\right\},$$

$$\frac{\lambda}{k M}\operatorname{cn}\left(\frac{u}{M}, \lambda\right) = \operatorname{cn} u \left\{1 + 2\Sigma \frac{\operatorname{cn}\frac{4sK}{n}}{1 - k^2 \operatorname{sn}^2\frac{4sK}{n}\operatorname{sn}^2 u}\right\} = \operatorname{cn} u \left\{1 + 2\Sigma \frac{(-)^s \operatorname{cn}\frac{2sK}{n}}{1 - k^2 \operatorname{sn}^2\frac{2sK}{n}\operatorname{sn}^2 u}\right\},$$

$$\frac{1}{M}\operatorname{dn}\left(\frac{u}{M}, \lambda\right) = \operatorname{dn} u \left\{1 + 2\Sigma \frac{\operatorname{dn}\frac{4sK}{u}}{1 - k^2 \operatorname{sn}^2\frac{4sK}{n}\operatorname{sn}^2 u}\right\} = \operatorname{dn} u \left\{1 + 2\Sigma \frac{\operatorname{dn}\frac{2sK}{n}}{1 - k^2 \operatorname{sn}^2\frac{2sK}{n}\operatorname{sn}^2 u}\right\},$$

$$\frac{\lambda'}{k' M}\operatorname{tn}\left(\frac{u}{M}, \lambda\right) = \operatorname{tn} u \left\{1 + 2\Sigma \frac{\operatorname{dn}\frac{4sK}{n}\operatorname{cn}^2 u}{\operatorname{cn}^2\frac{4sK}{n} - \operatorname{dn}^2\frac{4sK}{n}\operatorname{sn}^2 u}\right\} = \operatorname{tn} u \left\{1 + 2\Sigma \frac{\operatorname{dn}\frac{2sK}{n}\operatorname{cn}^2 u}{\operatorname{cn}^2\frac{2sK}{n} - \operatorname{dn}^2\frac{2sK}{n}\operatorname{sn}^2 u}\right\}.$$

Second Transformation, $\omega = \dfrac{iK'}{n}$ *(to a larger modulus* λ_1*).*

356. Write in the general formulæ $\omega = \dfrac{iK'}{n}$: the formulæ in the first instance present themselves in an imaginary form: these are given as well in the 4ω- as in the 2ω- form. For the conversion observe that the series

$$\operatorname{sn}\left(u + \frac{4iK'}{n}\right),\ \operatorname{sn}\left(u + \frac{8iK'}{n}\right) \dots \operatorname{sn}\left(u + \frac{2(n-1)iK'}{n}\right)$$

is in a different order

$$= \operatorname{sn}\left(u - \frac{2iK'}{n}\right),\ \operatorname{sn}\left(u + \frac{4iK'}{n}\right),\ \operatorname{sn}\left(u - \frac{6iK'}{n}\right) \dots$$
$$\operatorname{sn}\left(u \pm \frac{(n-1)iK'}{n}\right);$$

the series

$$\operatorname{cn}\left(u + \frac{4iK'}{n}\right),\ \operatorname{cn}\left(u + \frac{8iK'}{n}\right) \dots \operatorname{cn}\left(u + \frac{2(n-1)iK'}{n}\right)$$

is in a different order

$$= -\operatorname{cn}\left(u - \frac{2iK'}{n}\right),\ \operatorname{cn}\left(u + \frac{4iK'}{n}\right),\ -\operatorname{cn}\left(u - \frac{6iK'}{n}\right) \dots$$
$$\pm \operatorname{cn}\left(u \pm \frac{(n-1)iK'}{n}\right);$$

and the series

$$\operatorname{dn}\left(u + \frac{4iK'}{n}\right),\ \operatorname{dn}\left(u + \frac{8iK'}{n}\right) \dots \operatorname{dn}\left(u + \frac{2(n-1)iK'}{n}\right)$$

is in a different order

$$= -\operatorname{dn}\left(u - \frac{2iK'}{n}\right),\ \operatorname{dn}\left(u + \frac{4iK'}{n}\right),\ -\operatorname{dn}\left(u - \frac{6iK'}{n}\right) \dots$$
$$\pm \operatorname{dn}\left(u \pm \frac{(n-1)iK'}{n}\right).$$

357. There is a further change of form to be made in some of the formulæ. We have $k \operatorname{sn} v = \dfrac{1}{\operatorname{sn}(v + iK')}$, and thence

$$-k \operatorname{sn} \frac{2siK'}{n} = \frac{1}{\operatorname{sn}\left(iK' - \dfrac{2siK'}{n}\right)} = \frac{1}{\operatorname{sn} \dfrac{(n-2s)iK'}{n}}.$$

Putting for a moment $n-2s=2t-1$, we see that s having the positive integer values 1 to $\frac{1}{2}(n-1)$, t has the same series of values in a reverse order; whence finally writing s instead of t, $-k\,\mathrm{sn}\,\dfrac{2siK'}{n}$ has the same values as $\dfrac{1}{\mathrm{sn}\,\dfrac{(2s-1)\,iK'}{n}}$, or any the series

$$-k\,\mathrm{sn}\,\frac{2iK'}{n},\quad -k\,\mathrm{sn}\,\frac{4iK'}{n},\ldots\quad -k\,\mathrm{sn}\,\frac{(n-1)\,iK'}{n}$$

is in the reverse order

$$\frac{1}{\mathrm{sn}\,\dfrac{iK'}{n}},\qquad \frac{1}{\mathrm{sn}\,\dfrac{3iK'}{n}},\ldots\qquad \frac{1}{\mathrm{sn}\,\dfrac{(n-2)\,iK'}{n}};$$

and similarly $k\,\mathrm{sn}\left(K-\dfrac{2siK'}{n}\right)$ has the same values as

$$\frac{1}{\mathrm{sn}\left(K-\dfrac{(2s-1)\,iK'}{n}\right)}.$$

We have moreover

$$k\,\mathrm{cn}\,\frac{2siK'}{n}=i\,\mathrm{dn}\,\frac{(n-2s)\,iK'}{n}\div\mathrm{sn}\,\frac{(n-2s)\,iK'}{n}$$

$$\mathrm{dn}\,\frac{2siK'}{n}=\ \mathrm{cn}\,\frac{(n-2s)\,iK'}{n}\div\mathrm{sn}\,\frac{(n-2s)\,iK'}{n},$$

which may be similarly transformed by putting therein

$$n-2s=2t-1,$$

and finally s instead of t, as above.

In all the formulæ s has the different integer values from 1 to $\frac{1}{2}(n-1)$, and s' the different integer values from $-\frac{1}{2}(n-1)$ to $+\frac{1}{2}(n-1)$: or we may in the 4ω-formulæ consider s' as having the different integer values from 0 to $(n-1)$.

358. *Second Transformation, to a larger modulus λ_1 (imaginary form)*

$$\lambda_1 = k^n \left[\operatorname{sn}\left(K - \frac{4siK'}{n}\right)\right]^4, \qquad = k^n \left[\operatorname{sn}\left(K - \frac{2siK'}{n}\right)\right]^4,$$

$$\lambda_1' = k'^n + \left[\operatorname{dn}\frac{4siK'}{n}\right]^4, \qquad = k'^n + \left[\operatorname{dn}\frac{2siK'}{n}\right]^4,$$

$$M_1 = (-)^{\frac{1}{2}(n-1)} \left[\operatorname{sn}\left(K - \frac{4siK'}{n}\right)\right]^2 + \left[\operatorname{sn}\frac{4siK'}{n}\right]^2, \qquad = (-)^{\frac{1}{2}(n-1)} \left[\operatorname{sn}\left(K - \frac{2siK'}{n}\right)\right]^2 + \left[\operatorname{sn}\frac{2siK'}{n}\right]^2,$$

$$\operatorname{sn}\left(\frac{u}{M_1}, \lambda_1\right) = \frac{\operatorname{sn} u}{M_1}\left[1 - \frac{\operatorname{sn}^2 u}{\operatorname{sn}^2 \frac{4siK'}{n}}\right](+), \qquad = \frac{\operatorname{sn} u}{M_1}\left[1 - \frac{\operatorname{sn}^2 u}{\operatorname{sn}^2 \frac{2siK'}{n}}\right](+),$$

$$\operatorname{cn}\left(\frac{u}{M_1}, \lambda_1\right) = \operatorname{cn} u\left[1 - \frac{\operatorname{sn}^2 u}{\operatorname{sn}^2\left(K - \frac{4siK'}{n}\right)}\right](+), \qquad = \operatorname{cn} u\left[1 - \frac{\operatorname{sn}^2 u}{\operatorname{sn}^2\left(K - \frac{2siK'}{n}\right)}\right](+),$$

$$\operatorname{dn}\left(\frac{u}{M_1}, \lambda_1\right) = \operatorname{dn} u\left[1 - k^2 \operatorname{sn}^2\left(K - \frac{4siK'}{n}\right)\operatorname{sn}^2 u\right](+), \qquad = \operatorname{dn} u\left[1 - k^2 \operatorname{sn}^2\left(K - \frac{2siK'}{n}\right)\operatorname{sn}^2 u\right](+),$$

$$= \operatorname{dn} u\left[1 - \frac{\operatorname{sn}^2 u}{\operatorname{sn}^2\left(K - \frac{(2s-1)\,iK'}{n}\right)}\right](+),$$

$$\text{denom.} = \left[1 - k^2 \operatorname{sn}^2 \frac{4siK'}{n} \operatorname{sn}^2 u\right], \qquad = \left[1 - k^2 \operatorname{sn}^2 \frac{2siK'}{n} \operatorname{sn}^2 u\right],$$

$$= \left[1 - \frac{\operatorname{sn}^2 u}{\operatorname{sn}^2 \frac{(2s-1)\,iK'}{n}}\right].$$

$$1-\operatorname{sn}\left(\frac{u}{M_1},\lambda_1\right)=(1-\operatorname{sn}u)\left[1-\frac{\operatorname{sn}u}{\operatorname{sn}\left(K-\frac{4siK'}{n}\right)}\right]^s(+), \qquad =(1-\operatorname{sn}u)\left[1-\frac{\operatorname{sn}u}{\operatorname{sn}\left(K-\frac{2siK'}{n}\right)}\right]^s(+),$$

$$1+\operatorname{sn}\left(\frac{u}{M_1},\lambda_1\right)=(1+\operatorname{sn}u)\left[1+\frac{\operatorname{sn}u}{\operatorname{sn}\left(K-\frac{4siK'}{n}\right)}\right]^s(+), \qquad =(1+\operatorname{sn}u)\left[1+\frac{\operatorname{sn}u}{\operatorname{sn}\left(K-\frac{2siK'}{n}\right)}\right]^s(+),$$

$$1-\lambda_1\operatorname{sn}\left(\frac{u}{M_1},\lambda_1\right)=(1-k\operatorname{sn}u)\left[1-k\operatorname{sn}\left(K-\frac{4siK'}{n}\right)\operatorname{sn}u\right]^s(+), \qquad =(1-k\operatorname{sn}u)\left[1-k\operatorname{sn}\left(K-\frac{2siK'}{n}\right)\operatorname{sn}u\right]^s(+),$$

$$=(1-k\operatorname{sn}u)\left[1-\frac{\operatorname{sn}u}{\operatorname{sn}\left(K-\frac{(2s-1)iK'}{n}\right)}\right]^s(+),$$

$$1+\lambda_1\operatorname{sn}\left(\frac{u}{M_1},\lambda_1\right)=(1+k\operatorname{sn}u)\left[1+k\operatorname{sn}\left(K-\frac{4siK'}{n}\right)\operatorname{sn}u\right]^s(+), \qquad =(1+k\operatorname{sn}u)\left[1+k\operatorname{sn}\left(K-\frac{2siK'}{n}\right)\operatorname{sn}u\right]^s(+),$$

$$=(1+k\operatorname{sn}u)\left[1+\frac{\operatorname{sn}u}{\operatorname{sn}\left(K-\frac{(2s-1)iK'}{n}\right)}\right]^s(+),$$

$$\operatorname{sn}\left(\frac{u}{M_1},\lambda_1\right)=\sqrt{\frac{k^n}{\lambda_1}}\left[\operatorname{sn}\left(u+\frac{4s'iK'}{n}\right)\right], \qquad =\sqrt{\frac{k^n}{\lambda_1}}\left[\operatorname{sn}\left(u+\frac{2s'iK'}{n}\right)\right],$$

$$\operatorname{cn}\left(\frac{u}{M_1},\lambda_1\right)=\sqrt{\frac{\lambda_1'}{\lambda_1 k'^n}}\left[\operatorname{cn}\left(u+\frac{4s'iK'}{n}\right)\right], \qquad =\sqrt{\frac{\lambda_1' k^n}{\lambda_1 k'^n}}\left[\operatorname{cn}\left(u+\frac{2s'iK'}{n}\right)\right],$$

$$\operatorname{dn}\left(\frac{u}{M_1},\lambda_1\right)=\sqrt{\frac{\lambda_1'}{k'^n}}\left[\operatorname{dn}\left(u+\frac{4s'iK'}{n}\right)\right]. \qquad =\sqrt{\frac{\lambda_1'}{k'^n}}\left[\operatorname{dn}\left(u+\frac{2s'iK'}{n}\right)\right].$$

denom. as above.

$$\frac{\lambda_1}{kM_1}\,\mathrm{sn}\left(\frac{u}{M_1},\lambda_1\right)=\mathrm{sn}\,u\left\{1+2\Sigma\frac{\mathrm{cn}\frac{4siK'}{n}\,\mathrm{dn}\frac{4siK'}{n}}{1-k^2\mathrm{sn}^2\frac{4siK'}{n}\,\mathrm{sn}^2u}\right\}=\mathrm{sn}\,u\left\{1+2\Sigma\frac{\mathrm{cn}\frac{2siK'}{n}\,\mathrm{dn}\frac{2siK'}{n}}{1-k^2\mathrm{sn}^2\frac{2siK'}{n}\,\mathrm{sn}^2u}\right\}$$

$$=\mathrm{sn}\,u\left\{1-\frac{2}{k}\Sigma\frac{\mathrm{cn}\frac{(2s-1)\,iK'}{n}\,\mathrm{dn}\frac{(2s-1)\,iK'}{n}}{\mathrm{sn}^2\frac{(2s-1)\,iK'}{n}-\mathrm{sn}^2u}\right\}$$

$$(-)^{\frac{1}{2}(n-1)}\frac{\lambda_1}{kM_1}\,\mathrm{cn}\left(\frac{u}{M_1},\lambda_1\right)=\mathrm{cn}\,u\left\{1+2\Sigma\frac{\mathrm{cn}\frac{4siK'}{n}}{1-k^2\mathrm{sn}^2\frac{4siK'}{n}\,\mathrm{sn}^2u}\right\}=\mathrm{cn}\,u\left\{1+2\Sigma\frac{(-)^s\,\mathrm{cn}\frac{2siK'}{n}}{1-k^2\mathrm{sn}^2\frac{2siK'}{n}\,\mathrm{sn}^2u}\right\}$$

$$=\mathrm{cn}\,u\left\{1+\frac{(-)^{\frac{1}{2}(n-1)}2}{ik}\Sigma\frac{(-)^s\,\mathrm{sn}\frac{(2s-1)iK'}{n}\,\mathrm{dn}\frac{(2s-1)iK'}{n}}{\mathrm{sn}^2\frac{(2s-1)iK'}{n}-\mathrm{sn}^2u}\right\}$$

$$(-)^{\frac{1}{2}(n-1)}\frac{1}{M_1}\operatorname{dn}\left(\frac{u}{M_1},\lambda_1\right)=\operatorname{dn}u\left\{1+2\Sigma\frac{\operatorname{dn}\frac{4siK'}{n}}{1-k^2\operatorname{sn}^2\frac{4siK'}{n}\operatorname{sn}^2u}\right\}$$

$$=\operatorname{dn}u\left\{1+2\Sigma\frac{(-)^s\operatorname{dn}\frac{2siK'}{n}}{1-k^2\operatorname{sn}^2\frac{2siK'}{n}\operatorname{sn}^2u}\right\}$$

$$=\operatorname{dn}u\left\{1+\frac{(-)^{\frac{1}{2}(n-1)}}{i}2\Sigma\frac{(-)^s\operatorname{sn}\frac{(2s-1)iK'}{n}\operatorname{cn}\frac{(2s-1)iK'}{n}}{\operatorname{sn}^2\frac{(2s-1)iK'}{n}-\operatorname{sn}^2u}\right\}$$

$$\frac{\lambda_1'}{k'_1M_1}\operatorname{tn}\left(\frac{u}{M_1},\lambda_1\right)=\operatorname{tn}u\left\{1+2\Sigma\frac{\operatorname{dn}\frac{4siK'}{n}\operatorname{cn}^2u}{\operatorname{cn}^2\frac{4siK'}{n}-\operatorname{dn}^2\frac{4siK'}{n}\operatorname{sn}^2u}\right\}$$

$$=\operatorname{tn}u\left\{1+2\Sigma\frac{(-)^s\operatorname{dn}\frac{2siK'}{n}\operatorname{cn}^2u}{\operatorname{cn}^2\frac{2siK'}{n}-\operatorname{dn}^2\frac{2siK'}{n}\operatorname{sn}^2u}\right\}.$$

The Second Transformation under a real form.

359. The formulæ may be presented in a real form by means of the transformations,

$$\mathrm{sn}\,(iu, k) = \frac{i\,\mathrm{sn}\,(u, k')}{\mathrm{cn}\,(u, k')},$$

$$\mathrm{cn}\,(iu, k) = \frac{1}{\mathrm{cn}\,(u, k')},$$

$$\mathrm{dn}\,(iu, k) = \frac{\mathrm{dn}\,(u, k')}{\mathrm{cn}\,(u, k')},$$

$$\mathrm{sn}\,(K - iu, k) = \frac{\mathrm{cn}\,(iu, k)}{\mathrm{dn}\,(iu, k)}, = \frac{1}{\mathrm{dn}\,(u, k')}.$$

Writing for shortness sn′, cn′, dn′, to denote the functions to the modulus k', we have for instance

$$\lambda_1 = k^n + \left[\mathrm{dn}'\,\frac{2sK'}{n}\right]^4,$$

$$\lambda_1' = k'^n \left[\mathrm{sn}'\left(K' - \frac{2sK'}{n}\right)\right]^4,$$

$$M_1 = \left[\mathrm{sn}'\left(K - \frac{2sK'}{n}\right)\right]^2 + \left[\mathrm{sn}'\,\frac{2sK'}{n}\right]^2,$$

$$\mathrm{sn}\left(\frac{u}{M_1}, \lambda_1\right) = \frac{\mathrm{sn}\,u}{M_1}\left[1 + \frac{\mathrm{sn}^2 u}{\mathrm{tn}'^2\,\frac{2sK'}{n}}\right]$$

$$\div \left[1 + k^2\,\mathrm{tn}'^2\,\frac{2sK'}{n}\,\mathrm{sn}^2 u\right], \text{ \&c.;}$$

but I do not think it worth while to give the entire series of equations.

Two relations of the Complete Functions.

Art. Nos. 360 and 361.

360. In the first transformation, taken in the 2ω-form, (observe that this is essential)

$$\mathrm{sn}\left(\frac{u}{M}, \lambda\right) = \frac{\mathrm{sn}\,u}{M}\left[1 - \frac{\mathrm{sn}^2 u}{\mathrm{sn}^2\,\frac{2sK}{n}}\right] \quad (+),$$

On the left-hand side the least real and positive value for which $\operatorname{sn}\left(\frac{u}{M}, \lambda\right)$ vanishes is $\frac{u}{M} = 2\Lambda$, and on the right-hand side it is $u = \frac{2K}{n}$: hence we have $M\Lambda = \frac{K}{n}$ or $\frac{K}{nM} = \Lambda$.

361. In the second transformation

$$\operatorname{sn}\left(\frac{u}{M_1}, \lambda_1\right) = \frac{\operatorname{sn} u}{M_1}\left[1 + \frac{\operatorname{sn}^2 u}{\operatorname{tn}^2 \frac{2siK'}{n}}\right] \quad (+).$$

On the left-hand side the least real and positive value for which $\operatorname{sn}\left(\frac{u}{M_1}, \lambda_1\right)$ vanishes is $\frac{u}{M_1} = 2\Lambda_1$, and on the right-hand side (since here the only factor which can vanish is $\operatorname{sn} u$) it is $u = 2K$: hence $M_1\Lambda_1 = K$ or $\frac{K}{M_1} = \Lambda_1$.

Observe these equations, $\frac{K}{nM} = \Lambda$ and $\frac{K}{M_1} = \Lambda_1$.

The Complementary and Supplementary Transformations.

Art. Nos. 362 to 367.

The first complementary transformation.

362. Start from the first transformation: this may be presented in the form

$$\operatorname{tn}\left(\frac{u}{M}, \lambda\right) = (-)^{\frac{1}{2}(n-1)}\sqrt{\frac{k'^n}{\lambda'}}\left[\operatorname{tn}\left(u + \frac{2s'K}{n}\right)\right]. \qquad s' = -\tfrac{1}{2}(n-1) \text{ to } +\tfrac{1}{2}(n-1).$$

Writing herein iu instead of u, and recollecting that

$$\operatorname{tn}(iu, k) = i \operatorname{sn}(u, k'),$$

the equation becomes

$$\operatorname{sn}\left(\frac{u}{M}, \lambda'\right) = \sqrt{\frac{k'^n}{\lambda'}}\left[\operatorname{sn}\left(u - \frac{2s'iK}{n}, k'\right)\right]$$

$$= \sqrt{\frac{k'^n}{\lambda'}} \operatorname{sn}(u, k')\left[\operatorname{sn}\left(u + \frac{2siK}{n}, k'\right) \operatorname{sn}\left(u - \frac{2siK}{n}, k'\right)\right],$$

$$s = 1 \text{ to } \tfrac{1}{2}(n-1);$$

which is $=(-)^{\frac{1}{2}(n-1)}\left[\operatorname{sn}\left(\frac{2siK}{n}, k'\right)\right]^2\sqrt{\frac{k'^n}{\lambda'}}\operatorname{sn}u\left[1-\frac{\operatorname{sn}^2(u, k')}{\operatorname{sn}^2\left(\frac{2siK}{n}, k'\right)}\right]$

$$\div\left[1-k'^2\operatorname{sn}^2\left(\frac{2siK}{n}, k'\right)\operatorname{sn}^2(u, k')\right];$$

or, since clearly the outside multiplier must be $=\frac{1}{M}$,* this is

$$\operatorname{sn}\left(\frac{u}{M}, \lambda'\right)=\frac{\operatorname{sn}(u, k')}{M}\left[1+\frac{\operatorname{sn}^2(u, k')}{\operatorname{tn}^2\left(\frac{2sK}{n}, k\right)}\right]$$

$$\div\left[1+k'^2\operatorname{tn}^2\left(\frac{2sK}{n}, k\right)\operatorname{sn}^2(u, k')\right].$$

This is the first complementary transformation giving $\operatorname{sn}\left(\frac{u}{M}, \lambda'\right)$ in terms of $\operatorname{sn}(u, k')$. Observe that its form is analogous to the second transformation.

The second complementary transformation.

363. Start from the second transformation; this is

$$\operatorname{tn}\left(\frac{u}{M_1}, \lambda_1\right)=\sqrt{\frac{k'^n}{\lambda_1'}}\left[\operatorname{tn}\left(u+\frac{2s'iK'}{n}\right)\right]. \qquad \begin{array}{r} s'=-\frac{1}{2}(n-1) \\ \text{to} +\frac{1}{2}(n-1). \end{array}$$

* The formula is $\frac{1}{M}=(-)^{\frac{1}{2}(n-1)}\left[\operatorname{sn}'\frac{2siK}{n}\right]^2\sqrt{\frac{k'^n}{\lambda'}}$, which of course may be verified directly: we have $\operatorname{sn}'\left(\frac{2siK}{n}\right)=\frac{i\operatorname{sn}\frac{2sK}{n}}{\operatorname{cn}\cdot\frac{2sK}{n}}$, and thence

$$\left[\operatorname{sn}'\frac{2siK}{n}\right]^2=(-)^{\frac{1}{2}(n-1)}\frac{\left[\operatorname{sn}\frac{2sK}{n}\right]^2}{\left[\operatorname{cn}\frac{2sK}{n}\right]^2}:\ \text{also}\ \sqrt{\frac{k'^n}{\lambda'}}=\left[\operatorname{dn}\frac{2sK}{n}\right]^2,$$

whence the formula becomes

$$\frac{1}{M}=\frac{\left[\operatorname{sn}\frac{2sK}{n}\right]^2}{\left[\operatorname{cn}\frac{2sK}{n}\right]^2}\left[\operatorname{dn}\frac{2sK}{n}\right]^2,\ =\left[\operatorname{sn}\frac{2sK}{n}\right]^2\div\left[\operatorname{sn}\left(K-\frac{2sK}{n}\right)\right]^2,$$

which is right.

Writing herein iu instead of u, and recollecting that, as before,

$$\text{tn}(iu, k) = i\,\text{sn}(u, k'),$$

the equation becomes

$$\text{sn}\left(\frac{u}{M_1}, \lambda_1'\right) = (-)^{\frac{1}{2}(n-1)}\sqrt{\frac{k'^n}{\lambda_1'}}\left[\text{sn}\left(u + \frac{2s'iK'}{n}, k'\right)\right]$$

$$= (-)^{\frac{1}{2}(n-1)}\sqrt{\frac{k'^n}{\lambda_1'}}\,\text{sn}(u, k')$$

$$\times\left[\text{sn}\left(u + \frac{2sK'}{n}, k'\right)\text{sn}\left(u - \frac{2sK'}{n}, k'\right)\right],$$

$$s = 1 \text{ to } \tfrac{1}{2}(n-1);$$

$$= \sqrt{\frac{k'^n}{\lambda_1'}}\left[\text{sn}\frac{2sK'}{n}\right]^2 \text{sn}(u, k')\left[1 - \frac{\text{sn}^2(u, k')}{\text{sn}^2\left(\frac{2sK'}{n}, k'\right)}\right]$$

$$+\left[1 - k'^2\,\text{sn}^2\left(\frac{2sK'}{n}, k'\right)\text{sn}^2(u, k')\right];$$

where the outside multiplier must be $= \dfrac{1}{M_1}$. The formula is therefore

$$\text{sn}\left(\frac{u}{M_1}, \lambda_1'\right) = \frac{\text{sn}(u, k')}{M_1}\left[1 - \frac{\text{sn}^2(u, k')}{\text{sn}^2\left(\frac{2sK'}{n}, k'\right)}\right]$$

$$+\left[1 - k'^2\,\text{sn}^2\left(\frac{2sK'}{n}, k'\right)\text{sn}^2(u, k')\right];$$

which is the second complementary transformation giving $\text{sn}\left(\dfrac{u}{M_1}, \lambda_1'\right)$ in terms of $\text{sn}(u, k')$: observe that its form is analogous to the first transformation.

364. Writing the first complementary transformation in the form

$$\text{sn}\left(\frac{u}{M}, \lambda'\right) = \frac{\text{sn}(u, k')}{M}\left[1 + \frac{\text{sn}^2(u, k')}{\text{tn}^2\left(\frac{2sK}{n}, k\right)}\right], \quad (+)$$

and considering the least real positive value of u for which the two sides respectively vanish: these are on the left-hand side

$\frac{u}{M}=2\Lambda'$, and on the right-hand side $u=2K'$: hence we have $M\Lambda'=K'$ or $\Lambda'=\frac{K'}{M}$.

Similarly from the second complementary transformation,

$$\operatorname{sn}\left(\frac{u}{M_1},\lambda_1'\right)=\frac{\operatorname{sn}(u,k')}{M_1}\left[1-\frac{\operatorname{sn}^2(u,k')}{\operatorname{sn}^2\left(\frac{2sK'}{n},k'\right)}\right]; \quad (+)$$

the least real positive values for which the two sides vanish are $\frac{u}{M_1}=2\Lambda_1'$ and $u=\frac{2K'}{n}$, whence $M_1\Lambda_1'=\frac{K'}{n}$ or $\Lambda_1'=\frac{K'}{nM_1}$.

365. We have thus obtained the equations

$$\Lambda'=\frac{K'}{M},\qquad \Lambda_1'=\frac{K'}{nM_1},$$

to be taken along with the foregoing equations, No. 361,

$$\Lambda=\frac{K}{nM}\text{ and }\Lambda_1=\frac{K}{M_1}.$$

Eliminating M and M_1, we obtain

$$\frac{\Lambda'}{\Lambda}=n\frac{K'}{K},\quad \frac{K'}{K}=n\frac{\Lambda_1'}{\Lambda_1}:$$

the first of which is an equation between λ and k, and the second is the same equation between k and λ_1: and it thus appears that λ is the same function of k that k is of λ_1. The equations show that λ is less than k, and λ_1 greater than k.

The first supplementary transformation.

366. In the second transformation

$$\operatorname{sn}\left(\frac{u}{M_1},\lambda_1\right)=\frac{\operatorname{sn}u}{M_1}\left[1-\frac{\operatorname{sn}^2u}{\operatorname{sn}^2\frac{2siK'}{n}}\right]+\left[1-\frac{\operatorname{sn}^2u}{\operatorname{sn}^2\frac{(2s-1)iK'}{n}}\right],$$

change k into λ, and therefore λ_1 into k: writing for a moment N_1 as the new value of M_1 the formula becomes

$$\operatorname{sn}\left(\frac{u}{N_1},k\right)=\frac{\operatorname{sn}(u,\lambda)}{N_1}\left[1-\frac{\operatorname{sn}^2(u,\lambda)}{\operatorname{sn}^2\left(\frac{2si\Lambda'}{n},\lambda\right)}\right]$$

$$+\left[1-\frac{\operatorname{sn}^2(u,\lambda)}{\operatorname{sn}^2\left(\frac{(2s-1)i\Lambda'}{n},\lambda\right)}\right].$$

Change also u into $\frac{u}{M}$; then observing that

$$M_1=\frac{K}{\Lambda_1},\text{ and therefore } N_1=\frac{\Lambda}{K}=\frac{1}{nM}\text{ or } n=\frac{1}{MN_1},$$

the equation becomes

$$\operatorname{sn}(nu,k)=nM\operatorname{sn}\left(\frac{u}{M},\lambda\right)\left[1-\frac{\operatorname{sn}^2\left(\frac{u}{M},\lambda\right)}{\operatorname{sn}^2\left(\frac{2si\Lambda'}{n},\lambda\right)}\right]$$

$$+\left[1-\frac{\operatorname{sn}^2\left(\frac{u}{M},\lambda\right)}{\operatorname{sn}^2\left(\frac{(2s-1)i\Lambda'}{n},\lambda\right)}\right],$$

which is the first supplementary transformation.

Combining herewith the first transformation,

$$\operatorname{sn}\left(\frac{u}{M},\lambda\right)=\frac{\operatorname{sn}u}{M}\left[1-\frac{\operatorname{sn}^2u}{\operatorname{sn}^2\frac{2sK}{n}}\right]+\left[1-k^2\operatorname{sn}^2\frac{2sK}{n}\operatorname{sn}^2u\right],$$

we see that the two together lead to an expression of $\operatorname{sn}(nu,k)$ in terms of $\operatorname{sn}(u,k)$.

The second supplementary transformation.

367. In the first transformation

$$\operatorname{sn}\left(\frac{u}{M},\lambda\right)=\frac{\operatorname{sn}u}{M}\left[1-\frac{\operatorname{sn}^2u}{\operatorname{sn}^2\frac{2sK}{n}}\right]+\left[1-k^2\operatorname{sn}^2\frac{2sK}{n}\operatorname{sn}^2u\right],$$

change k into λ_1, and therefore λ into k: writing for a moment N as the new value of M, the formula becomes

$$\operatorname{sn}\left(\frac{u}{N},k\right)=\frac{\operatorname{sn}(u,\lambda_1)}{N}\left[1-\frac{\operatorname{sn}^2(u,\lambda_1)}{\operatorname{sn}^2\left(\frac{2s\Lambda_1}{n},\lambda_1\right)}\right]$$

$$+\left[1-\lambda_1^2\operatorname{sn}^2\left(\frac{2s\Lambda_1}{n},\lambda_1\right)\operatorname{sn}^2(u,\lambda_1)\right];$$

change also u into $\frac{u}{M_1}$; then observing that

$M = \frac{K'}{\Lambda'}$, and therefore $N = \frac{\Lambda_1'}{K'}, = \frac{1}{nM_1}$, that is $n = \frac{1}{M_1 N}$,

the equation becomes

$$\mathrm{sn}(nu, k) = nM_1 \,\mathrm{sn}\left(\frac{u}{M_1}, \lambda_1\right) \left[1 - \frac{\mathrm{sn}^2\left(\frac{u}{M_1}, \lambda_1\right)}{\mathrm{sn}^2\left(\frac{2s\Lambda_1}{n}, \lambda_1\right)}\right]$$

$$\div \left[1 - \lambda_1^2 \,\mathrm{sn}^2\left(\frac{2s\Lambda_1}{n}, \lambda_1\right) \mathrm{sn}^2\left(\frac{u}{M_1}, \lambda_1\right)\right],$$

which is the second supplementary transformation.

Combining herewith the second transformation,

$$\mathrm{sn}\left(\frac{u}{M_1}, \lambda_1\right) = \frac{\mathrm{sn}(u, k)}{M_1} \left[1 - \frac{\mathrm{sn}^2 u}{\mathrm{sn}^2 \frac{2siK'}{n}}\right]$$

$$\div \left[1 - \frac{\mathrm{sn}^2 u}{\mathrm{sn}^2 \frac{(2s-1)iK'}{n}}\right];$$

we see that the two together lead to an expression of $\mathrm{sn}(nu, k)$ in terms of $\mathrm{sn}(u, k)$.

The Multiplication-formulæ. Art. No. 368.

369. For the actual determination of the multiplication-formulæ, observe that the first supplementary transformation may be written in the form

$$\mathrm{sn}(nu, k) = \sqrt{\frac{\lambda^n}{k}} \left[\mathrm{sn}\left(\frac{u}{M} + \frac{2s'i\Lambda'}{n}, \lambda\right)\right], \qquad s' = -\tfrac{1}{2}(n-1) \text{ to } +\tfrac{1}{2}(n-1);$$

or, what is the same thing,

$$\mathrm{sn}(nu, k) = \sqrt{\frac{\lambda^n}{k}} \left[\mathrm{sn}\left(\frac{u}{M} + \frac{2m'i\Lambda'}{n}, \lambda\right)\right], \qquad m' = -\tfrac{1}{2}(n-1) \text{ to } +\tfrac{1}{2}(n-1).$$

But the first transformation gives

$$\mathrm{sn}\left(\frac{u}{M}, \lambda\right) = (-)^{\frac{1}{2}(n-1)} \sqrt{\frac{k^n}{\lambda}} \left[\mathrm{sn}\left(u + \frac{2s'K}{n}\right)\right], \qquad s' = -\tfrac{1}{2}(n-1) \text{ to } +\tfrac{1}{2}(n-1);$$

or say

$$\operatorname{sn}\left(\frac{u}{M},\lambda\right)=(-)^{\frac{1}{2}(n-1)}\sqrt{\frac{k^n}{\lambda}}\left[\operatorname{sn}\left(u+\frac{2mK}{n}\right)\right],\qquad m=-\tfrac{1}{2}(n-1) \text{ to } +\tfrac{1}{2}(n-1);$$

and writing herein $u+\dfrac{2m'iK'}{n}$ for u, $\dfrac{u}{M}$ becomes

$$\frac{u}{M}+\frac{2m'iK'}{nM},=\frac{u}{M}+\frac{2m'i\Lambda'}{n},$$

and the formula is

$$(-)^{\frac{1}{2}(n-1)}\operatorname{sn}\left(\frac{u}{M}+\frac{2m'i\Lambda'}{n},\lambda\right)=\sqrt{\frac{k^n}{\lambda}}\left[\operatorname{sn}\left(u+\frac{2mK}{n}+\frac{2m'iK'}{n}\right)\right],$$

where on the right-hand side m has the last-mentioned values.

Giving herein to m' the different values from $-\frac{1}{2}(n-1)$ to $+\frac{1}{2}(n-1)$ and multiplying the results together, observing that

$$(-)^{\frac{1}{2}n(n-1)}=(-)^{\frac{1}{2}(n-1)},$$

we obtain

$$(-)^{\frac{1}{2}(n-1)}\left[\operatorname{sn}\left(\frac{u}{M}+\frac{2m'i\Lambda'}{n},\lambda\right)\right]=\sqrt{\frac{k^{n^2}}{\lambda^n}}\left\{\operatorname{sn}\left(u+\frac{2mK}{n}+\frac{2m'iK'}{n}\right)\right\},$$

or, the left-hand side being $=(-)^{\frac{1}{2}(n-1)}\sqrt{\dfrac{k}{\lambda^n}}\operatorname{sn}(nu,k)$, the formula is

$$\operatorname{sn} nu=(-)^{\frac{1}{2}(n-1)}k^{\frac{1}{2}(n^2-1)}\left\{\operatorname{sn}\left(u+\frac{2mK}{n}+\frac{2m'iK'}{n}\right)\right\},$$

where on the right-hand side the { } denote the double product obtained by giving to m, m' respectively the values $-\frac{1}{2}(n-1)$ to $+\frac{1}{2}(n-1)$, or say the values $0, \pm 1, \pm 2, \ldots \pm\frac{1}{2}(n-1)$.

And in the same way

$$\operatorname{cn} nu=\left(\frac{k}{k'}\right)^{\frac{1}{2}(n^2-1)}\left\{\operatorname{cn}\left(u+\frac{2mK}{n}+\frac{2m'iK'}{n}\right)\right\},$$

and

$$\operatorname{dn} nu=\left(\frac{1}{k'}\right)^{\frac{1}{2}(n^2-1)}\left\{\operatorname{dn}\left(u+\frac{2mK}{n}+\frac{2m'iK'}{n}\right)\right\},$$

which are the formulæ obtained Chap. IV.

CHAPTER XI.

THE q-FUNCTIONS: FURTHER THEORY OF THE FUNCTIONS H, Θ.

369. In the present Chapter we start with the transformation of the order n in the form of the first supplementary transformation, whereby the functions sn (nu, k), &c. are given in terms of sn $\left(\frac{u}{M}, \lambda\right)$: writing first $\frac{u}{n}$ for u, we make $n = \infty$, and (as will appear) we thus obtain the elliptic functions sn (u, k), &c., as fractions, the numerators and denominators being respectively obtained in terms of the circular functions of $\frac{\pi u}{2K}$, viz. as products depending on these functions, and involving also the quantity $e^{-\frac{\pi K'}{K}}$, which is put $= q$: the elliptic functions have been already in Chapter VI. expressed as fractions by means of the functions H, Θ: and identifying the two expressions, we obtain the expressions of these functions as series involving powers of q, or say as q-series.

Derivation of the q-formulæ. Art. Nos. 370 to 378.

370. The first supplementary transformation is

$$\text{sn}\,(nu, k) = nM\,\text{sn}\left(\frac{u}{M}, \lambda\right)\left[1 - \frac{\text{sn}^2\left(\frac{u}{M}, \lambda\right)}{\text{sn}^2\left(\frac{2si\Lambda'}{n}, \lambda\right)}\right]$$

$$\div \left[1 - \frac{\text{sn}^2\left(\frac{u}{M}, \lambda\right)}{\text{sn}^2\left(\frac{(2s-1)i\Lambda'}{n}, \lambda\right)}\right]:$$

we write herein $\frac{u}{n}$ for u, and make n infinite. This gives $\lambda = 0$, $\text{sn}(\theta, \lambda) = \sin\theta$, $\Lambda = \frac{1}{2}\pi$; whence $\left(\text{in virtue of } \Lambda = \frac{K}{nM} \text{ and } \Lambda' = \frac{K'}{M}\right)$,

$$nM = \frac{2K}{\pi}, \quad \frac{\Lambda'}{n} = \frac{\pi K'}{2K};$$

and the equation becomes

$$\text{sn}\, u = \frac{2K}{\pi} \sin\frac{\pi u}{2K} \left[1 - \frac{\sin^2 \frac{\pi u}{2K}}{\sin^2 \frac{si\pi K'}{K}}\right] \div \left[1 - \frac{\sin^2 \frac{\pi u}{2K}}{\sin^2 \frac{(2s-1)\, i\pi K'}{2K}}\right].$$

This is one of a group of formulæ obtained in the same manner.

371. The formulæ are

$$\text{sn}\, u = \frac{2K}{\pi} \sin\frac{\pi u}{2K} \left[1 - \frac{\sin^2 \frac{\pi u}{2K}}{\sin^2 \frac{mi\pi K'}{K}}\right] \qquad (+),$$

$$\text{cn}\, u = \cos\frac{\pi u}{2K} \left[1 - \frac{\sin^2 \frac{\pi u}{2K}}{\cos^2 \frac{mi\pi K'}{K}}\right] \qquad (+),$$

$$\text{dn}\, u = \left[1 - \frac{\sin^2 \frac{\pi u}{2K}}{\cos^2 \frac{(2m-1)\, i\pi K'}{2K}}\right] \qquad (+),$$

$$1 - \text{sn}\, u = \left(1 - \sin\frac{\pi u}{2K}\right) \left[1 - \frac{\sin \frac{\pi u}{2K}}{\cos \frac{mi\pi K'}{K}}\right]^2 \qquad (+),$$

$$1 + \text{sn}\, u = \left(1 + \sin\frac{\pi u}{2K}\right) \left[1 + \frac{\sin \frac{\pi u}{2K}}{\cos \frac{mi\pi K'}{K}}\right]^2 \qquad (+),$$

$$1 - k\,\mathrm{sn}\,u = \left[1 - \frac{\mathrm{sn}\,\dfrac{\pi u}{2K}}{\cos\dfrac{(2m-1)\,i\pi K'}{2K}}\right]^2 \quad (+),$$

$$1 + k\,\mathrm{sn}\,u = \left[1 + \frac{\mathrm{sn}\,\dfrac{\pi u}{2K}}{\cos\dfrac{(2m-1)\,i\pi K'}{2K}}\right]^2 \quad (+),$$

$$\text{denom.} = \left[1 - \frac{\mathrm{sn}^2\,\dfrac{\pi u}{2K}}{\mathrm{sn}^2\,\dfrac{(2m-1)\,i\pi K'}{2K}}\right];$$

372. We obtain in like manner another group of formulæ, in which also m has the values 1, 2, 3... to infinity,

$$\mathrm{sn}\,u = -\frac{\pi}{kK}\sin\frac{\pi u}{2K}\,\Sigma\left(\frac{\cos\dfrac{(2m-1)\,i\pi K'}{2K}}{\sin^2\dfrac{(2m-1)\,i\pi K'}{2K} - \sin^2\dfrac{\pi u}{2K}}\right),$$

$$\mathrm{cn}\,u = \frac{i\pi}{kK}\cos\frac{\pi u}{2K}\,\Sigma\left((-)^{m-1}\frac{\sin\dfrac{(2m-1)\,i\pi K'}{2K}}{\sin^2\dfrac{(2m-1)\,i\pi K'}{2K} - \sin^2\dfrac{\pi u}{2K}}\right);$$

$$\mathrm{dn}\,u = 1 + \frac{i\pi}{K}\sin^2\frac{\pi u}{2K}\,\Sigma\left(\frac{(-)^{m-1}\cot\dfrac{(2m-1)\,i\pi K'}{K}}{\sin^2\dfrac{(2m-1)\,i\pi K'}{K} - \sin^2\dfrac{\pi u}{2K}}\right).$$

The deduction of this last formula presents some peculiarity: writing $\pm$ instead of $(-)^{\frac{1}{2}(n-1)}$, the formula originally presents itself in the form

$$\pm\,\mathrm{dn}\,u = \frac{\pi}{2K} \pm \frac{\pi}{iK}\,\Sigma\left(\frac{(-)^m\sin\dfrac{(2m-1)\,i\pi K'}{K}\cos\dfrac{(2m-1)\,i\pi K'}{K}}{\sin^2\dfrac{(2m-1)\,i\pi K'}{K} - \sin^2\dfrac{\pi u}{2K}}\right);$$

viz. the upper or lower sign must here be taken according as the number of terms in the series is even or odd. To get rid of this variable sign, write in the equation $u=0$, the equation becomes

$$\pm 1 = \frac{\pi}{2K} \pm \frac{\pi}{iK} \Sigma \left(\frac{(-)^m \sin \frac{(2m-1) i\pi K'}{K} \cos \frac{(2m-1) i\pi K'}{K}}{\sin^2 \frac{(2m-1) i\pi K'}{K}} \right);$$

and subtracting this from the general formula, each side of the equation is affected with the same sign $\pm$, which sign may therefore be omitted: whence, observing that in general

$$\frac{\sin a \cos a}{\sin^2 a - \sin^2 x} - \frac{\sin a \cos a}{\sin^2 a}$$

$$= \sin a \cos a \frac{\sin^2 x}{\sin^2 a (\sin^2 a - \sin^2 x)} = \frac{\cot a \sin^2 x}{\sin^2 a - \sin^2 x},$$

it is at once seen that we thus obtain the result first written down.

All the formulæ assume a more convenient form by writing therein $u = \frac{2Kx}{\pi}$, viz. we have thus $\sin \frac{\pi u}{2K} = \sin x$, and consequently the elliptic functions $\operatorname{sn} \frac{2Kx}{\pi}$, &c. expressed in terms of $\sin x$.

373. Introducing now the quantity

$$q, = e^{-\frac{\pi K'}{K}},$$

we have

$$\sin \frac{mi\pi K'}{K} = \frac{1}{2i}(q^m - q^{-m}) = \frac{i(1-q^{2m})}{2q^m},$$

$$\cos \frac{mi\pi K'}{K} = \frac{1}{2}(q^m + q^{-m}) = \frac{1+q^{2m}}{2q^m},$$

$$1 - \frac{\sin^2 x}{\sin^2 \frac{mi\pi K'}{K}} = 1 + \frac{4q^{2m} \sin^2 x}{(1-q^{2m})^2} = \frac{1 - 2q^{2m} \cos 2x + q^{4m}}{(1-q^{2m})^2}, \text{&c.},$$

and in the resulting formulæ the right-hand sides contain as factors certain functions of q, which functions are afterwards determined as will presently appear. Supposing this done, the formulæ of the first group are

$$\operatorname{sn}\frac{2Kx}{\pi} = 2\ \frac{1}{\sqrt{k}}\sqrt[4]{q}\sin x \qquad [1-2q^{2m}\cos 2x+q^{4m}], \quad (+)$$

$$\operatorname{cn}\frac{2Kx}{\pi} = 2\sqrt{\frac{k'}{k}}\sqrt[4]{q}\cos x \qquad [1+2q^{2m}\cos 2x+q^{4m}], \quad (+)$$

$$\operatorname{dn}\frac{2Kx}{\pi} = \quad \sqrt{k'} \qquad [1+2q^{2m-1}\cos 2x+q^{4m-2}], \quad (+)$$

$$1-\operatorname{sn}\frac{2Kx}{\pi} = 2\sqrt{\frac{k'}{k}}\sqrt[4]{q}\,(1-\sin x)\,[1-2q^{m}\ \sin x+q^{2m}]^2, \quad (+)$$

$$1+\operatorname{sn}\frac{2Kx}{\pi} = 2\sqrt{\frac{k'}{k}}\sqrt[4]{q}\,(1+\sin x)\,[1+2q^{m}\ \sin x+q^{2m}]^2, \quad (+)$$

$$1-k\operatorname{sn}\frac{2Kx}{\pi} = \quad \sqrt{k'} \qquad [1-2q^{m-\frac{1}{2}}\sin x+q^{2m-1}]^2, \quad (+)$$

$$1+k\operatorname{sn}\frac{2Kx}{\pi} = \quad \sqrt{k'} \qquad [1+2q^{m-\frac{1}{2}}\sin x+q^{2m-1}]^2, \quad (+)$$

where

$$\text{denom.} = [1-2q^{2m-1}\cos 2x+q^{4m-2}];$$

and the formulæ of the second group are

$$\operatorname{sn}\frac{2Kx}{\pi} = \frac{2\pi}{kK}\sin x\,\Sigma\left\{\frac{q^{m-\frac{1}{2}}(1+q^{2m-1})}{1-2q^{2m-1}\cos 2x+q^{4m-2}}\right\},$$

$$\operatorname{cn}\frac{2Kx}{\pi} = \frac{2\pi}{kK}\cos x\,\Sigma\left\{(-)^{m-1}\frac{q^{m-\frac{1}{2}}(1-q^{2m-1})}{1-2q^{2m-1}\cos 2x+q^{4m-2}}\right\};$$

$$1-\operatorname{dn}\frac{2Kx}{\pi} = \frac{4\pi}{K}\sin^2 x\,\Sigma\left\{\frac{(-)^{m-1}q^{2m-1}\dfrac{1+q^{2m-1}}{1-q^{2m-1}}}{1-2q^{2m-1}\cos 2x+q^{4m-2}}\right\},$$

and to these Jacobi has joined an expression for $\operatorname{am}\frac{2Kx}{\pi}$, that is $\sin^{-1}\operatorname{sn}\frac{2Kx}{\pi}$: viz. the equation is

$$\sin^{-1}\left(\operatorname{sn}\frac{2Kx}{\pi}\right) = \pm x + 2\Sigma\,(-)^{m-1}\tan^{-1}\frac{(1+q^{2m-1})\tan x}{1-q^{2m-1}},$$

where the sign is $-$ or $+$, according as in the calculation of the series the number of terms taken is odd or even. Writing herein $k = 0$, the formula becomes

$$x = \pm x + 2\Sigma (-)^{m-1} \tan^{-1} (\tan x),$$

where the sign is $-$ or $+$ as before, a particular formula, the truth of which is evident at sight: and subtracting this from the general formula, we convert it into

$$\sin^{-1} \text{sn} \frac{2Kx}{\pi} = x + 2\Sigma \left\{(-)^{m-1} \tan^{-1}\left(\frac{1+q^{2m-1}}{1-q^{2m-1}} \tan x\right) - \tan^{-1} \tan x\right\},$$

or, what is the same thing,

$$\sin^{-1} \left(\text{sn} \frac{2Kx}{\pi}\right) = x + 2\Sigma (-)^{m-1} \tan^{-1}\left(\frac{q^{2m-1} \sin 2x}{1-q^{2m-1} \cos 2x}\right);$$

a form of the formula, free from the discontinuity, and in which the series is convergent. Differentiating in regard to x and using a formula $\frac{2K}{\pi} = 1 + 4\Sigma \frac{(-)^{m-1} q^{2m-1}}{1-q^{2m-1}}$, which will be proved further on, we obtain the foregoing expression for $1 - \text{dn} \frac{2Kx}{\pi}$; and conversely by the integration of this we obtain the last-mentioned formula for $\sin^{-1} \left(\text{sn} \frac{2Kx}{\pi}\right)$.

374. In completion of the investigation of the formulæ of No. 372, observe that writing

$$\frac{1}{A} = [1-q^{2m}]^2, \qquad (+)$$

$$\frac{1}{B} = [1+q^{2m}]^2, \qquad (+)$$

$$\frac{1}{C} = [1+q^{2m-1}]^2, \qquad (+)$$

where $\qquad \text{denom.} = [1-q^{2m-1}]^2;$

the formulæ obtained in the first instance are

$$\text{sn} \frac{2Kx}{\pi} = \frac{2AK}{\pi} \sin x \, [1 - 2q^{2m} \cos 2x + q^{4m}], \qquad (+)$$

$$\text{cn} \frac{2Kx}{\pi} = \quad B \quad \cos x \, [1 + 2q^{2m} \cos 2x + q^{4m}], \qquad (+)$$

$$\operatorname{dn}\frac{2Kx}{\pi} = C\,[1 + 2q^{2m-1}\cos 2x + q^{4m-2}], \qquad (+)$$

where denom. $= [1 - 2q^{2m-1}\cos 2x + q^{4m-2}]$.

Writing in the first and third of these $x = \frac{1}{2}\pi$, we find

$$1 = \frac{2AK}{\pi}\cdot\frac{C}{B}; \quad k' = C.C;$$

whence $$C = \sqrt{k'}, \quad \text{and} \quad B = \frac{2\sqrt{k'}AK}{\pi}.$$

375. To determine A, write for a moment U in place of e^{ix}, then we have

$$\operatorname{sn}\frac{2Kx}{\pi} = \frac{AK}{\pi}\,\frac{U - U^{-1}}{i}\,\frac{[1-q^{2m}U^2][1-q^{2m}U^{-2}]}{[1-q^{2m-1}U^2][1-q^{2m-1}U^{-2}]};$$

which, observing that

$$U - U^{-1} = U(1 - U^{-2}), \quad = -\frac{1}{U}(1 - U^2),$$

may be written

$$\operatorname{sn}\frac{2Kx}{\pi} = \frac{AK}{\pi i}\,\frac{U[1-q^{2m}U^2][1-q^{2m-2}U^{-2}]}{[1-q^{2m-1}U^2][1-q^{2m-1}U^{-2}]},$$

or $$= -\frac{AK}{\pi i}\,\frac{1}{U}\,\frac{[1-q^{2m-2}U^2][1-q^{2m}U^{-2}]}{[1-q^{2m-1}U^2][1-q^{2m-1}U^{-2}]}.$$

For x write $x + \frac{i\pi K'}{2K}$, $\operatorname{sn}\frac{2Kx}{\pi}$ becomes

$$\operatorname{sn}\left(\frac{2Kx}{\pi} + iK'\right), \quad = \frac{1}{k\operatorname{sn}\frac{2Kx}{\pi}},$$

U is changed into $q^{\frac{1}{2}}U$, and taking the second formula we have

$$\frac{1}{k\operatorname{sn}\frac{2Kx}{\pi}} = -\frac{AK}{\pi i}\,\frac{1}{q^{\frac{1}{2}}U}\,\frac{[1-q^{2m-1}U^2][1-q^{2m-1}U^{-2}]}{[1-q^{2m}U^2][1-q^{2m-2}U^{-2}]}.$$

Hence multiplying, we find

$$\frac{1}{k} = \left(\frac{AK}{\pi}\right)^2\cdot\frac{1}{\sqrt{q}}, \quad \text{whence} \quad \frac{AK}{\pi} = \frac{\sqrt[4]{q}}{\sqrt{k}};$$

and therefore

$$A = \frac{\pi}{K}\frac{\sqrt[4]{q}}{\sqrt{k}}, \quad B = 2\sqrt{k'}\frac{\sqrt[4]{q}}{\sqrt{k}}, \quad C = \sqrt{k'};$$

and substituting we have the foregoing formulæ for sn, cn, and dn of $\dfrac{2Kx}{\pi}$.

370. We also obtain various other q-formulæ.

Multiplying the expressions for B, C we have

$$\frac{2\sqrt[4]{q}\,.\,k'}{\sqrt{k}} = \frac{[1-q^{2m-1}]^4}{[1+q^m]^2};$$

and observing that

$$[1+q^m] = \frac{[1-q^{2m}]}{[1-q^m]}, \quad = \frac{1}{[1-q^{2m-1}]},$$

we find

$$[1-q^{2m-1}]^6 = \frac{2\sqrt[4]{q}\,k'}{\sqrt{k}},$$

and thence, using the foregoing value of A,

$$[1-q^{2m}]^6 = \frac{2kk'K^3}{\pi^3\sqrt{q}}:$$

which two formulæ give

$$[1-q^{2m-1}]^2[1-q^{2m}] = \sqrt{\frac{2k'K}{\pi}},$$

$$[1-q^m]^6 \qquad = \frac{4\sqrt{k}\,k'^2K^3}{\pi^3\sqrt[4]{q}};$$

and to these may be joined

$$[1+q^{2m-1}]^6 \qquad = \frac{2\sqrt[4]{q}}{\sqrt{kk'}},$$

$$[1+q^{2m}]^6 \qquad = \frac{k}{4\sqrt{k'}\sqrt{q}},$$

$$[1+q^m]^6 \qquad = \frac{\sqrt{k}}{2k'\sqrt[4]{q}}.$$

377. If for shortness we write:

$$\begin{aligned}
\alpha &= [1+q^{2m-1}], \qquad \text{whence } \alpha\beta = [1+q^m],\ \gamma\delta = [1-q^m],\\
\beta &= [1+q^{2m}], \qquad\quad \text{and } \alpha\beta\gamma = 1\,^*;\\
\gamma &= [1-q^{2m-1}],\\
\delta &= [1-q^{2m}];
\end{aligned}$$

then the foregoing formulæ give

$$\begin{aligned}
k &= 4\sqrt{q}\left(\frac{\beta}{\alpha}\right)^4,\\
k' &= \left(\frac{\gamma}{\alpha}\right)^4,\\
\frac{2K}{\pi} &= \left(\frac{\alpha\delta}{\beta\gamma}\right)^2,\\
\frac{2kK}{\pi} &= 4\sqrt{q}\left(\frac{\beta\delta}{\alpha\gamma}\right)^2,\\
\frac{2k'K}{\pi} &= \left(\frac{\gamma\delta}{\alpha\beta}\right)^2, \quad = \gamma^4\delta^2,\\
\frac{2\sqrt{k}K}{\pi} &= 2\sqrt[4]{q}\cdot\left(\frac{\delta}{\gamma}\right)^2,\\
\frac{2\sqrt{k'}K}{\pi} &= \left(\frac{\delta}{\beta}\right)^2.
\end{aligned}$$

The equation $k^2+k'^2=1$ gives $\gamma^8+16q\beta^8=\alpha^8$, or written at length

$$[(1-q)(1-q^3)(1-q^5)\ldots]^8 + 16q\,[(1+q^2)(1+q^4)(1+q^6)\ldots]^8$$
$$= [(1+q)(1+q^3)(1+q^5)\ldots]^8;$$

a remarkable identity.

* This is in fact the formula $[1+q^m] = \dfrac{1}{[1-q^{2m-1}]}$ proved No. 376: it occurs in Euler's Memoir, *De Partitione Numerorum* (1750), Op. Min. Coll. p. 93.

378. It will be noticed that we have obtained expressions for $\operatorname{sn}\dfrac{2Kx}{\pi}$, $\operatorname{cn}\dfrac{2Kx}{\pi}$, $\operatorname{dn}\dfrac{2Kx}{\pi}$, as rational fractions having a common denominator, the three numerators and the denominator being each of them a q-function, $\left(q = e^{-\frac{\pi K'}{K}}\right)$, involving circular functions of x respectively. Imagining x replaced by its value $\dfrac{\pi u}{2K}$, we have $\operatorname{sn} u$, $\operatorname{cn} u$, $\operatorname{dn} u$ expressed as rational fractions, the three numerators and the denominator being each of them a q-function involving circular functions of $\dfrac{\pi u}{2K}$. We have already obtained for $\operatorname{sn} u$, $\operatorname{cn} u$, $\operatorname{dn} u$ fractional expressions having a common denominator Θu, and in their numerators Hu, $H(u+K)$, $\Theta(u+K)$ respectively; and it thus appears that these functions must be, to proper factors près, multiples of the q-functions of $\dfrac{\pi u}{2K}$ respectively: viz. the factors to multiply the q-functions must be of the form AU, BU, CU, DU where U is an unknown function of u, but A, B, C, D are known constants or exponential factors; and the theorem at once suggests itself, that the two sets of functions differ only by these factors A, B, C, D, or what is the same thing, that U is a mere constant, which may be taken $=1$. But Jacobi in fact directly identifies $\Theta\dfrac{2Kx}{\pi}$ with a q-function of x $\left(\text{that is, } \Theta u \text{ with a } q\text{-function of } \dfrac{\pi u}{2K}\right)$, by an investigation of some complexity but of very great interest.

Θ, *H expressed as q-functions.* Art. Nos. 379 to 383.

379. We have

$$\sqrt{\frac{1 - k\operatorname{sn}\dfrac{2Kx}{\pi}}{1 + k\operatorname{sn}\dfrac{2Kx}{\pi}}} = \frac{[1 - 2q^{m-\frac{1}{2}}\sin x + q^{2m-1}]}{[1 + 2q^{m-\frac{1}{2}}\sin x + q^{2m-1}]}.$$

Taking the logarithm of each side, and expanding the logarithms of the factors on the right-hand side, after some obvious reductions we obtain

$$\log\sqrt{\frac{1-k\,\mathrm{sn}\,\dfrac{2Kx}{\pi}}{1+k\,\mathrm{sn}\,\dfrac{2Kx}{\pi}}} = \Sigma(-)^{m-1}\frac{4q^{m-\frac{1}{2}}\sin(2m-1)x}{(2m-1)(1-q^{2m-1})}:$$

or, writing the series at full length,

$$= \frac{4\sqrt{q}\sin x}{1-q} - \frac{4\sqrt{q^3}\sin 3x}{3(1-q^3)} + \frac{4\sqrt{q^5}\sin 5x}{5(1-q^5)} - \ldots$$

Differentiating each side in regard to x, we find without difficulty

$$\frac{2kK}{\pi}\frac{\mathrm{cn}\,\dfrac{2Kx}{\pi}}{\mathrm{dn}\,\dfrac{2Kx}{\pi}} = \frac{4\sqrt{q}\cos x}{1-q} - \frac{4\sqrt{q^3}\cos 3x}{1-q^3} + \frac{4\sqrt{q^5}\cos 5x}{1-q^5} - \&c.,$$

or, observing that the left hand is

$$\frac{2kK}{\pi}\,\mathrm{sn}\left(K - \frac{2Kx}{\pi}\right), \quad = \frac{2kK}{\pi}\,\mathrm{sn}\,\frac{2K}{\pi}\left(\frac{\pi}{2}-x\right),$$

and writing $\frac{1}{2}\pi - x$ in place of x, this is

$$\frac{2kK}{\pi}\,\mathrm{sn}\,\frac{2Kx}{\pi} = \frac{4\sqrt{q}\sin x}{1-q} + \frac{4\sqrt{q^3}\sin 3x}{1-q^3} + \frac{4\sqrt{q^5}\sin 5x}{1-q^5} + \&c.$$

380. It is this formula which leads to the identification just spoken of; viz. squaring the two sides we obtain after all reductions

$$\left(\frac{2kK}{\pi}\right)^2 \mathrm{sn}^2\,\frac{2Kx}{\pi} = \frac{4K}{\pi^2}(K-E)$$

$$-4\left\{\frac{2q\cos 2x}{1-q^2} + \frac{4q^2\cos 4x}{1-q^4} + \frac{6q^3\cos 6x}{1-q^6} + \ldots\right\},$$

or, multiplying by dx and integrating from $x=0$,

$$\left(\frac{2kK}{\pi}\right)^2\int \mathrm{sn}^2\,\frac{2Kx}{\pi}\,dx = \frac{4K^2}{\pi^2}\left(1-\frac{E}{K}\right)x$$

$$-4\left\{\frac{q\sin 2x}{1-q^2} + \frac{q^2\sin 4x}{1-q^4} + \frac{q^3\sin 6x}{1-q^6} + \ldots\right\};$$

whence, from the definition of Zu, *ante*, No. 131,

$$\frac{2K}{\pi} Z\left(\frac{2Kx}{\pi}\right) = 4\left\{\frac{q \sin 2x}{1-q^2} + \frac{q^2 \sin 4x}{1-q^4} + \frac{q^3 \sin 6x}{1-q^6} + \dots\right\};$$

and if we again multiply by dx and integrate from $x = 0$,

$$\frac{2K}{\pi}\int Z\left(\frac{2Kx}{\pi}\right) dx = \log \frac{(1-2q\cos 2x+q^2)(1-2q^3\cos 2x+q^6)\dots}{\{(1-q)(1-q^3)\dots\}^2};$$

that is,

$$\Theta\left(\frac{2Kx}{\pi}\right) = \frac{\sqrt{\dfrac{2k'K}{\pi}}}{\{(1-q)(1-q^3)\dots\}^2}$$
$$(1-2q\cos 2x+q^2)(1-2q^3\cos 2x+q^6)\dots;$$

or say

$$\Theta\left(\frac{2Kx}{\pi}\right) = \frac{\sqrt{\dfrac{2k'K}{\pi}}}{\prod_{m=1}^{\infty}[1-q^{2m-1}]^2}\prod_{m=1}^{\infty}[1-2q^{2m-1}\cos 2x+q^{4m-2}];$$

viz. we have obtained this value of $\Theta\dfrac{2Kx}{\pi}$ from the definition

$$\Theta u = \sqrt{\frac{2k'K}{\pi}}\, e^{-\frac{1}{2}\frac{E}{K}u^2+\int_0 du\int_0 du\,\mathrm{dn}^2 u}.$$

381. We have to prove the theorem for the squaring of the right-hand side of the equation

$$\frac{2kK}{\pi}\,\mathrm{sn}\,\frac{2Kx}{\pi} = \frac{4\sqrt{q}\sin x}{1-q} + \frac{4\sqrt{q^3}\sin 3x}{1-q^3} + \frac{4\sqrt{q^5}\sin 5x}{1-q^5} + \dots$$

Forming the square and reducing by the substitution

$$2\sin mx \sin nx = \cos(m-n)x - \cos(m+n)x,$$

the square is

$$= A + A'\cos 2x + A''\cos 4x + A'''\cos 6x \dots,$$

where

$$A = \frac{8q}{(1-q)^2} + \frac{8q^3}{(1-q^3)^2} + \dots;$$

and moreover

$$A^{(m)} = 8\,[2B^{(m)} - C^{(m)}],$$

where

$$B^{(n)} = \frac{q^{n+1}}{(1-q)(1-q^{2n+1})} + \frac{q^{n+3}}{1-q^3 \,.\, 1-q^{2n+3}} + \frac{q^{n+5}}{1-q^5 \,.\, 1-q^{2n+5}} + \dots$$

$$= \frac{q^n}{1-q^{2n}} \left\{ \begin{array}{l} \dfrac{q}{1-q} + \dfrac{q^3}{1-q^3} + \dfrac{q^5}{1-q^5} + \dots \\ -\dfrac{q^{2n+1}}{1-q^{2n+1}} - \dfrac{q^{2n+3}}{1-q^{2n+3}} - \dfrac{q^{2n+5}}{1-q^{2n+5}} + \dots \end{array} \right\}$$

$$= \frac{q^n}{1-q^{2n}} \left\{ \frac{q}{1-q} + \frac{q^3}{1-q^3} \dots + \frac{q^{2n-1}}{1-q^{2n-1}} \right\},$$

and

$$C^{(n)} = \frac{q^n}{1-q \,.\, 1-q^{2n-1}} + \frac{q^n}{1-q^3 \,.\, 1-q^{2n-3}} + \frac{q^n}{1-q^5 \,.\, 1-q^{2n-5}}$$

$$\dots + \frac{q^n}{1-q^{2n-1} \,.\, 1-q}$$

$$= \frac{q^n}{1-q^{2n}} \left\{ \begin{array}{llll} \dfrac{q}{1-q} & + \dfrac{q^3}{1-q^3} & \dots + \dfrac{q^{2n-1}}{1-q^{2n-1}} \\ + \dfrac{q^{2n-1}}{1-q^{2n-1}} & + \dfrac{q^{2n-3}}{1-q^{2n-3}} & \dots + \dfrac{q}{1-q} \\ +1 & +1 & \dots +1 \end{array} \right\}$$

$$= \frac{nq^n}{1-q^{2n}} + \frac{2q^n}{1-q^{2n}} \left\{ \frac{q}{1-q} + \frac{q^3}{1-q^3} \dots + \frac{q^{2n-1}}{1-q^{2n-1}} \right\};$$

whence $$A^{(n)} = \frac{-8nq^n}{1-q^{2n}};$$

viz. each coefficient (except A, which is an infinite series) has this finite expression, and we have

$$\left(\frac{2kK}{\pi}\right)^2 \operatorname{sn}^2 \frac{2Kx}{\pi} = A - 4\left\{ \frac{2q \cos 2x}{1-q^2} + \frac{4q^2 \cos 4x}{1-q^4} + \frac{6q^3 \cos 6x}{1-q^6} + \dots \right\}.$$

382. To find the value of

$$A_1 = 8\left\{ \frac{q}{(1-q)^2} + \frac{q^3}{(1-q^3)^2} + \&\text{c.} \right\}.$$

multiply by dx and integrate from 0 to $\frac{1}{2}\pi$, we have

$$\left(\frac{2kK}{\pi}\right)^2 \int_0^{\frac{1}{2}\pi} \mathrm{sn}^2 \frac{2Kx}{\pi}\, dx = \tfrac{1}{2}\pi A\,;$$

or what is the same thing,

$$A = \frac{4K}{\pi^2} k^2 \int_0^K \mathrm{sn}^2 u\, du\,;$$

viz. from the equation

$$ZK = 0 = K\left(1 - \frac{E}{K}\right) - k^2 \int_0^K \mathrm{sn}^2 u\, du,$$

we have
$$A = \frac{4K}{\pi^2}(K - E),$$

and the proof is thus completed.

383. Write for shortness

$$\frac{\sqrt{\frac{2k'K}{\pi}}}{\Pi[1-q^{2m-1}]^2} = G,$$

where, *ante*, No. 357,

$$\Pi[1-q^{2m-1}]^6 = \frac{2\sqrt[4]{q}\,k'}{\sqrt{k}}\,;$$

then the relation obtained is

$$\Theta\left(\frac{2Kx}{\pi}\right) = G\,.\,\Pi[1 - 2q^{2m-1}\cos 2x + q^{4m-2}],$$

which is the required expression for Θ, leading as mentioned above to the corresponding expressions for the other functions.

In fact, comparing the forms

$\operatorname{sn}\frac{2Kx}{\pi} = \frac{1}{\sqrt{k}}\, 2\sqrt[4]{q}\sin x\,[1-2q^{2m}\cos 2x+q^{4m}],\quad (+)$	$\operatorname{sn}\frac{2Kx}{\pi} = \frac{1}{\sqrt{k}}\, H\left(\frac{2Kx}{\pi}\right),\quad (+)$
$\operatorname{cn}\frac{2Kx}{\pi} = \sqrt{\frac{k'}{k}}\, 2\sqrt[4]{q}\cos x\,[1+2q^{2m}\cos 2x+q^{4m}],\quad (+)$	$\operatorname{cn}\frac{2Kx}{\pi} = \sqrt{\frac{k'}{k}}\, H\frac{2K}{\pi}\left(x+\frac{\pi}{2}\right),\quad (\div)$
$\operatorname{dn}\frac{2Kx}{\pi} = \sqrt{k'}\quad [1+2q^{2m-1}\cos 2x+q^{4m-2}],\quad (+)$	$\operatorname{dn}\frac{2Kx}{\pi} = \sqrt{k'}\;\Theta\frac{2K}{\pi}\left(x+\frac{\pi}{2}\right),\quad (\div)$
$\text{denom.} = \quad [1-2q^{2m-1}\cos 2x+q^{4m-2}],\quad (+)$	$\text{denom.} = \Theta\left(\frac{2Kx}{\pi}\right),$

we obtain

$$H\left(\frac{2Kx}{\pi}\right) = G\,.\,2\sqrt[4]{q}\sin x\,[1-2q^{2m}\cos 2x+q^{4m}],$$

$$H\frac{2K}{\pi}\left(x+\frac{\pi}{2}\right) = G\,.\,2\sqrt[4]{q}\cos x\,[1+2q^{2m}\cos 2x+q^{4m}],$$

$$\Theta\left(\frac{2Kx}{\pi}\right) = G\,.\quad [1-2q^{2m-1}\cos 2x+q^{4m-2}],$$

$$\Theta\frac{2K}{\pi}\left(x+\frac{\pi}{2}\right) = G\,.\quad [1+2q^{2m-1}\cos 2x+q^{4m-2}].$$

or omitting the second and fourth equations which are included in the first and third respectively,

$$H\left(\frac{2Kx}{\pi}\right) = G\,.\,2\sqrt[4]{q}\sin x\,[1-2q^{2n}\cos 2x+q^{4n}],$$

$$\Theta\left(\frac{2Kx}{\pi}\right) = G\,. \qquad\qquad [1-2q^{2n-1}\cos 2x+q^{4n-2}].$$

New developments of the functions H, Θ.
Art. Nos. 384 to 386.

384. We have identically

$$[1-2q^{2n-1}\cos 2x+q^{4n-2}]$$
$$=\frac{1}{[1-q^{2n}]}\{1-2q\cos 2x+2q^4\cos 4x-2q^9\cos 6x+\ldots\},$$

$$2\sqrt[4]{q}\sin x\,[1-2q^{2n}\cos 2x+q^{4n}]$$
$$=\frac{1}{[1-q^{2n}]}\{2\sqrt[4]{q}\sin x-2\sqrt[4]{q^9}\sin 3x+2\sqrt[4]{q^{25}}\sin 5x-\ldots\};$$

and hence observing that

$$\frac{G}{[1-q^{2n}]},\ =\frac{\sqrt{\dfrac{2k'K}{\pi}}}{[1-q^{2n-1}]^2.[1-q^{2n}]},\ =1,$$

we find

$$\Theta\left(\frac{2Kx}{\pi}\right)=1-2q\cos 2x+2q^4\cos 4x-2q^9\cos 6x+\ldots,$$

$$H\left(\frac{2Kx}{\pi}\right)=2\sqrt[4]{q}\sin x-2\sqrt[4]{q^9}\sin 3x+2\sqrt[4]{q^{25}}\sin 5x-\&c.,$$

which are the expressions of these two functions developed in cosines and sines of multiples of x.

385. For the direct proof of the identities we require the development of $[1+q^{2n-1}z]$, that is of $(1+qz)(1+q^3z)(1+q^5z)\ldots$ in powers of z. Assuming it

$$=1+Az+Bz^2+Cz^3+\ldots,$$

if for z we write q^2z, and multiply by $1+qz$, the result is

$= 1 + qz$ into $(1+q^3z)(1+q^5z)\ldots$; viz. this is the original function. We thus have

$$(1 + Az + Bz^2 + Cz^3 + \ldots)$$

$$= (1+qz)(1 + Aq^2z + Bq^4z^2 + Cq^6z^3 + \ldots);$$

that is, $A(1-q^2) = q, \quad B(1-q^4) = q^3A, \quad C(1-q^6) = q^5B, \ldots$

and thence

$$[1+q^{2m-1}z] = 1 + \frac{qz}{1-q^2} + \frac{q^4z^2}{1-q^2 \,.\, 1-q^4} + \frac{q^9z^3}{1-q^2 \,.\, 1-q^4 \,.\, 1-q^6} + \ldots$$

In a very similar manner it is shown that

$$\frac{1}{[1-q^mz]} = 1 + \frac{q}{1-q}\,\frac{z}{1-qz} + \frac{q^4}{1-q \,.\, 1-q^2}\,\frac{z^2}{1-qz \,.\, 1-q^2z}$$

$$+ \frac{q^9}{1-q \,.\, 1-q^2 \,.\, 1-q^3}\,\frac{z^3}{1-qz \,.\, 1-q^2z \,.\, 1-q^3z} + \ldots$$

386. Starting with the equation

$$[1+q^{2m-1}z] = 1 + \frac{qz}{1-q^2} + \frac{q^4z^2}{1-q^2 \,.\, 1-q^4} + \frac{q^9z^3}{1-q^2 \,.\, 1-q^4 \,.\, 1-q^6} + \ldots,$$

we have similarly

$$[1+q^{2m-1}z^{-1}] = 1 + \frac{qz^{-1}}{1-q^2} + \frac{q^4z^{-2}}{1-q^2 \,.\, 1-q^4} + \frac{q^9z^{-3}}{1-q^2 \,.\, 1-q^4 \,.\, 1-q^6} + \ldots,$$

and these two are to be multiplied together; the product will be an infinite series of the form

$$B_0 + B_1(z + z^{-1}) + B_2(z^2 + z^{-2}) + \ldots,$$

where $B_0, B_1, B_2 \ldots$ are functions of q given in the first instance as infinite series, which however admit of summation by means of the last formula in No. 385, viz.

$$\frac{1}{[1-q^mz]} = 1 + \frac{q}{1-q}\,\frac{z}{1-qz} + \frac{q^4}{1-q \,.\, 1-q^2}\,\frac{z^2}{1-qz \,.\, 1-q^2z}$$

$$+ \frac{q^9}{1-q \,.\, 1-q^2 \,.\, 1-q^3}\,\frac{z^3}{1-qz \,.\, 1-q^2z \,.\, 1-q^3z};$$

this, putting therein q^2 instead of q and $z = q^{2m}$, becomes

$$\frac{1}{[1-q^{2n+2m}]} = 1 + \frac{q^2}{1-q^2}\,\frac{q^{2n}}{1-q^{2n+2}} + \frac{q^8}{1-q^2.\,1-q^4}\,\frac{q^{4n}}{1-q^{2n+2}.\,1-q^{2n+4}}$$

$$+\frac{q^{18}}{1-q^2.\,1-q^4.\,1-q^6}\,\frac{q^{6n}}{1-q^{2n+2}.\,1-q^{2n+4}.\,1-q^{2n+6}}+\ldots,$$

where of course $[1-q^{2n+2m}]$ denotes $1-q^{2n}.\,1-q^{2n+2}.\,1-q^{2n+4}\ldots$

387. Effecting the multiplication of the first-mentioned two expressions, we have

$$[1-q^{2m-1}(z+z^{-1})+q^{4m-2}] = B_0 + B_1(z+z^{-1}) + B_2(z^2+z^{-2}) + \&c.,$$

where in general

$$B_n = \frac{q^{n^2}}{1-q^2.\,1-q^4\ldots1-q^{2n}}$$

$$\left\{1+\frac{q^2}{1-q^2}\,\frac{q^{2n}}{1-q^{2n+2}}+\frac{q^8}{1-q^2.\,1-q^4}\,\frac{q^{4n}}{1-q^{2n+2}.\,1-q^{2n+4}}+\ldots\right\},$$

that is by the last formula

$$=\frac{q^{n^2}}{1-q^2.\,1-q^4\ldots1-q^{2n}}\cdot\frac{1}{[1-q^{2n+2m}]},\quad =\frac{q^{n^2}}{[1-q^{2m}]};$$

and we have consequently

$$[1-q^{2m-1}(z+z^{-1})+q^{4m-2}]$$
$$=\frac{1}{[1-q^{2m}]}\{1+q(z+z^{-1})+q^4(z^2+z^{-2})+q^9(z^3+z^{-3})+\ldots\}.$$

388. This equation, writing therein $-e^{2ix}$ for z, becomes

$$[1-2q^{2m-1}\cos 2x+q^{4m-2}]$$
$$=\frac{1}{[1-q^{2m}]}\{1-2q\cos 2x+2q^4\cos 4x-2q^9\cos 6x+\ldots\};$$

and if in the same equation we write qz for z it becomes

$$(1+z^{-1})[1+q^{2m}z][1+q^{2m}z^{-1}]$$
$$=\frac{1}{[1-q^{2m}]}\frac{1}{\sqrt{z}}\{z^{\frac{1}{2}}+z^{-\frac{1}{2}}+q^2(z^{\frac{3}{2}}+z^{-\frac{3}{2}})+q^6(z^{\frac{5}{2}}+z^{-\frac{5}{2}})+\ldots\};$$

viz. putting here $-e^{2ix}$ for z, or say $z^{\frac{1}{2}}=\frac{1}{i}.e^{ix}$, we have

$$\sin x\,[1-2q^{2m}\cos 2x+q^{4m}]$$
$$=\frac{1}{[1-q^{2m}]}\{\sin x-q^2\sin 3x+q^6\sin 5x-q^{12}\sin 5x+\ldots\};$$

or what is the same thing,

$$2\sqrt[4]{q}\sin x\,[1-2q^{2m}\cos 2x+q^{4m}]$$

$$=\frac{1}{[1-q^{2m}]}\,[2\sqrt[4]{q}\sin x-2\sqrt[4]{q^9}\sin 3x+2\sqrt[4]{q^{25}}\sin 5x-\dots].$$

Double factorial expressions of H, Θ. General theory.
Art. Nos. 389 to 395.

389. Reverting to the expressions of $\Theta\left(\frac{2Kx}{\pi}\right)$, $H\left(\frac{2Kx}{\pi}\right)$ as q-products, and writing $x=\frac{\pi u}{2K}$, the q-products thus identified with the functions Hu and Θu respectively, presented themselves at the commencement of this Chapter as mere constant multiples of the expressions

$$\sin\frac{\pi u}{2K}\left[1-\frac{\sin^2\frac{\pi u}{2K}}{\sin^2\frac{si\pi K'}{K}}\right],\quad\left[1-\frac{\sin^2\frac{\pi u}{2K}}{\sin^2\frac{(2s-1)i\pi K'}{K}}\right],$$

$$(s=1\text{ to }\infty),$$

so that Hu, Θu, are constant multiples of these expressions respectively; and since $\Theta 0=\sqrt{\frac{2k'K}{\pi}}$, it follows that the complete values are

$$Hu=\sqrt{k}\sqrt{\frac{2k'K}{\pi}}\cdot\frac{2K}{\pi}\cdot\sin\frac{\pi u}{2K}\left[1-\frac{\sin^2\frac{\pi u}{2K}}{\sin^2\frac{si\pi K'}{K}}\right],$$

$$\Theta u=\sqrt{\frac{2k'K}{\pi}}\left[1-\frac{\sin^2\frac{\pi u}{2K}}{\sin^2\frac{(2s-1)i\pi K'}{2K}}\right].$$

It is important to examine the meaning of these formulæ. Consider the function which enters into the expression of Hu; or writing for greater convenience m' in place of s, say

$$\sin\frac{\pi u}{2K}\left[1-\frac{\sin^2\frac{\pi u}{2K}}{\sin^2\frac{m'i\pi K'}{K}}\right], \qquad (m'=1 \text{ to } \infty).$$

390. Observing that in general

$$\sin^2 u - \sin^2 a = \sin(u+a)\sin(u-a),$$

this (disregarding for the moment a constant factor) is

$$=\left[\sin\left(\frac{\pi u}{2K}+\frac{m'i\pi K'}{K}\right)\right],$$

$$=\left[\sin\frac{\pi}{2K}(u+2m'iK')\right],$$

where m' has every positive or negative integer value (zero included) from $-\infty$ to $+\infty$; say from $-\mu'$ to $+\mu'$, $\mu'=\infty$.

Now we have

$$\sin x = x\left[1-\frac{x^2}{s^2\pi^2}\right], \quad (s=1 \text{ to } \infty),$$

which writing $\frac{\pi u}{2K}$ for x, becomes

$$\sin\frac{\pi u}{2K}=\frac{\pi}{2K}u\left[1-\frac{u^2}{4s^2K^2}\right],$$

or disregarding a constant factor,

$$\sin\frac{\pi u}{2K}=[u+2mK],$$

where m has every positive or negative integer value, zero included, from $-\infty$ to $+\infty$; say from $-\mu$ to μ, $\mu=\infty$.

Assuming for a moment that it is allowable to write herein $u+2m'iK$ in place of u, we have

$$\sin\frac{\pi}{2K}(u+2m'iK')=[u+2mK+2m'iK'],$$

m as above, and consequently the numerator is

$$=[u+2mK+2m'iK'],$$

m, m' each extending from $-\infty$ to $+\infty$ as above. As regards the omitted constant factor, it is clear that $\operatorname{sn} u \div u$ reduces itself to unity for u indefinitely small, and the formula thus becomes

$$\frac{2K}{\pi}\sin\frac{\pi u}{2K}\left[1-\frac{\sin^2\frac{\pi u}{2K}}{\sin^2\frac{s i\pi K'}{K}}\right]=u\left\{1+\frac{u}{2mK+2m'iK'}\right\},$$

m, m' as before, excepting that the set of values $m=0$, $m'=0$ (having been taken account of in the factor u) is to be omitted.

391. But when in the sine-formula we write $u+2m'iK'$ in place of u, we assume that $u+2m'iK'$ is indefinitely small in regard to the extreme values $\pm\mu$ of m (viz. the infinite product $x\left(1-\frac{x^2}{\pi^2}\right)\left(1-\frac{x^2}{4\pi^2}\right)\ldots$ taken to the term $1-\frac{x^2}{s^2\pi^2}$, approximates to $\sin x$ only on the assumption that $\frac{x}{s\pi}$ is indefinitely small): of course this is so when m' is finite, but m' acquires the values $\pm\mu'$; in order to sustain the assumption we must suppose that μ' is indefinitely small as regards μ; or say that $\mu'\div\mu=0$.

Hence in the last-mentioned equation the limits of the doubly-infinite product are $m=-\mu$ to $m=+\mu$; $m'=-\mu'$ to $m'=+\mu'$; μ, μ' each infinite; but $\mu'\div\mu=0$. Putting for shortness $2mK+2m'iK'=(m, m')$ the equation is

$$\frac{2K}{\pi}\sin\frac{\pi u}{2K}\left[1-\frac{\sin^2\frac{\pi u}{2K}}{\sin^2\frac{s i\pi K'}{K}}\right]=u\left\{1+\frac{u}{(m, m')}\right\},$$

which is one of a group of four formulæ.

392. Writing for shortness as in Nos. 39 and 117,

$$\begin{aligned}
(m, m') &= 2mK + 2m'iK',\\
(\overline{m}, m') &= (2m+1)K + 2m'iK',\\
(m, \overline{m}') &= 2mK + (2m'+1)iK',\\
(\overline{m}, \overline{m}') &= (2m+1)K + (2m'+1)iK',
\end{aligned}$$

these are

$$\frac{2K}{\pi}\sin\frac{\pi u}{2K}\left[1-\frac{\sin^2\frac{\pi u}{2K}}{\sin^2\frac{s i\pi K'}{K}}\right] = u\left\{1+\frac{u}{(m, m')}\right\},$$

$$\left[1-\frac{\sin^2\frac{\pi u}{2K}}{\cos^2\frac{s i\pi K'}{K}}\right] = \left\{1+\frac{u}{(\overline{m}, m')}\right\},$$

$$\left[1-\frac{\sin^2\frac{\pi u}{2K}}{\cos^2\frac{(2s-1)i\pi K'}{K}}\right] = \left\{1+\frac{u}{(\overline{m}, \overline{m}')}\right\},$$

$$\left[1-\frac{\sin^2\frac{\pi u}{2K}}{\sin^2\frac{(2s-1)i\pi K'}{K}}\right] = \left\{1+\frac{u}{(m, \overline{m}')}\right\},$$

where on the right-hand side the limits are to be taken so that (m, m'), &c. may have equal positive and negative values, or say, as regards

m, from $m = -\mu$ to $+\mu$,
$\overline{m}$ " " $= -\mu - 1$ " $+\mu$,
m' " $m' = -\mu'$ " $+\mu'$,
$\overline{m}'$ " " $= -\mu' - 1$ " $+\mu'$;

viz. m, m' have all positive and negative integer values between these limits (both inclusive) respectively: but as regards (m, m') the combination $(0, 0)$, (which is separately taken account of in the exterior factor u), is to be omitted: μ, μ' are each infinite, but $\mu' + \mu = 0$.

393. The values of the Jacobian functions H, Θ thus are, as mentioned No. 39,

$$Hu = \sqrt{k}\sqrt{\frac{2k'K}{\pi}}\,u\left\{1+\frac{u}{(m, m')}\right\},$$

$$\Theta u = \sqrt{\frac{2k'K}{\pi}}\left\{1+\frac{u}{(m, \overline{m}')}\right\}.$$

limits as just mentioned. It is on account of the unsymmetrical condition $\mu' + \mu = 0$ in regard to the limits that H, Θ have a perfect periodicity as regards $4K$, but only an imperfect periodicity as regards $4iK'$; viz. as regards this quantity the functions are only periodic to an exponential factor près. The resulting expressions of the elliptic functions are, as mentioned No. 123,

$$\operatorname{sn} u = u \left\{1 + \frac{u}{(m, m')}\right\}, \qquad (+)$$

$$\operatorname{cn} u = \left\{1 + \frac{u}{(\overline{m}, m')}\right\}, \qquad (+)$$

$$\operatorname{dn} u = \left\{1 + \frac{u}{(\overline{m}, \overline{m'})}\right\}, \qquad (+)$$

$$\text{denom.} = \left\{1 + \frac{u}{(m, \overline{m'})}\right\};$$

where as regards $4iK'$, although the numerators and denominator are not separately periodic, they acquire by the change equal factors, and thus the quotients are periodic as well in regard to $4iK'$ as to $4K$.

394. We may state the general theory thus: consider the doubly infinite product

$$\left\{1 + \frac{u}{a + m\Omega + m'\Omega'}\right\},$$

where m, m' have within infinite limits every positive or negative integer value whatever.

To avoid difficulties, it is assumed *first* that Ω, Ω' are incommensurable, (for if they had a greatest common measure Δ the function would be an infinite power of the single product $\left[1 + \frac{u}{a + t\Delta}\right]$) t a positive or negative integer; *secondly*, that the ratio $\Omega : \Omega'$ is imaginary, for if it were real there would be an infinity of factors for which $m\Omega + m'\Omega'$ is indefinitely near to any given real value whatever. The function $a + m\Omega + m'\Omega'$ can at most vanish for a single set of values of m, m'; viz. it will do this if $a = -\lambda\Omega - \lambda'\Omega'$, λ, λ' being positive or negative integers; and we must in this case replace the product by

$$u\left[1 + \frac{u}{a + m\Omega + m'\Omega'}\right],$$

and exclude from the product the combination of values $m = \lambda$, $m' = \lambda'$; but this makes no real difference in the theory, and we need only attend to the other case, that in which $a + m\Omega + m'\Omega'$ does not vanish for any integer values of m, m'. *Thirdly*, that the limits are such that to each given value of $a + m\Omega + m'\Omega'$ there corresponds an equal and opposite value; or what is the same thing, regarding m, m' as rectangular co-ordinates, then that the product is extended to all integer values of m, m' lying within a closed curve having a centre at the real point given by the equation $a + m\Omega + m'\Omega' = 0$, (viz. if $a = -\lambda\Omega - \lambda'\Omega'$, then the co-ordinates of the centre are $m = \lambda$, $m' = \lambda'$). Say this is the "bounding curve," we may regard the linear magnitude of this curve as proportional to a parameter C, in such wise that C being indefinitely large, each radius vector of the curve (measured from the centre) is indefinitely large. Upon the foregoing suppositions, regarding the bounding curve as given in its form (for instance, if it be a circle, or a square, or again a rectangle with its sides in a given ratio, &c.), then as C increases and ultimately becomes infinite, the product in question

$$\left[1 + \frac{u}{a + m\Omega + m'\Omega'}\right],$$

tends to and ultimately attains a certain definite value; but this value is dependent on the form of the bounding curve.

395. There is, however, a relation between the values of the product for different forms of the bounding curve; viz. this is

$$\Pi_1 = e^{-\frac{1}{2}(A_1 - A_2)u^2}\,\Pi_2,$$

where A_1, A_2 denote the values of the integral

$$\iint \frac{dm\,dm'}{(a + m\Omega + m'\Omega')^2}$$

taken for the two forms of the bounding curve respectively.

In particular let the bounding curve be the rectangle $m = \lambda \pm \mu$, $m' = \lambda' \pm \mu'$, and let the product, when $\mu' + \mu = 0$, be called Π_0, and when $\mu' + \mu = \infty$ be called Π_∞; then if Π be the product for any other given form of the bounding curve, we have

$$\Pi = e^{\frac{1}{2}B_0 u^2}\Pi_0, \quad = e^{\frac{1}{2}B_\infty u^2}\Pi_\infty,$$

where B_0, B_∞ are constants depending on the form of the bounding curve; and observing that Π_0 has the period 2Ω, or say $\Pi_0(u+2\Omega) = \Pi_0 u$, while Π_∞ has the period $2\Omega'$, or say $\Pi_\infty(u+2\Omega') = \Pi_\infty u$, we obtain

$$\Pi(u+2\Omega) = e^{\frac{1}{2}B_0(u+2\Omega)^2}\Pi_0 = e^{2B_0\Omega(u+\Omega)}\Pi u,$$

$$\Pi(u+2\Omega') = e^{\frac{1}{2}B_\infty(u+2\Omega')^2}\Pi_\infty = e^{2B_\infty\Omega'(u+\Omega')}\Pi u',$$

which shows that the function Πu is not perfectly, but to an exponential factor près, periodic in regard to the two quantities 2Ω and $2\Omega'$ respectively. See as to this theory my papers, *Camb. and Dub. Math. Jour.* t. IV. 1845, pp. 257—277, and *Liouville*, t. X. 1845, pp. 385—420.

Transformation of the function Π, Θ. (Only the first transformation is here considered.) Art. No. 396.

396. The equation

$$\Theta u = \sqrt{\frac{2kK}{\pi}}\,\Pi\left[1 - \frac{\sin^2\dfrac{\pi u}{2K}}{\sin^2\dfrac{(2m-1)i\pi K'}{2K}}\right], \quad (m = 1 \text{ to } \infty),$$

putting therein $\dfrac{u}{M}$, λ for u, k, and attending to the relations $\Lambda = \dfrac{K}{nM}$, $\Lambda' = \dfrac{K'}{M}$, becomes

$$\Theta\left(\frac{u}{M}, \lambda\right) = \sqrt{\frac{2\lambda\Lambda}{\pi}}\,\Pi\left[1 - \frac{\sin^2\dfrac{n\pi u}{2K}}{\sin^2\dfrac{(2m-1)ni\pi K'}{2K}}\right], \quad (m = 1 \text{ to } \infty),$$

and we may hence deduce an equation of the form

$$\Theta\left(\frac{u}{M}, \lambda\right) = A\,\Pi[\Theta(u + 2s'K)]. \qquad (s' = -\tfrac{1}{2}(n-1) \text{ to } +\tfrac{1}{2}(n-1).)$$

In fact, disregarding constant factors we have

$$\Theta\left(\frac{u}{M}, \lambda\right) = \left[\sin\frac{n\pi}{2K}(u + 2m - 1\, iK')\sin\frac{n\pi}{2K}(u - \overline{2m-1}\, iK')\right],$$

or, what is the same thing,

$$\Theta\left(\frac{u}{M}, \lambda\right) = \left[\sin\frac{n\pi}{2K}(u + \overline{2m-1}\, iK')\right],$$

if m has now all positive or negative integer values from $-\infty$ to ∞: and similarly

$$\Theta u = \left[\sin\frac{\pi}{2K}(u + \overline{2m-1}\, iK')\right];$$

and if in this last equation, we write for u, $u + 2s'K$, giving to s' all the values from $-\frac{1}{2}(n-1)$ to $+\frac{1}{2}(n-1)$, and multiply the resulting expressions; then by aid of a known trigonometrical formula, we see that $\Theta\left(\frac{u}{M}, \lambda\right)$ has a value of the form in question. Writing $u = 0$, we obtain at once the value of A, and the equation becomes

$$\Theta\left(\frac{u}{M}, \lambda\right) = \left[\Theta\left(u + \frac{2s'K}{n}\right)\right]\Theta(0, \lambda) \div \Theta 0\left[\Theta^2\left(\frac{2sK}{n}\right)\right];$$

and similarly

$$H\left(\frac{u}{M}, \lambda\right) = \left[H\left(u + \frac{2s'K}{n}\right)\right] \cdot \frac{\sqrt{\lambda}}{M\sqrt{k}}\Theta(0, \lambda) \div \Theta 0\left[-H^2\left(\frac{2sK}{n}\right)\right],$$

which are formulæ for the transformation of the functions H, Θ.

306*. The theory of the transformation of the functions Hu, Θu might be derived from the double factorial expressions given in No. 303: it is, however, somewhat difficult to carry out the process, and I propose only to give a general idea of it. Disregarding a constant factor we have

$$\Theta\left(\frac{u}{M}, \lambda\right) = \left\{1 + \frac{u}{2mM\Lambda + (2m' + 1)\, iM\Lambda'}\right\};$$

which, substituting for $M\Lambda$ and $M\Lambda'$ their values $\dfrac{K}{n}$ and K', and writing for convenience l in place of m, becomes

$$\Theta\left(\frac{u}{M},\lambda\right)=\left\{1+\frac{u}{\frac{2l}{n}K+(2m'+1)iK'}\right\},$$

where on the right-hand side l, m' have every integer value whatever from $-\infty$ to $+\infty$: grouping these according to the remainder of the division l by n, we separate the right-hand side into n factors, each of which is a Θ-function with the original periods K, K'; thus writing $l=mn+s'$, where s' has any one of the values $0, 1, \ldots n-1$, and m has any integer value whatever from $-\infty$ to $+\infty$, the factor corresponding to a given value of s' is

$$=\left\{1+\frac{u+\frac{2s'K}{n}}{2mK+(2m'+1)iK'}\right\}$$

$$\div\left\{1+\frac{\frac{2s'K}{n}}{2mK+(2m'+1)iK'}\right\};$$

viz. disregarding the constant divisor, the factor is

$$=\Theta\left(u+\frac{2s'K}{n}\right),$$

and we thus have $\Theta\left(\dfrac{u}{M},\lambda\right)$ expressed as a constant multiple of the product of the n factors $\Theta\left(u+\dfrac{2s'K}{n}\right)$: and in like manner $H\left(\dfrac{u}{M},\lambda\right)$ is a constant multiple of the product of the n factors $H\left(u+\dfrac{2s'K}{n}\right)$; s' having in each case the values $0, 1, 2, \ldots n-1$, as above.

Numerator and Denominator functions.

Art. Nos. 397 and 398.

397. The results obtained No. 396 may be written

$$\Theta\left(\frac{u}{M}, \lambda\right) = \Theta u \left[\Theta\left(u + \frac{2sK}{n}\right)\Theta\left(u - \frac{2sK}{n}\right)\right]$$

into constant factor as above,

$$H\left(\frac{u}{M}, \lambda\right) = Hu \left[H\left(u + \frac{2sK}{n}\right)H\left(u - \frac{2sK}{n}\right)\right]$$

into constant factor as above.

We then have

$$\Theta\left(u + \frac{2sK}{n}\right)\Theta\left(u - \frac{2sK}{n}\right) \div \Theta^2 \frac{2sK}{n}$$

$$= \frac{\Theta^2 u}{\Theta^2 0}\left(1 - \frac{H^2 u}{\Theta^2 u}\,\frac{H^2 \frac{2sK}{n}}{\Theta^2 \frac{2sK}{n}}\right)$$

$$= \frac{\Theta^2 u}{\Theta^2 0}\left(1 - k^2 \operatorname{sn}^2 u \operatorname{sn}^2 \frac{2sK}{n}\right),$$

and

$$H\left(u + \frac{2sK}{n}\right)H\left(u - \frac{2sK}{n}\right) \div \left(-H^2 \frac{2sK}{n}\right)$$

$$= \frac{\Theta^2 u}{\Theta^2 0}\left(1 - \frac{H^2 u}{\Theta^2 u}\cdot\frac{\Theta^2 \frac{2sK}{n}}{H^2 \frac{2sK}{n}}\right)$$

$$= \frac{\Theta^2 u}{\Theta^2 0}\left(1 - \frac{\operatorname{sn}^2 u}{\operatorname{sn}^2 \frac{2sK}{n}}\right);$$

and we thence obtain

$$\Theta^{n-1}0\,\Theta\left(\frac{u}{M}, \lambda\right) \div \Theta^n u = \frac{\Theta(0, \lambda)}{\Theta 0}\left[1 - k^2 \operatorname{sn}^2 u \operatorname{sn}^2 \frac{2sK}{n}\right],$$

$$\Theta^{n-1}0\,H\left(\frac{u}{M}, \lambda\right) \div \Theta^n u = \frac{\sqrt{\lambda}}{M}\,\frac{\Theta(0, \lambda)}{\Theta 0} \operatorname{sn} u \left[1 - \frac{\operatorname{sn}^2 u}{\operatorname{sn}^2 \frac{2sK}{n}}\right].$$

398. Now in the function, see No. 355,

$$\sqrt{\lambda}\,\mathrm{sn}\left(\frac{u}{M},\lambda\right), =\frac{\sqrt{\lambda}\,\mathrm{sn}\,u}{M}\left[1-\frac{\mathrm{sn}^2 u}{\mathrm{sn}^2\frac{2sK}{n}}\right]+\left[1-k^2\,\mathrm{sn}^2 u\,\mathrm{sn}^2\frac{2sK}{n}\right],$$

multiplying the original numerator and denominator each by $\frac{\Theta^n(0,\lambda)}{\Theta^n 0}$, so that the denominator shall for $u=0$ reduce itself to $\frac{\Theta^n(0,\lambda)}{\Theta^n 0}$, $=\left\{\frac{\lambda'\Lambda}{k'K}\right\}^{\frac{1}{2}n}$, the numerator and denominator are $\Theta\left(\frac{u}{M},\lambda\right)$, $H\left(\frac{u}{M},\lambda\right)$, each multiplied by $\Theta^{n-1}(0,\lambda)$ and divided by $\Theta^n u$; and in like manner the numerators of $\sqrt{\lambda}\,\mathrm{sn}\left(\frac{u}{M},\lambda\right)$, $\sqrt{\frac{\lambda}{\lambda'}}\,\mathrm{cn}\left(\frac{u}{M},\lambda\right)$, $\frac{1}{\sqrt{\lambda'}}\,\mathrm{dn}\left(\frac{u}{M},\lambda\right)$, and the common denominator (constant factor as just mentioned) are

$$H\left(\frac{u}{M},\lambda\right),\ H\left(\frac{u}{M}+\Lambda,\lambda\right),\ \Theta\left(\frac{u}{M}+\Lambda,\lambda\right),\ \Theta\left(\frac{u}{M},\lambda\right),$$

each multiplied by $\Theta^{n-1}(0,\lambda)$ and divided by $\Theta^n u$; viz. writing $\Theta_1 0$ in place of $\Theta(0,\lambda)$ we have thus the theorem stated Chap. IX. No. 306.

CHAPTER XII.

REDUCTION OF A DIFFERENTIAL EXPRESSION $\frac{Rdx}{\sqrt{X}}$.

399. In the present Chapter, working out the steps of the processes referred to, Art. Nos. 1 to 11, we show how the differential expression $\frac{Rdx}{\sqrt{X}}$ is by the substitutions $\frac{p+qx}{1+x}$ for x, and $\frac{a+bx^2}{c+dx^2}$ for x^2, reduced successively to the forms

$$\frac{Rdx}{\sqrt{\pm(1\pm mx^2)(1\pm nx^2)}}, \quad \frac{Rdx}{\sqrt{1-x^2 . 1-k^2x^2}},$$

but for greater clearness we consider the substitutions under the forms $x=\frac{p+qy}{1+y}$, and $x^2=\frac{a+by^2}{c+dy^2}$.

Reduction to the form $\frac{Rdx}{\sqrt{\pm(1\pm mx^2)(1\pm nx^2)}}$.

Art. Nos. 400 to 407.

400. We start with an expression

$$\frac{Rdx}{\sqrt{X}},$$

where R is a rational function of x, X a quartic function with real coefficients, and which is therefore the product of two factors $\zeta+2\eta x+\theta x^2$, $\lambda+2\mu x+\nu x^2$, with real coefficients: the values of x are real, and such that X is positive or $\sqrt{X}$ real.

401. Writing $x = \dfrac{p+qy}{1+y}$, we have

$$dx = \frac{(q-p)\,dy}{(1+y)^2},$$

and the two factors of X become respectively

$$\frac{1}{(1+y)^2}\,[\zeta(1+y)^2 + 2\eta(1+y)(p+qy) + \theta(p+qy)^2],$$

$$\frac{1}{(1+y)^2}\,[\lambda(1+y)^2 + 2\mu(1+y)(p+qy) + \nu(p+qy)^2],$$

so that representing for a moment the functions in [ ] by M, N respectively, the differential expression becomes

$$= \frac{(q-p)\,R\,dy}{\sqrt{MN}},$$

where MN is a quartic function of y.

402. To make the odd powers of y disappear in the function MN, we write

$$\zeta + \eta(p+q) + \theta pq = 0,$$

$$\lambda + \mu(p+q) + \nu pq = 0,$$

for p, q being thus determined, then

$$M, = \zeta + 2\eta p + \theta p^2 + (\zeta + 2\eta q + \theta q^2)\,y^2,$$

$$N, = \lambda + 2\mu p + \nu p^2 + (\lambda + 2\mu q + \nu q^2)\,y^2,$$

will be functions of y^2 only.

The two equations give $p+q$ and pq rationally, but in order that the resulting values of p, q may be real, the values of $p+q$ and pq must be such that $(p+q)^2 - 4pq, = (p-q)^2$, is positive.

403. If the roots of the equation $X = 0$ are not all real, that is, if they are either all imaginary or else two real and two imaginary, we may take the equation $\lambda + 2\mu x + \nu x^2 = 0$ to have its two roots imaginary, and write therefore $\lambda\nu > \mu^2$. But this

being so, the second of the two equations in p, q written in the form

$$\frac{4\lambda}{\nu}+\frac{4\mu}{\nu}(p+q)+(p+q)^2-(p-q)^2=0,$$

gives

$$(p-q)^2=\left(p+q+\frac{2\mu}{\nu}\right)^2+\frac{4(\lambda\nu-\mu^2)}{\nu^2};$$

so that $p+q$ being real, $(p-q)^2$ is positive, or p and q are real.

404. If the roots of the equation $X=0$ are all real, let their values be $\alpha, \beta, \gamma, \delta$; then assuming

$$\zeta+2\eta x+\theta x^2=\theta(x-\alpha)(x-\beta), \text{ and } \lambda+2\mu x+\nu x^2=\nu(x-\gamma)(x-\delta),$$

the equations in p, q become

$$\alpha\beta-\tfrac{1}{2}(\alpha+\beta)(p+q)+pq=0,$$
$$\gamma\delta-\tfrac{1}{2}(\gamma+\delta)(p+q)+pq=0;$$

whence

$$p+q=\frac{2(\alpha\beta-\gamma\delta)}{\alpha+\beta-\gamma-\delta},$$

$$pq=\frac{\alpha\beta(\gamma+\delta)-\gamma\delta(\alpha+\beta)}{(\alpha+\beta-\gamma-\delta)},$$

and thence

$$\tfrac{1}{4}(p-q)^2=\frac{(\alpha-\gamma)(\alpha-\delta)(\beta-\gamma)(\beta-\delta)}{(\alpha+\beta-\gamma-\delta)^2},$$

which is positive if we take for α, β the greatest two roots or the least two roots, or the two extreme roots, or the two mean roots; viz. we thus have $(p-q)^2$ positive, and therefore p and q real.

405. The rational function R is the sum of an even function and an odd function of y: the differential expression is thus divided into two parts; that containing the odd function may be integrated by circular and logarithmic functions (as at once appears by making therein the substitution $\sqrt{y}$ in place of y), and there remains for consideration only the part depending

on the even function of y, or, what is the same thing, we may take R to be an even function of y, that is, a rational function of x^2.

406. [It may be remarked that in the case where the function X has four real factors, say that the value is $X=(x-\alpha)(x-\beta)(x-\gamma)(x-\delta)$, then writing $y^2=\dfrac{x-\alpha}{x-\beta}$, or, what is the same thing, $x=\dfrac{\alpha-\beta y^2}{1-y^2}$, we have

$$x-\alpha=(\alpha-\beta)\,y^2\,(\div),\qquad\qquad x-\beta=(\alpha-\beta)\;(+),$$

$$x-\gamma=\alpha-\gamma-(\beta-\gamma)\,y^2\,(+),\; x-\delta=\alpha-\delta-(\beta-\delta)\,y^2\,(+),$$

where denominator is $=1-y^2$; also $dx=2(\alpha-\beta)\,y\,dy+(1-y^2)^2$, and we thence find

$$\frac{dx}{\sqrt{x-\alpha\,.\,x-\beta\,.\,x-\gamma\,.\,x-\delta}}=\frac{2dy}{\sqrt{\alpha-\gamma-(\beta-\gamma)\,y^2\,.\,\alpha-\delta-(\beta-\delta)\,y^2}},$$

where the radical on the right-hand is in the required form; but in the case where the expression is $\dfrac{R\,dx}{\sqrt{X}}$, we have thus in place of R a function of y^2, so that no part of the integral is directly reducible to circular or logarithmic functions, and the form of the result would appear to be more complicated than if we had begun by the linear substitution upon x.]

407. Restoring x in place of y, the conclusion is that the original differential expression may be replaced by one of the form

$$\frac{R\,dx}{\sqrt{MN}},$$

where R is a rational function of x^2, and M and N are each of them a real function of the form $A+Bx^2$ *.

* The above is the investigation given in Legendre's Chap. II., and the result is as stated: but Legendre in his following Chap. III. only assumes that the radical is reduced to the form $\sqrt{\alpha+\beta x^2+\gamma x^4}$, and he considers (as his first case) that in which the equation $\alpha+\beta x^2+\gamma x^4=0$ gives imaginary values of x^2, that is where

The function MN may have the several forms

$$\pm(1 \pm mx^2)(1 \pm nx^2),$$

where m and n are positive; but we may assume that the signs are not such as to make the function $= -(1 + mx^2)(1 + nx^2)$; in fact X assumed to be positive for at least some real value of the original x, cannot be by a real substitution transformed into an essentially negative function.

Reduction to the standard form $\dfrac{R\,dx}{\sqrt{1-x^2 \,.\, 1-k^2x^2}}$.

Art. Nos. 408 to 412.

408. Retaining for convenience MN to signify

$$\pm(1 \pm mx^2)(1 \pm nx^2),$$

we have to show that $\dfrac{R\,dx}{\sqrt{MN}}$ can by the substitution $x^2 = \dfrac{a + by^2}{c + dy^2}$ be transformed into $\dfrac{S\,dy}{\sqrt{1-y^2 \,.\, 1-k^2y^2}}$, where S is a rational function of y^2, and k^2 is positive and < 1: and since by the substitution in question R is changed into a rational function of y^2, the theorem will it is clear hold good if only we have

$$\frac{\lambda\,dx}{\sqrt{MN}} = \frac{dy}{\sqrt{1-y^2 \,.\, 1-k^2y^2}},$$

where λ is a constant. On account of the definite form of the expression on the right-hand side, it is rather more

$\alpha + \beta x^2 + \gamma x^4$ is of the form $\lambda^2 + 2\lambda\mu x^2 \cos\theta + \mu^2 x^4$, a case which he further considers in Chap. XI. The idea seems to be, that since in the case in question there are no odd powers of x, the transformation $x = \dfrac{p + qy}{1 + y}$ of Chap. II. is unnecessary; if, however, we do make this substitution, we obtain under the radical sign a new quartic function without odd powers: the substitution is found to be $x = \dfrac{1-y}{1+y}\sqrt{\dfrac{\lambda}{\mu}}$ (mentioned in Chap. XI.), and the radical is thereby reduced to the form $m^2(1 + p^2y^2)(1 + q^2y^2)$, which is the fourth case of Chap. III.

convenient, writing the relation between x^2, y^2 in the form $y^2 = \dfrac{-a+cx^2}{b-dx^2}$, to transform this into the form $\dfrac{dx}{\sqrt{MN}}$ on the left-hand side. The last-mentioned equation gives

$$y\,dy = \frac{(bc-ad)\,x\,dx}{(b-dx^2)^2},$$

$$\frac{1}{y} = \frac{\sqrt{b-dx^2}}{\sqrt{-a+cx^2}},$$

$$\frac{1}{\sqrt{1-y^2}} = \frac{\sqrt{b-dx^2}}{\sqrt{b+a-(d+c)\,x^2}},$$

$$\frac{1}{\sqrt{1-k^2y^2}} = \frac{\sqrt{b-dx^2}}{\sqrt{b+k^2a-(d+k^2c)\,x^2}};$$

and we thence have $\dfrac{dy}{\sqrt{1-y^2\,.\,1-k^2y^2}}$

$$= \frac{(bc-ad)\,x\,dx}{\sqrt{b-dx^2\,.\,-a+cx^2\,.\,b+a-(d+c)\,x^2\,.\,b+k^2a-(d+k^2c)\,x^2}}.$$

Here in the denominator one of the four factors must reduce itself to a constant, and another of them to a multiple of x^2, in order that the second side may be of the required form $\dfrac{\lambda\,dx}{\sqrt{MN}}$.

409. For instance, if $b+a=0$, $d+k^2c=0$, that is, $b=-a$, $k^2=-\dfrac{d}{c}$, then the relation between y^2, x^2 is $y^2 = \dfrac{a-cx^2}{a+dx^2}$, and the differential formula, after some easy reductions, becomes

$$\sqrt{\frac{a}{c}}\,\frac{dy}{\sqrt{1-y^2\,.\,1-k^2y^2}} = \frac{dx}{\sqrt{1+\dfrac{d}{a}x^2\,.\,1-\dfrac{c}{a}x^2}};$$

or writing for greater convenience $a=1$, then we have

$y^2 = \frac{1-cx^2}{1+dx^2}$, leading to the differential formula

$$\frac{1}{\sqrt{c}}\frac{dy}{\sqrt{1-y^2 . 1-k^2y^2}} = \frac{dx}{\sqrt{1-cx^2 . 1+dx^2}},$$

where $k^2 = -\frac{d}{c}$; this implies c positive, d negative and in absolute magnitude less than c; and we have thus a formula applicable to the case $MN = (1-mx^2)(1-nx^2)$; viz. assuming $m > n$, we may write $c = m$, $d = -n$, and the relation then is $y^2 = \frac{1-mx^2}{1-nx^2}$, or, what is the same thing, $x^2 = \frac{1-y^2}{m-ny^2}$, giving

$$\frac{1}{\sqrt{m}}\frac{dy}{\sqrt{1-y^2 . 1-k^2y^2}} = \frac{dx}{\sqrt{1-mx^2 . 1-nx^2}},$$

where $k^2 = \frac{n}{m}$. A more simple formula giving this same relation is $x^2 = \frac{y^2}{m}$.

410. We thus obtain transformations applicable to the several forms of MN, viz. numbering the cases as in Legendre's Chap. III., but for the reason appearing in the foot-note p. 312, omitting his first case, we have

2°. $MN = \quad (1+mx^2)(1-nx^2)$, $\qquad x < \frac{1}{n}$,

3°. $MN = -(1+mx^2)(1-nx^2)$, $\qquad x > \frac{1}{n}$,

4°. $MN = \quad (1+mx^2)(1+nx^2)$, $\quad m > n$,

5°. $MN = \quad (1-mx^2)(1-nx^2)$, $\quad m > n$, $\quad x$ from 0 to $\frac{1}{\sqrt{m}}$, or from $\frac{1}{\sqrt{n}}$ to ∞,

6°. $MN = -(1-mx^2)(1-nx^2)$, $\quad m > n$, $\quad x$ from $\frac{1}{\sqrt{m}}$ to $\frac{1}{\sqrt{n}}$;

and writing for shortness $Y = 1 - y^2 . 1 - k^2y^2$, the formulæ are

2°. $k^2 = \dfrac{m}{m+n}$, $x^2 = \dfrac{1}{n}(1-y^2)$, $\dfrac{dx}{\sqrt{MN}} = -\dfrac{k}{\sqrt{m}}\dfrac{dy}{\sqrt{Y}}$,

or else $x^2 = \dfrac{y^2}{m+n-my^2}$, $\dfrac{dx}{\sqrt{MN}} = \dfrac{k}{\sqrt{m}}\dfrac{dy}{\sqrt{Y}}$,

3°. $k^2 = \dfrac{n}{m+n}$, $x^2 = \dfrac{1}{n(1-y^2)}$, $\dfrac{dx}{\sqrt{MN}} = \dfrac{k}{\sqrt{n}}\dfrac{dy}{\sqrt{Y}}$,

4°. $k^2 = \dfrac{m-n}{n}$, $x^2 = \dfrac{1}{m}\dfrac{y^2}{1+y^2}$, $\dfrac{dx}{\sqrt{MN}} = \dfrac{1}{\sqrt{m}}\dfrac{dy}{\sqrt{Y}}$,

5°. $k^2 = \dfrac{n}{m}$, $x^2 = \dfrac{y^2}{m}$, $\dfrac{dx}{\sqrt{MN}} = \dfrac{1}{\sqrt{m}}\dfrac{dy}{\sqrt{Y}}$,

6°. $k^2 = \dfrac{m-n}{n}$, $x^2 = \dfrac{1}{m-(m-n)y^2}$, $\dfrac{dx}{\sqrt{MN}} = \dfrac{1}{\sqrt{m}}\dfrac{dy}{\sqrt{Y}}$.

411. It is to be added that if in the expression $\dfrac{dy}{\sqrt{Y}}$ we have $y > \dfrac{1}{k}$, in which case writing $Y = (y^2-1)(k^2y^2-1)$ the radical is still real, then assuming $y = \dfrac{1}{kz}$, we have $\dfrac{dy}{\sqrt{Y}} = -\dfrac{dz}{\sqrt{Z}}$, where $Z = (1-z^2)(1-k^2z^2)$, and as y passes from $\dfrac{1}{k}$ to ∞, z passes from 1 to 0. Hence, replacing y or z by the original letter x, the conclusion is that in every case the differential expression $\dfrac{dx}{\sqrt{\pm(1\pm mx^2)(1\pm nx^2)}}$ can by a real substitution $\dfrac{A+Bx^2}{C+Dx^2}$ in place of x^2 be reduced to the form

$$\frac{dx}{\sqrt{(1-x^2)(1-k^2x^2)}},$$

where the variable x extends between the limits 0 and 1.

Further investigations. Art. Nos. 412 to 417.

412. Reverting to the investigation, Art. Nos. 400—407, but abandoning the condition that the transformation shall be real, it is clear that we can by such a transformation reduce the differential expression to the form

$$\frac{Rdx}{\sqrt{1-x^2 . 1-k^2x^2}},$$

where, however, k^2 is not of necessity real, or if real and positive not of necessity less than 1: it is interesting to inquire further into this question, and to show how the modulus k of this in general abnormal form is determined. The process in fact was by a substitution $\frac{p+qx}{1+x}$ in place of x, or say by a linear transformation performed upon x, to transform the quartic function X into a quartic function Y containing only the even powers of the variable; the solution of this problem depends on a cubic equation which is solved rationally when we know any decomposition of the function X, Y into quadric factors; and it was in order to have such rational solution of the cubic equation, and with a view to obtain real transformations that we commenced by assuming the function X to be decomposed into factors of the form $(\zeta + 2\eta x + \theta x^2)(\lambda + 2\mu x + \nu x^2)$ or $\theta(x-\alpha)(x-\beta)(x-\gamma)(x-\delta)$. But analytically it is more elegant to deal with the undecomposed quartic function, as was done by me in a paper in the *Camb. and Dub. Math. Journal*, t. I. (1846), pp. 70—73, and I here reproduce the investigation.

413. Let the two quartic functions

$$P = (a, b, c, d, e)(x, y)^4,$$

$$P' = (a', b', c', d', e')(x', y')^4,$$

be linear transformations one of the other, say the second is derived from the first by the substitution

$$x, y = \lambda x' + \mu y', \lambda_1 x' + \mu_1 y'.$$

Then writing $m = \lambda\mu_1 - \lambda_1\mu$ for the determinant of substitution, we have

$$\frac{x\,dy - y\,dx}{\sqrt{P}} = \frac{m\,(x'dy' - y'dx')}{\sqrt{P'}},$$

or writing
$$u = \frac{y}{x}, \quad u' = \frac{y'}{x'},$$

and therefore

$$\frac{x\,dy - y\,dx}{\sqrt{P}} = \frac{du}{\sqrt{U}}, \quad \frac{x'dy' - y'dx'}{\sqrt{P'}} = \frac{du'}{\sqrt{U'}},$$

where
$$U = (a,\ b,\ c,\ d,\ e)\,(1,\ u)^4,$$
$$U' = (a',\ b',\ c',\ d',\ e')\,(1,\ u')^4,$$

the differential equation becomes

$$\frac{du}{\sqrt{U}} = m\,\frac{du'}{\sqrt{U'}},$$

viz. we have this from the transformation $u = \dfrac{\lambda + \mu u'}{\lambda_1 + \mu_1 u'}$.

The functions P, P' are obtained the one from the other by the foregoing linear substitution; viz. if I, J are the invariants of P, viz.

$$I = ae - 4bd + 3c^2,$$
$$J = ace - ad^2 - b^2e + 2bcd - c^3,$$

and by I', J' the corresponding invariants of P'; then we have between the coefficients of the functions and the coefficients of transformation the relations

$$I' = m^4 I, \quad J' = m^6 J, \text{ whence } \frac{J'^2}{I'^3} = \frac{J^2}{I^3}.$$

414. Supposing now

$$U' = a'\,(1 + pu'^2)\,(1 + qu'^2),$$

or
$$b' = 0, \quad d' = 0, \quad 6c' = a'\,(p+q), \quad e' = a'pq,$$

we have $$I' = \tfrac{1}{12}a'^2(p^2+q^2+14pq),$$

$$J' = \tfrac{1}{216}a'^3(p+q)(34pq-p^2-q^2);$$

and thence $$\frac{(p+q)^2(34pq-p^2-q^2)^2}{(p^2+q^2+14pq)^3} = \frac{27J^2}{I^3},$$

or, as this may also be written,

$$\frac{108pq(p-q)^4}{(p^2+q^2+14pq)^3} = 1 - \frac{27J^2}{I^3},$$

which determines the relation between p and q. Also

$$\frac{m}{\sqrt{a'}} = \left(\frac{p^2+q^2+14pq}{12I}\right)^{\frac{1}{4}},$$

so that the differential equation is

$$\frac{du}{\sqrt{U}} = \left(\frac{p^2+q^2+14pq}{12I}\right)^{\frac{1}{4}} \cdot \frac{du'}{\sqrt{1+pu'^2 \,.\, 1+qu'^2}}.$$

415. If in particular $p=-1$, then writing also $-q$ in place of q, this is

$$\frac{du}{\sqrt{U}} = \left(\frac{q^2+14q+1}{12I}\right)^{\frac{1}{4}} \cdot \frac{du'}{\sqrt{1-u'^2 \,.\, 1-qu'^2}},$$

where q is determined by the equation

$$\frac{108q(1-q)^4}{(q^2+14q+1)^3} = 1 - \frac{27J^2}{I^3}.$$

Writing for shortness

$$1 - \frac{27J^2}{I^3} = \frac{27}{4M},$$

the equation in q becomes

$$(q^2+14q+1)^3 - 16Mq(q-1)^4 = 0,$$

or, as this may be written,

$$\left(q+\frac{1}{q}+14\right)^3-16M\left(q^{\frac{1}{2}}-q^{-\frac{1}{2}}\right)^4=0;$$

viz. writing $q^{\frac{1}{2}}-q^{-\frac{1}{2}}=\dfrac{4}{\sqrt{\theta-1}}$, this is $\theta^3-M(\theta-1)=0$, which determines θ, and then

$$q=\frac{7+\theta+4\sqrt{\theta+3}}{\theta-1}.$$

416. Suppose $q=\alpha$ is one of the values of q, the equation becomes

$$\frac{(q^2+14q+1)^3}{q(q-1)^4}=\frac{(\alpha^2+14\alpha+1)^3}{\alpha(\alpha-1)^4}$$

$$=\frac{(\beta^{16}+14\beta^8+1)^3}{\beta^8(\beta^8-1)^4}\ \text{ if } \alpha=\beta^8.$$

Now if $q=\left(\dfrac{1-\beta}{1+\beta}\right)^4$, then

$$q^2+14q+1=\frac{16(\beta^8+14\beta^4+1)}{(1+\beta)^8},\quad q-1=\frac{-8\beta(1+\beta^2)}{(1+\beta)^4},$$

which satisfy the equation: hence also identically

$$(q^2+14q+1)^3-q(q-1)^4\,\frac{(\beta^{16}+14\beta^8+1)^3}{\beta^8(\beta^8-1)^4}$$

$$=(q-\beta^8)\left(q-\frac{1}{\beta^8}\right)\left\{q-\left(\frac{1-\beta}{1+\beta}\right)^4\right\}\left\{q-\left(\frac{1+\beta}{1-\beta}\right)^4\right\}$$

$$\times\left\{q-\left(\frac{1-\beta i}{1+\beta i}\right)^4\right\}\left\{q-\left(\frac{1+\beta i}{1-\beta i}\right)^4\right\};$$

or the values of q take the form

$$\beta^8,\ \frac{1}{\beta^8},\ \left(\frac{1-\beta}{1+\beta}\right)^4,\ \left(\frac{1+\beta}{1-\beta}\right)^4,\ \left(\frac{1-\beta i}{1+\beta i}\right)^4,\ \left(\frac{1+\beta i}{1-\beta i}\right)^4.$$

(Compare Abel, *Œuvres*, t. I, p. 310); viz. when by a linear substitution performed on the variable u, we reduce the expression $\frac{du}{\sqrt{U}}$ to the form

$$\frac{du'}{\sqrt{1-u'^2 \cdot 1-qu'^2}},$$

the squared modulus q of the resulting form is determined by a sextic equation depending upon $\frac{I^3}{J^2}$, the absolute invariant of the original quartic function U, and such that its several roots can be expressed in terms of any one of them in the manner just appearing.

417. Jacobi, in the *Fund. Nova*, pp. 6—17, treats the question of reduction in a somewhat different manner, giving it as an illustration of his general theory of transformation, explained *ante*, Chap. VII. He proposes to transform the expression

$$\frac{dy}{\sqrt{\pm(y-\alpha)(y-\beta)(y-\gamma)(y-\delta)}},$$

into the form
$$\frac{dx}{M\sqrt{1-x^2 \cdot 1-k^2x^2}},$$

by a substitution
$$y=\frac{a+a'x+a''x^2}{b+b'x+b''x^2}.$$

Writing for shortness U, V for the numerator and denominator of this fraction respectively, it follows from the general theory that there will be such a transformation if only

$$(U-\alpha V)(U-\beta V)(U-\gamma V)(U-\delta V)$$
$$=K(1-x^2)(1-k^2x^2)(1+mx)^2(1+nx)^2,$$

K, m, n being constants; and he is thence led to assume

$$U-\alpha V=A(1-x)(1-kx),$$
$$U-\beta V=B(1+x)(1+kx),$$
$$U-\gamma V=C(1+mx)^2,$$
$$U-\delta V=D(1+nx)^2,$$

where A, B, C, D are constants, one of which may be assumed at pleasure. Having found $C+D$ equal to a fraction, he assumes C and D equal to the numerator and denominator of this fraction respectively, viz.

$$C=\sqrt{\alpha-\gamma\,.\,\beta-\gamma},\quad D=\sqrt{\alpha-\delta\,.\,\beta-\delta};$$

and he then further finds

$$A=-\frac{\sqrt{\alpha-\gamma\,.\,\alpha-\delta}}{\gamma-\delta}\,[\sqrt{\alpha-\gamma\,.\,\beta-\delta}-\sqrt{\alpha-\delta\,.\,\beta-\gamma}],$$

$$B=\frac{\sqrt{\beta-\gamma\,.\,\beta-\delta}}{\gamma-\delta}\,[\sqrt{\alpha-\gamma\,.\,\beta-\delta}-\sqrt{\alpha-\delta\,.\,\beta-\gamma}],$$

$$\sqrt{k}=\frac{\sqrt[4]{\alpha-\gamma\,.\,\beta-\delta}-\sqrt[4]{\alpha-\delta\,.\,\beta-\gamma}}{\sqrt[4]{\alpha-\gamma\,.\,\beta-\delta}+\sqrt[4]{\alpha-\delta\,.\,\beta-\gamma}},$$

$$M=\tfrac{1}{4}\,[\sqrt[4]{\alpha-\gamma\,.\,\beta-\delta}-\sqrt[4]{\alpha-\delta\,.\,\beta-\gamma}]^2,$$

which completes the system of values.

Moreover, $m=-n=\sqrt{k}$, and consequently the expression of y in terms of x may be written

$$\frac{y-\gamma}{y-\delta}=\frac{\sqrt{\alpha-\gamma\,.\,\beta-\gamma}}{\sqrt{\alpha-\delta\,.\,\beta-\delta}}\left(\frac{1+x\sqrt{k}}{1-x\sqrt{k}}\right)^2.$$

The results, adapted to give real transformations, and for the different limits of the integrals, are given in the tables I., II., III., IV., pp. 12 to 17.

416. It is somewhat remarkable that Jacobi fails to remark, as coming under his form $y=\dfrac{a+a'x+a''x^2}{b+b'x+b''x^2}$, the before-mentioned transformation $y=\dfrac{\alpha-\beta x^2}{1-x^2}$, which in fact reduces the expression $\dfrac{dy}{\sqrt{y-\alpha\,.\,y-\beta\,.\,y-\gamma\,.\,y-\delta}}$ to the form

$$\frac{2dx}{\sqrt{\alpha-\gamma-(\beta-\gamma)x^2\,.\,\alpha-\delta-(\beta-\delta)x^2}},$$

which is $=\dfrac{dx}{M\sqrt{1-x^2\,.\,1-k^2x^2}}$; and this is the more singular, that the transformation at which he arrives may be separated into the two transformations $\frac{y-\gamma}{y-\delta}=z^2$, that is $y=\frac{\gamma-\delta z^2}{1-z^2}$, which is a transformation of the form in question, and the further linear transformation $z=m\left(\frac{1+x\sqrt{k}}{1-x\sqrt{k}}\right)$, which is unnecessary, in so far as the quartic function of z is already a function without odd powers, at once reducible to the standard form $1-z^2.1-k^2z^2$. At the conclusion of his investigation Jacobi remarks that the inverse substitution $x=\frac{a+a'y+a''y^2}{b+b'y+b''y^2}$ leads also to very elegant results. I have not investigated the formulæ.

It may be added that, applying the transformation

$$y=\frac{a+a'x+a''x^2}{b+b'x+b''x^2}$$

to a differential expression $\frac{dx}{\sqrt{1-x^2.\overline{1-k^2x^2}}}$ of the standard form, so as to obtain a new expression of the like form with a different modulus, there are in all eighteen such transformations, viz., six wherein the equation of transformation is of the form

$$y=\frac{a+a''x^2}{b+b''x^2},$$

four where it is of the form

$$y=\frac{a'x}{b+b''x^2},$$

and eight where it is of the form

$$y=m\frac{a+a'x+a''x^2}{a-a'x+a''x^2}.$$

See as to this Abel's letter to Legendre (1828), *Œuvres*, t. II. p. 256.

CHAPTER XIII.

QUADRIC TRANSFORMATION OF THE ELLIPTIC INTEGRALS OF THE FIRST AND SECOND KIND: THE ARITHMETICO-GEOMETRICAL MEAN.

419. WRITING for greater convenience ϕ_1 in place of θ, and k_1 in place of λ, it has already been shown, *ante* No. 243, that if $k_1 = \frac{1-k'}{1+k'}$ and $k_1 \sin\phi_1 = \sin(2\phi - \phi_1)$, then

$$\frac{d\phi}{\Delta(k, \phi)} = \tfrac{1}{2}(1+k_1)\frac{d\phi_1}{\Delta(k_1, \phi_1)},$$

or what is the same thing, $F(k, \phi) = \frac{1}{2}(1+k_1)\,F(k_1, \phi_1)$.

But there is as regards this transformation a peculiar convenience in adopting, instead of the standard form of radical $\sqrt{1-k^2\sin^2\phi}$, a new form $\sqrt{a^2\cos^2\phi + b^2\sin^2\phi}$ (where a is taken to be $>b$); and I write in the present Chapter

$$F(a, b, \phi) = \int \frac{d\phi}{\sqrt{a^2\cos^2\phi + b^2\sin^2\phi}},$$

$$E(a, b, \phi) = \int d\phi\sqrt{a^2\cos^2\phi + b^2\sin^2\phi},$$

where the integrals are taken from zero.

Obviously

$$\sqrt{a^2\cos^2\phi + b^2\sin^2\phi} = \sqrt{a^2 - (a^2 - b^2)\sin^2\phi}, \; = a\sqrt{1-k^2\sin^2\phi}$$

if $k^2 = 1 - \frac{b^2}{a^2}$ $\left(\text{whence also } k' = \frac{b}{a}\right)$; and the two functions are thus $= a^{-1}F(k, \phi)$ and $aE(k, \phi)$ respectively.

Geometrical Investigation of the formulæ of Transformation.

Art. Nos. 420 to 423.

420. I reproduce the original geometrical investigation of Landen's transformation in the new notation, as follows:

Taking in the figure P a point on the circle, O the centre, Q any other point on the diameter AB,

$$QA = a,\quad QB = b,\quad \angle AQP = \phi_1,\quad \angle ABP = \phi,$$

and therefore $$\angle AOP = 2\phi,$$

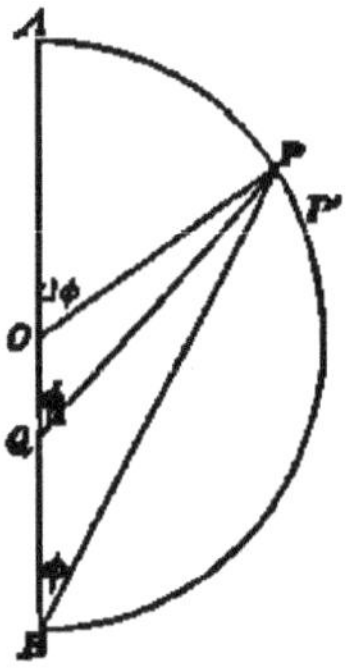

we write

$$a_1 = \tfrac{1}{2}(a+b),\ b_1 = \sqrt{ab},\ c_1 = \tfrac{1}{2}(a-b);$$

we have then

$$OA = OB = OP = a_1;\quad OQ = a_1 - b,\ = \tfrac{1}{2}(a-b),\ = c_1;$$

$$QP \sin\phi_1 = \qquad a_1 \sin 2\phi,$$

$$QP \cos\phi_1 = c_1 + a_1 \cos 2\phi;$$

and $$QP^2,\ = c_1^2 + 2c_1a_1 \cos 2\phi + a_1^2,$$

$$= \tfrac{1}{2}(a^2+b^2) + \tfrac{1}{2}(a^2-b^2)\cos 2\phi,$$

$$= \tfrac{1}{2}(a^2+b^2)(\cos^2\phi+\sin^2\phi) + \tfrac{1}{2}(a^2-b^2)(\cos^2\phi-\sin^2\phi),$$

$$= a^2\cos^2\phi + b^2\sin^2\phi;$$

that is,
$$\sin\phi_1 = \frac{a_1 \sin 2\phi}{\sqrt{a^2\cos^2\phi + b^2\sin^2\phi}};$$

$$\cos\phi_1 = \frac{c_1 + a_1\cos 2\phi}{\sqrt{a^2\cos^2\phi + b^2\sin^2\phi}};$$

and thence

$$a_1^2\cos^2\phi_1 + b_1^2\sin^2\phi_1 = \frac{a_1^2(a\cos^2\phi + b\sin^2\phi)^2}{a^2\cos^2\phi + b^2\sin^2\phi}.$$

421. We hence find

$$\sin(2\phi - \phi_1) = \frac{\tfrac{1}{2}(a-b)\sin 2\phi}{\sqrt{a^2\cos^2\phi + b^2\sin^2\phi}};$$

$$\cos(2\phi - \phi_1) = \frac{a\cos^2\phi + b\sin^2\phi}{\sqrt{a^2\cos^2\phi + b^2\sin^2\phi}};$$

and then further

$$\cos(2\phi - \phi_1) = \frac{1}{a_1}\sqrt{a_1^2\cos^2\phi + b_1^2\sin^2\phi}.$$

Considering the point P' consecutive to P, we have

$$PQ\,d\phi_1 = PP'\sin P'PQ, = 2a_1 d\phi\cos(2\phi-\phi_1);$$

viz. substituting for PQ its value, we have

$$2d\phi\sqrt{a_1^2\cos^2\phi + b_1^2\sin^2\phi} = \sqrt{a^2\cos^2\phi + b^2\sin^2\phi}\,d\phi_1;$$

that is,

$$\frac{2d\phi}{\sqrt{a^2\cos^2\phi + b^2\sin^2\phi}} = \frac{d\phi_1}{\sqrt{a_1^2\cos^2\phi_1 + b_1^2\sin^2\phi_1}},$$

the required differential relation: and by integration

$$F(a, b, \phi) = \tfrac{1}{2}F(a_1, b_1, \phi_1).$$

422. Moreover

$$\sin^2\phi_1 = \frac{4a_1^2\sin^2\phi\cos^2\phi}{a^2\cos^2\phi + b^2\sin^2\phi}.$$

Write for a moment

$$X = a^2\cos^2\phi + b^2\sin^2\phi,$$

then

$$X - a^2 = (b^2 - a^2)\sin^2\phi,$$

$$X - b^2 = (a^2 - b^2)\cos^2\phi,$$

$$(X - a^2)(X - b^2) = -4(a-b)^2 a_1^2\sin^2\phi\cos^2\phi;$$

and therefore

$(X - a^2)(X - b^2) + X(a-b)^2\sin^2\phi_1 = 0,$ that is

$$X^2 + X[-(a^2+b^2)(\sin^2\phi_1 + \cos^2\phi_1) + (a-b)^2\sin^2\phi_1] + a^2b^2 = 0,$$

or, what is the same thing,

$$\begin{aligned}
&(X - \{\tfrac{1}{2}(a^2+b^2)\cos^2\phi_1 + ab\sin^2\phi_1\})^2 \\
&= \tfrac{1}{4}(a^2+b^2)^2\cos^4\phi_1 + (a^2+b^2)\,ab\cos^2\phi_1\sin^2\phi_1 + a^2b^2\sin^4\phi_1 \\
&\qquad - a^2b^2\cos^4\phi_1 - 2a^2b^2\cos^2\phi_1\sin^2\phi_1 - a^2b^2\sin^4\phi_1 \\
&= \tfrac{1}{4}(a^2-b^2)^2\cos^4\phi_1 + ab(a-b)^2\cos^2\phi_1\sin^2\phi_1 \\
&= 4c_1^2\cos^2\phi_1(a_1^2\cos^2\phi_1 + b_1^2\sin^2\phi_1);
\end{aligned}$$

viz. restoring for X its value, we have

$$\begin{aligned}
&a^2\cos^2\phi + b^2\sin^2\phi \\
&= \tfrac{1}{2}(a^2+b^2)\cos^2\phi_1 + ab\sin^2\phi_1 + 2c_1\cos\phi_1\sqrt{a_1^2\cos^2\phi_1 + b_1^2\sin^2\phi_1} \\
&= 2(a_1^2\cos^2\phi_1 + b_1^2\sin^2\phi_1) - b_1^2 + 2c_1\cos\phi_1\sqrt{a_1^2\cos^2\phi_1 + b_1^2\sin^2\phi_1},
\end{aligned}$$

which is another form of the integral equation.

423. Write this in the form

$$(a^2\cos^2\phi + b^2\sin^2\phi) \div \sqrt{a_1^2\cos^2\phi_1 + b_1^2\sin^2\phi_1} =$$

$$2\left\{\sqrt{a_1^2\cos^2\phi_1 + b_1^2\sin^2\phi_1} - \frac{\tfrac{1}{2}b_1^2}{\sqrt{a_1^2\cos^2\phi_1 + b_1^2\sin^2\phi_1}} + c_1\cos\phi_1\right\};$$

then combining with

$$\frac{2d\phi}{\sqrt{a^2\cos^2\phi+b^2\sin^2\phi}}=\frac{d\phi_1}{\sqrt{a_1^2\cos^2\phi_1+b_1^2\sin^2\phi_1}},$$

we have

$$d\phi\sqrt{a^2\cos^2\phi+b^2\sin^2\phi}=$$

$$d\phi_1\left\{\sqrt{a_1^2\cos^2\phi_1+b_1^2\sin^2\phi_1}-\frac{\frac{1}{2}b_1^2}{\sqrt{a_1^2\cos^2\phi_1+b_1^2\sin^2\phi_1}}+c_1\cos\phi_1\right\};$$

and thence by integration

$$E(a,b,\phi)=E(a_1,b_1,\phi_1)-\tfrac{1}{2}b_1^2F(a_1,b_1,\phi)+c_1\sin\phi_1,$$

and *ante*, No. 421, we have

$$F(a,b,\phi)=\tfrac{1}{2}F(a_1,b_1,\phi_1),$$

which are the required transformation equations corresponding to the relation

$$\sin\phi_1=\frac{a_1\sin 2\phi}{\sqrt{a^2\cos^2\phi+b^2\sin^2\phi}}.$$

Reduction to Standard Form of Radical.

Art. Nos. 424 to 425.

424. The two angles correspond to each other as follows,

$$\phi=0,\qquad \phi_1=0,$$

$$\phi=\tan^{-1}\frac{\sqrt{a}}{\sqrt{b}},\qquad \phi_1=\tfrac{1}{2}\pi,$$

$$\phi=\tfrac{1}{2}\pi,\qquad \phi_1=\pi,$$

viz. ϕ passing from 0 to $\frac{1}{2}\pi$, ϕ_1 passes from 0 to π; and for $\phi=\frac{1}{2}\pi$ the functions of ϕ_1 are consequently the doubles of the complete functions; we thus obtain

$$E(a,b)=2E(a_1,b_1)-b_1^2F(a,b),$$
$$F(a,b)=F(a_1,b_1),$$

where $E(a,b)$, &c., denote the complete functions.

425. Recollecting that $k^2 = 1 - \frac{b^2}{a^2}$, whence $k' = \frac{b}{a}$; and assuming also $k_1^2 = 1 - \frac{b_1^2}{a_1^2}$, we have

$$k_1^2 = \frac{1}{a_1^2}(a_1^2 - b_1^2) = \frac{(a-b)^2}{(a+b)^2} = \frac{(1-k')^2}{(1+k')^2},$$

that is $k_1 = \frac{1-k'}{1+k'}$ as before, and the formulæ become

$$E(k, \phi) = \frac{a_1}{a} E(k_1, \phi_1) - \tfrac{1}{2}\frac{b_1^2}{a_1 a} F(k_1, \phi_1) + \frac{c_1}{a}\sin\phi_1,$$

$$F(k, \phi) = \tfrac{1}{2}\frac{a}{a_1} F(k_1, \phi);$$

or, what is the same thing,

$$E(k, \phi) = \tfrac{1}{2}(1+k') E(k_1, \phi_1) - \frac{k'}{1+k'} F(k_1, \phi_1) + \tfrac{1}{2}(1-k')\sin\phi_1,$$

$$F(k, \phi) = \frac{1}{(1+k')} F(k_1, \phi_1),$$

where

$$\sin\phi_1 = \frac{\tfrac{1}{2}(1+k')\sin 2\phi}{\sqrt{1-k^2\sin^2\phi}};$$

but it is convenient, in the first instance at any rate, to retain the formulæ in their original form.

Continued Repetition of the Transformation.

Arts. Nos. 426 to 429.

426. In the same manner as a_1, b_1, c_1 were derived from a, b, we may from a_1, b_1 derive a_2, b_2, c_2, and so on indefinitely: viz.

$$\begin{array}{lll} a_1 = \tfrac{1}{2}(a+b), & b_1 = \sqrt{ab}, & c_1 = \tfrac{1}{2}(a-b), \\ a_2 = \tfrac{1}{2}(a_1+b_1), & b_2 = \sqrt{a_1 b_1}, & c_2 = \tfrac{1}{2}(a_1-b_1), \\ a_3 = \tfrac{1}{2}(a_2+b_2), & b_3 = \sqrt{a_2 b_2}, & c_3 = \tfrac{1}{2}(a_2-b_2); \\ \vdots & \vdots & \vdots \end{array}$$

it is easy to see that, as n increases, a_n and b_n will approach (and that very rapidly) one and the same determinate limit, which from the mode of obtaining it from the two original quantities (a, b), is said to be the "arithmetico-geometrical mean" of these quantities, and is represented by $M(a, b)$: and of course c_n will rapidly approach the limiting value zero.

But for $a_n = b_n$ we have

$$F(a_n, b_n, \phi) = a_n^{-1}\phi, \quad E(a_n, b_n, \phi) = a_n\phi\,;$$

and in particular

$$F(a_n, b_n) = a_n^{-1}\,.\,\tfrac{1}{2}\pi, \quad E(a_n, b_n) = a_n\,.\,\tfrac{1}{2}\pi.$$

427. Considering first the complete function $F(a, b)$, we have

$$F(a, b) = F(a_1, b_1)\,\ldots = F(a_n, b_n) = \tfrac{1}{2}\pi \div M(a, b),$$

viz. the complete function is given as $\frac{1}{2}\pi$ into the reciprocal of the arithmetico-geometrical mean of a, b.

428. Considering next the incomplete function $F(a, b, \phi)$, the equations

$$\sin\phi_1 = \frac{a_1 \sin 2\phi}{\sqrt{a^2\cos^2\phi + b^2\sin^2\phi}}, \quad \sin\phi_2 = \frac{a_1 \sin 2\phi}{\sqrt{a_1^2\sin^2\phi_1 + b_1^2\sin^2\phi_1}}, \text{ \&c.}$$

show without difficulty that as n increases, ϕ_n continually approaches a value, $= 2^n$ into a determinate magnitude, say $M(a, b, \phi)$: in fact n being large and therefore a_{n-1}, b_{n-1}, a_n approximately equal, we have very nearly $\sin\phi_n = \sin 2\phi_{n-1}$, that is $\phi_n = 2\phi_{n-1}$: the limit in question $M(a, b, \phi)$ is of course to be calculated from a, b, ϕ by means of the equation itself $\phi_n = 2^n M(a, b, \phi)$: and it is to be remarked that for $\phi = \frac{1}{2}\pi$, the value of ϕ_n is $= 2^n\,.\,\frac{1}{2}\pi$, so that $M(a, b, \frac{1}{2}\pi) = \frac{1}{2}\pi$.

The equations $F(a, b, \phi) = \frac{1}{2}F(a_1, b_1, \phi_1) = \ldots$ then give

$$F(a, b, \phi)\,\ldots = \frac{1}{2^n}F(a_n, b_n, \phi_n) = \frac{1}{2^n}\,.\,\frac{1}{a_n}\phi_n = \frac{M(a, b, \phi)}{M(a, b)}.$$

Or if we choose to combine this with

$$F(a, b) = \tfrac{1}{2}\pi + M(a, b),$$

then

$$F(a, b, \phi) = \frac{2}{\pi} M(a, b, \phi)\, F(a, b).$$

429. Considering next the E-formula, this may be written

$$[E(a, b, \phi) - a^2 F(a, b, \phi)] = [E(a_1, b_1, \phi_1) - a_1^2 F(a_1, b_1, \phi_1)]$$
$$+ F(a_1, b_1, \phi_1)(a_1^2 - \tfrac{1}{2}a^2 - \tfrac{1}{2}b_1^2) + c_1 \sin\phi_1,$$

where in the second line the coefficient of $F(a_1, b_1, \phi_1)$ is $-\tfrac{1}{4}(a^2 - b^2), = -a_1c_1$, or the equation is

$$[E(a, b, \phi) - a^2 F(a, b, \phi)] = [E(a_1, b_1, \phi_1) - a_1^2 F(a_1, b_1, \phi_1)]$$
$$- a_1c_1 F(a_1, b_1, \phi_1) + c_1 \sin\phi_1.$$

And hence observing that as n increases

$$E(a_n, b_n, \phi_n) - a_n^2 F(a_n, b_n, \phi_n)$$

continually approaches to zero, we obtain

$$E(a, b, \phi) - a^2 F(a, b, \phi) = -\{2a_1c_1 + 4a_2c_2 + 8a_3c_3 \ldots\} F(a, b, \phi)$$
$$+ c_1 \sin\phi_1 + c_2 \sin\phi_2 + c_3 \sin\phi_3 + \ldots$$

Or substituting for $F(a, b, \phi)$ its value

$$E(a, b, \phi) = \{a^2 - 2a_1c_1 - 4a_2c_2 - \ldots\} \frac{M(a, b, \phi)}{M(a, b)}$$
$$+ (c_1 \sin\phi_1 + c_2 \sin\phi_2 + c_3 \sin\phi_3 + \ldots),$$

and in particular if $\phi = \tfrac{1}{2}\pi$, then $\phi_1 = \pi$, $\phi_2 = 2\pi$, &c.,

$$M(a, b, \tfrac{1}{2}\pi) = \tfrac{1}{2}\pi$$

as before, and the equation becomes

$$E(a, b) = \{a^2 - 2a_1c_1 - 4a_2c_2 - \ldots\}\, \tfrac{1}{2}\pi \div M(a, b).$$

Reduction to Standard Form of Radical.

Art. Nos. 430 to 431.

430. Introducing the modulus k, viz. writing $b = ak'$, and ultimately $a = 1$, the formula for $F(a, b)$ becomes

$$F_1(k) = \tfrac{1}{2}\pi \div M(1, k');$$

viz. the complete function is given as $= \tfrac{1}{2}\pi$ into the reciprocal of the arithmetico-geometrical mean of 1 and the complementary modulus.

Gauss has given the formula

$$\frac{1}{M(1+x, 1-x)} = 1 + \frac{1^2}{2^2}x^2 + \frac{1^2 \cdot 3^2}{2^2 \cdot 4^2}x^4 + \dots$$

Writing this under the form

$$\frac{1}{M(1+k_1, 1-k_1)} = 1 + \frac{1}{2^2}k_1^2 + \frac{1^2 \cdot 3^2}{2^2 \cdot 4^2}k_1^4 + \dots$$

we at once connect it with the formula last obtained: viz. the right-hand side is

$$= \frac{2}{\pi}F_1 k_1 = \frac{2}{\pi}\frac{1}{1+k_1}F_1 k = \frac{1}{(1+k_1)M(1, k')}.$$

Or the equation is

$$M(1+k_1, 1-k_1) = (1+k_1)\,M(1, k') = (1+k_1)\,M\left(1, \frac{1-k_1}{1+k_1}\right);$$

which is obviously true, since in general

$$M(a, b) = \theta\, M\left(\frac{a}{\theta}, \frac{b}{\theta}\right).$$

The formula for $F(a, b, \phi)$ gives in like manner

$$F(k, \phi) = \frac{M(1, k', \phi)}{M(1, k')},$$

which is a formula for the numerical calculation of the function $F(k, \phi)$.

431. Proceeding next to the function E, we have

$$E(k,\phi)=\left\{1-\frac{2a_1c_1}{a^2}-\frac{4a_2c_2}{a^2}-\dots\right\}\frac{M(1,k',\phi)}{M(1,k')}$$
$$+\left(\frac{c_1}{a}\sin\phi_1+\frac{c_2}{a}\sin\phi_2+\frac{c_3}{a}\sin\phi_3+\dots\right).$$

Hence forming the equations

$$\frac{a_1c_1}{a^2}=\tfrac{1}{4}k^2,\quad \frac{a_2c_2}{a_1c_1}=\tfrac{1}{4}k_1,\quad \frac{a_3c_3}{a_2c_2}=\tfrac{1}{4}k_2\dots$$

$$\frac{c_1}{a}=\frac{k_1}{1+k_1},$$

$$\frac{c_2}{a_1}=\frac{k_2}{1+k_2},\quad \frac{a_1}{a}=\frac{1}{1+k_1},$$

$$\frac{c_3}{a_2}=\frac{k_3}{1+k_3},\quad \frac{a_2}{a_1}=\frac{1}{1+k_2},\quad \frac{a_1}{a}=\frac{1}{1+k_1},\ \&c$$

the equation becomes

$$E(k,\phi)=\left[1-\tfrac{1}{2}k^2\left(1+\tfrac{1}{2}k_1+\tfrac{1}{4}k_1k_2+\tfrac{1}{8}k_1k_2k_3+\dots\right)\right]\frac{M(1,k',\phi)}{M(1,k')}$$
$$+\frac{k_1}{1+k_1}\sin\phi_1+\frac{k_2}{(1+k_1)(1+k_2)}\sin\phi_2+\frac{k_3}{(1+k_1)(1+k_2)(1+k_3)}\sin\phi_3$$
$$+\ \&c.,$$

or observing that

$$\frac{1}{1+k_1}=\frac{k}{2\sqrt{k_1}},\quad\text{that is}\quad \frac{1}{1+k_1}=\frac{k}{2\sqrt{k_1}},$$

$$\frac{1}{1+k_2}=\frac{k_1}{2\sqrt{k_2}},\quad ,,\quad \frac{1}{1+k_1\,.\,1+k_2}=\frac{k\sqrt{k_1}}{4\sqrt{k_2}},$$

$$\frac{1}{1+k_3}=\frac{k_2}{2\sqrt{k_3}},\quad ,,\quad \frac{1}{1+k_1\,.\,1+k_2\,.\,1+k_3}=\frac{k\sqrt{k_1k_2}}{8\sqrt{k_3}},$$

&c.

the last line may be written

$$+k\left[\tfrac{1}{2}\sqrt{k_1}\sin\phi_1+\tfrac{1}{4}\sqrt{k_1k_2}\sin\phi_2+\tfrac{1}{8}\sqrt{k_1k_2k_3}\sin\phi_3+\dots\right].$$

In particular, if $\phi=\frac{1}{2}\pi$, the equation becomes

$$E_1k=\left[1-\tfrac{1}{2}k^2\left(1+\tfrac{1}{2}k_1+\tfrac{1}{4}k_1k_2+\tfrac{1}{8}k_1k_2k_3\dots\dots\right)\right]\tfrac{1}{2}\pi\div M(1,k');$$

or if we please,

$$E_1k=\left[1-\tfrac{1}{2}k^2\left(1+\tfrac{1}{2}k_1+\tfrac{1}{4}k_1k_2+\tfrac{1}{8}k_1k_2k_3+\dots\right)\right]F_1k.$$

Application to Integrals of the third kind.

Art. No. 432.

432. The transformation is applicable to elliptic integrals of the third kind, but the results are not of any particular interest. Writing down the equations

$$\frac{2d\phi}{\sqrt{a^2\cos^2\phi+b^2\sin^2\phi}}=\frac{d\phi_1}{\sqrt{a_1^2\cos^2\phi_1+b_1^2\sin^2\phi_1}},$$

$$\frac{a^2\cos^2\phi+b^2\sin^2\phi}{(a^2\cos^2\phi+b^2\sin^2\phi)(\cos^2\phi+\sin^2\phi)+4n_1a_1^2\sin^2\phi\cos^2\phi}$$

$$=\frac{1}{1+n_1\sin^2\phi_1},$$

the expression on the left-hand of this last equation is

$$=\frac{A}{a\cos^2\phi+bX\sin^2\phi}+\frac{BX}{aX\cos^2\phi+b\sin^2\phi},$$

where

$$a^2\cos^4\phi+b^2\sin^4\phi+(a^2+b^2+4n_1a_1^2)\sin^2\phi\cos^2\phi$$

$$=(a\cos^2\phi+bX\sin^2\phi)\left(a\cos^2\phi+\frac{b}{X}\sin^2\phi\right);$$

that is

$$X+\frac{1}{X}=\frac{1}{ab}(a^2+b^2+4n_1a_1^2)$$

$$=\frac{2}{b_1^2}(a_1^2+c_1^2+2n_1a_1^2);$$

whence

$$\left(X-\frac{1}{X}\right)^2=\frac{4}{b_1^4}\{(a_1^2+c_1^2+2n_1a_1^2)^2-b_1^4\}$$

$$=\frac{16a_1^2}{b_1^4}(1+n_1)(c_1^2+n_1a_1^2);$$

that is $$X = \frac{1}{b_1^2}\left[a_1^2 + c_1^2 + 2n_1 a_1^2 + 2a_1\sqrt{(1+n_1)(c_1^2+n_1 a_1^2)}\right],$$

$$\frac{1}{X} = \frac{1}{b_1^2}\left[a_1^2 + c_1^2 + 2n_1 a_1^2 - 2a_1\sqrt{(1+n_1)(c_1^2+n_1 a_1^2)}\right];$$

and then

$$A = \frac{(aX-b)X}{X^2-1}, \quad B = -\frac{a-bX}{X^2-1},$$

and we have

$$\left\{\frac{(aX-b)X}{X^2-1}\,\frac{1}{a\cos^2\phi+bX\sin^2\phi} - \frac{(a-bX)X}{X^2-1}\,\frac{1}{aX\cos^2\phi+b\sin^2\phi}\right\}\times$$

$$\frac{2d\phi}{\sqrt{a^2\cos^2\phi+b^2\sin^2\phi}} =$$

$$\frac{d\phi_1}{(1+n_1\sin^2\phi_1)\sqrt{a_1^2\cos^2\phi_1+b_1^2\sin^2\phi_1}},$$

whence, integrating, the function

$$\int\frac{d\phi_1}{(1+n_1\sin^2\phi_1)\sqrt{a_1^2\cos^2\phi_1+b_1^2\sin^2\phi_1}}$$

is expressed as the sum of the two elliptic integrals of the third kind having a common modulus but different parameters.

Numerical instance for complete Functions E_1, F_1, and for an incomplete F. Art. No. 433.

433. As a numerical instance take (as in Legendre's example, t. I. p. 91),

$a=1$, $b=\frac{1}{2}\sqrt{2+\sqrt{3}}=\cos 75^\circ$ (whence $k=\sin 75^\circ$), $\tan\phi=\frac{\sqrt{2}}{\sqrt{3}}$;

we have

	a	b	c	k	k'	ϕ
(0)	1·000,0000	0·258,8190		0·965,9258	0·258,8190	47° 8′ 31″
(1)	0·629,4095	0·508,7426	·870,5905	0·588,7908	0·808,2856	62° 36′ 8″
(2)	0·569,0761	0·565,8688	·060,3834	0·106,0200	0·994,3686	119° 55′ 46″
(3)	0·567,4724	0·567,4701	·001,6037	0·002,8260	0·999,9959	240° 0′ 0″
(4)	0·567,4712	0·567,4712	·000,0011	0·000,0020	0·999,9999	480° 0′ 0″

first as to the complete functions we have

$$F_1 = \frac{\pi}{2} \cdot \frac{1}{a_4} \quad = 2{\cdot}768{,}064.$$

$$\begin{aligned} \tfrac{1}{2}\left(1 - \frac{E_1}{F_1}\right) = \quad & a_1c_1 & = {\cdot}233{,}2532 \\ & + 2a_2c_2 & {\cdot}068{,}0686 \\ & + 4a_3c_3 & {\cdot}003{,}6402 \\ & + 8a_4c_4 & {\cdot}000{,}0051 \\ & & = {\cdot}305{,}5671 \end{aligned}$$

agreeing with Legendre's values $F_1 = 2{\cdot}768{,}0631$, $E_1 = 1{\cdot}076{,}4051$, and thence $\tfrac{1}{2}\left(1 - \frac{E_1}{F_1}\right) = {\cdot}305{,}5671$.

Also, we have $F(k, \phi) = \frac{1}{2^4}\frac{1}{a_4} \cdot \phi_4$, or since $\frac{1}{2^4}\phi_4 = 30^\circ = \tfrac{1}{6}\pi$, this is $F(k, \phi) = \tfrac{1}{3}F_1 = 0{\cdot}9226877$: it is in fact easily verified that the assumed value of ϕ is such as to give exactly

$$F(k, \phi) = \tfrac{1}{3}F_1.$$

The notion of the arithmetico-geometrical mean was established by Gauss in the memoir "Determinatio Attractionis &c." Comm. Gott. Rec. t. IV. (1818), but his later researches in relation to the subject were not published until after his death, *Werke*, t. IV. pp. 361—403; a table is given p. 403, of the values of the arithmetico-geometrical mean $M(1, \sin\theta)$ and of its logarithm, $\theta = 0^\circ$ to 90° at intervals of $30'$.

CHAPTER XIV.

THE GENERAL DIFFERENTIAL EQUATION $\dfrac{dx}{\sqrt{X}} = \dfrac{dy}{\sqrt{Y}}$.

Integration of the differential equation.
Art. Nos. 434 to 436.

434. In the present Chapter, writing

$$X = a + bx + cx^2 + dx^3 + ex^4,$$

$$Y = a + by + cy^2 + dy^3 + ey^4,$$

I consider the differential equation

$$\frac{dx}{\sqrt{X}} + \frac{dy}{\sqrt{Y}} = 0.$$

435. A direct process for finding the algebraical integral as follows was given by Lagrange.

Assume $\dfrac{dx}{dt} = \sqrt{X}$, and therefore $\dfrac{dy}{dt} = -\sqrt{Y}$;

then
$$2\frac{d^2x}{dt^2} = b + 2cx + 3dx^2 + 4ex^3,$$

$$2\frac{d^2y}{dt^2} = b + 2cy + 3dy^2 + 4ey^3;$$

and if $p = x + y$, $q = x - y$, then

$$\frac{d^2p}{dt^2} = \frac{d^2x}{dt^2} + \frac{d^2y}{dt^2} = b + cp + \tfrac{3}{4}d(p^2 + q^2) + \tfrac{1}{2}e(p^3 + 3pq^2),$$

$$\frac{dp}{dt}\frac{dq}{dt} = X - Y = bq + cpq + \tfrac{1}{4}dq(3p^2 + q^2) + \tfrac{1}{2}epq(p^2 + q^2);$$

whence
$$q\frac{d^2p}{dt^2} - \frac{dp}{dt}\frac{dq}{dt} = \tfrac{1}{2}dq^3 + epq^3,$$

or
$$\frac{2}{q^3}\frac{d^2p}{dt^2}\frac{dp}{dt} - \frac{2}{q^4}\left(\frac{dp}{dt}\right)^2\frac{dq}{dt} = (d + 2ep)\frac{dp}{dt}.$$

which is integrable as it stands and gives

$$\frac{1}{q^2}\left(\frac{dp}{dt}\right)^2 = C + dp + ep^2;$$

or substituting for q, $\frac{dp}{dt}$ and p their values

$$\left(\frac{\sqrt{X} - \sqrt{Y}}{x-y}\right)^2 = C + d(x+y) + e(x+y)^2,$$

which is the general integral, C being the constant of integration.

436. To further develope this result observe that we have

$$X + Y - 2\sqrt{XY} = C(x-y)^2 + d(x^3 - x^2y - xy^2 + y^3) + e(x^4 - 2x^2y^2 + y^4),$$

that is

$$2\sqrt{XY} = 2a + b(x+y) + c(x^2+y^2) - C(x-y)^2 + dxy(x+y) + 2ex^2y^2;$$

or say

$$\sqrt{XY} = a + \tfrac{1}{2}b(x+y) + \tfrac{1}{2}c(x^2+y^2) - \tfrac{1}{2}C(x-y)^2 + \tfrac{1}{2}dxy(x+y) + ex^2y^2\,^*:$$

whence squaring and transposing

$$[a + \tfrac{1}{2}b(x+y) + \tfrac{1}{2}c(x^2+y^2) - \tfrac{1}{2}C(x-y)^2 + \tfrac{1}{2}dxy(x+y) + ex^2y^2]^2 - (a + bx + cx^2 + dx^3 + ex^4)(a + by + cy^2 + dy^3 + ey^4) = 0;$$

* Write $x = \sin\phi$, $y = \sin\psi$, $(a, b, c, d, e) = (1, 0, -1-k^2, 0, k^2)$, the equation becomes

$$\cos\phi\cos\psi\Delta\phi\Delta\psi = 1 - \tfrac{1}{2}(1+k^2)(\sin^2\phi + \sin^2\psi) - \tfrac{1}{2}C(\sin\phi - \sin\psi)^2 + k^2\sin^2\phi\sin^2\psi;$$

and to introduce μ instead of C we must write

$$\cos\mu\Delta\mu = 1 - \tfrac{1}{2}(1+k^2)\sin^2\mu - \tfrac{1}{2}C\sin^2\mu,$$

that is

$$\tfrac{1}{2}C\sin^2\mu = 1 - \tfrac{1}{2}(1+k^2)\sin^2\mu - \cos\mu\Delta\mu.$$

The equation thus is

$$1 - \tfrac{1}{2}(1+k^2)(\sin^2\phi + \sin^2\psi) + k^2\sin^2\phi\sin^2\psi - \cos\phi\cos\psi\Delta\phi\Delta\psi = \frac{(\sin\phi - \sin\psi)^2}{\sin^2\mu}[1 - \tfrac{1}{2}(1+k^2)\sin^2\mu - \cos\mu\Delta\mu];$$

this is of course a form of the addition equation, and could be verified as such by substituting for $\cos\mu$, $\sin\mu$, $\Delta\mu$ their values in terms of ϕ, ψ: but the form is not a convenient one.

viz. this is

$$\begin{array}{lll}
\tfrac{1}{4}C^2(x-y)^4, & & \\
\multicolumn{3}{l}{-\,C(x-y)^2\,[a+\tfrac{1}{2}b(x+y)+\tfrac{1}{2}c(x^2+y^2)+\tfrac{1}{2}dxy(x+y)+ex^2y^2],} \\
+a^2\,.\,1 & -1 & =0, \\
+ab\,.\,x+y & -x-y & =0, \\
+ac\,.\,x^2+y^2 & -x^2-y^2 & =0, \\
+ad\,.\,xy(x+y) & -x^3-y^3 & =-(x-y)^2(x+y), \\
+ae\,.\,2x^2y^2 & -x^4-y^4 & =-(x-y)^2(x+y)^2, \\
+b^2\,.\,\tfrac{1}{4}(x+y)^2 & -xy & =+\tfrac{1}{4}(x-y)^2, \\
+bc\,.\,\tfrac{1}{2}(x+y)(x^2+y^2) & -x^2y-xy^2 & =+\tfrac{1}{2}(x-y)^2(x+y), \\
+bd\,.\,\tfrac{1}{2}xy(x+y)^2 & -xy^3-x^3y & =-\tfrac{1}{2}xy(x-y)^2, \\
+be\,.\,x^2y^2(x+y) & -xy^4-x^4y & =-xy(x-y)^2(x+y), \\
+c^2\,.\,\tfrac{1}{4}(x^2+y^2)^2 & -x^2y^2 & =+\tfrac{1}{4}(x-y)^2(x+y)^2, \\
+cd\,.\,\tfrac{1}{2}xy(x+y)(x^2+y^2) & -x^3y^2-x^2y^3 & =+\tfrac{1}{2}(x-y)^2xy(x+y), \\
+ce\,.\,x^2y^2(x^2+y^2) & -x^4y^2-x^2y^4 & =0, \\
+d^2\,.\,\tfrac{1}{4}x^2y^2(x+y)^2 & -x^3y^3 & =+\tfrac{1}{4}(x-y)^2x^2y^2, \\
+de\,.\,x^3y^3(x+y) & -x^4y^3-x^3y^4 & =0, \\
+e^2\,.\,x^4y^4 & -x^4y^4 & =0, \qquad\qquad =0\,;
\end{array}$$

viz. the whole equation divides by $(x-y)^2$. Omitting this factor it is

$$\begin{array}{l}
\tfrac{1}{4}C^2(x-y)^2 \\
\quad -\,C\,[a+\tfrac{1}{2}b(x+y)+\tfrac{1}{2}c(x^2+y^2)+\tfrac{1}{2}dxy(x+y)+ex^2y^2], \\
\quad -\,ad(x+y) \\
\quad -\,ae(x+y)^2 \\
\quad +\,\tfrac{1}{4}b^2 \\
\quad +\,\tfrac{1}{2}bc(x+y) \\
\quad -\,\tfrac{1}{2}bd\,xy \\
\quad -\,be\,xy(x+y) \\
\quad +\,\tfrac{1}{4}c^2(x+y)^2 \\
\quad +\,\tfrac{1}{2}cd\,xy(x+y) \\
\quad +\,\tfrac{1}{4}d^2x^2y^2 \qquad\qquad\qquad\qquad =0,
\end{array}$$

or what is the same thing, it is

$$
\begin{array}{lllllll}
 & (& -\ Ca & & +\tfrac{1}{4}b^2 &) & \\
+(x+y) & (& -\tfrac{1}{2}Cb & -\ ad & +\tfrac{1}{2}bc &) & \\
+(x^2+y^2) & (\ \tfrac{1}{4}C^2 & -\tfrac{1}{2}Cc & -\ ae & & +\tfrac{1}{4}c^2\ ) & \\
+xy & (-\tfrac{1}{2}C^2 & & -2ae & -\tfrac{1}{2}bd & +\tfrac{1}{2}c^2\ ) & \\
+xy(x+y) & (& -\tfrac{1}{2}Cd & & -\ be & +\tfrac{1}{2}cd) & \\
+x^2y^2 & (& -\ Ce & & & +\tfrac{1}{4}d^2\ ) & =0.
\end{array}
$$

This may be written

$$
\begin{aligned}
&(\mathfrak{a}+2\mathfrak{h}x+\mathfrak{g}x^2)\\
+\,2y\,&(\mathfrak{h}+2\mathfrak{b}x+\mathfrak{f}x^2)\\
+\,y^2\,&(\mathfrak{g}+2\mathfrak{f}x+\mathfrak{c}x^2)=0,
\end{aligned}
$$

where the several coefficients have the values

$$
\begin{aligned}
\mathfrak{a} &= \quad b^2-4aC,\\
\mathfrak{b} &= -2ae-\tfrac{1}{2}bd+\tfrac{1}{2}c^2-\tfrac{1}{2}C^2,\\
\mathfrak{c} &= \quad d^2-4eC,\\
\mathfrak{f} &= \quad cd-2be-Cd,\\
\mathfrak{g} &= -4ae+c^2-2Cc+C^2,\\
\mathfrak{h} &= \quad bc-2ad-Cb.
\end{aligned}
$$

The result shows that the complete integral of the differential equation is an equation $u=0$, where u is a symmetric quadriquadric function of (x, y); that is, a symmetric function, quadric in regard to each variable separately.

Further development of the theory. Art. Nos. 437 to 446.

437. This may be verified almost instantaneously: starting from

$$
\begin{aligned}
u = \quad &(\mathfrak{a}+2\mathfrak{h}x+\mathfrak{g}x^2),\\
+\,2y\,&(\mathfrak{h}+2\mathfrak{b}x+\mathfrak{f}x^2),\\
+\,y^2\,&(\mathfrak{g}+2\mathfrak{f}x+\mathfrak{c}x^2)=0,
\end{aligned}
$$

we may write

$$u=A+2By+Cy^2=A'+2B'x+C'x^2=0,$$

A, B, C being given quadric functions of x, and A', B', C' the same quadric functions of y.

Then differentiating

$$\frac{du}{dx}dx + \frac{du}{dy}dy = 0.$$

But

$$\frac{du}{dy} = 2(Cy + B) = 2\sqrt{B^2 - AC},$$

$$\text{since } u = 0 \text{ gives } (Cy+B)^2 = B^2 - AC,$$

$$\frac{du}{dx} = 2(C'x + B') = 2\sqrt{B'^2 - A'C'},$$

$$\text{,, } (C'x + B')^2 = B'^2 - A'C',$$

and the differential equation thus is

$$\frac{dx}{\sqrt{B'^2 - A'C'}} + \frac{dy}{\sqrt{B^2 - AC}} = 0.$$

This will coincide with

$$\frac{dx}{\sqrt{X}} + \frac{dy}{\sqrt{Y}} = 0,$$

if only the quadric functions A, B, C are determined so that $B^2 - AC = \theta X$ (which of course implies $B'^2 - A'C' = \theta Y$). We have in all six disposable quantities a, b, c, f, g, h, that is five ratios; and the equation in question

$$(\mathrm{h} + 2\mathrm{b}x + \mathrm{f}x^2)^2 - (\mathrm{a} + 2\mathrm{h}x + \mathrm{g}x^2)(\mathrm{g} + 2\mathrm{f}x + \mathrm{c}x^2) = \theta X,$$

establishes four relations between the five ratios, and thus leaves one indeterminate ratio serving as a constant of integration: we in fact satisfy the equations by means of the before-mentioned values of a, b, c, f, g, h, which contain the arbitrary constant C; viz. we then have

$$(\mathrm{h} + 2\mathrm{b}x + \mathrm{f}x^2)^2 - (\mathrm{a} + 2\mathrm{h}x + \mathrm{g}x^2)(\mathrm{g} + 2\mathrm{f}x + \mathrm{c}x^2)$$

$$= \left(\frac{\mathrm{h}^2 - \mathrm{ag}}{a} \text{ or } \frac{\mathrm{f}^2 - \mathrm{cg}}{e}\right) X$$

$$= 4\{ad^2 + b^2e - bcd + \{-4ae + bd + (C - c)^2\} C\} X.$$

As a partial verification, observe that the equation

$$\frac{\mathrm{h}^2 - \mathrm{ag}}{a} = \frac{\mathrm{f}^2 - \mathrm{cg}}{e} \text{ or } e\mathrm{h}^2 - a\mathrm{f}^2 = \mathrm{g}(e\mathrm{a} - a\mathrm{c}), = (eb^2 - ad^2)\,\mathrm{g},$$

is satisfied identically.

438. Regard u as a function of C; we have

$$\begin{array}{lr|ll} u = b^2 & +2C & -2a & +C^2.(x-y)^2 \\ +(2bc-4ad)(x+y) & & -b(x+y) & \\ +(c^2-4ae)(x^2+y^2) & & -c(x^2+y^2) & \\ +(-8ae-2bd+2c^2)xy & & & \\ +(2cd-4be)xy(x+y) & & -dxy(x+y) & \\ +\quad d^2 \quad x^2y^2 & & -2e\,x^2y^2 & \end{array}$$

say this is

$$u = \lambda + 2\mu C + \nu C^2;$$

then we have

$$\mu^2 - \lambda\nu = 4a^2 \ldots + 4e^2x^4y^4, = 4XY:$$

viz. calculating $\mu^2 - \lambda\nu$, it will be found to have this value.

439. Now starting with the equation $u = 0$, and treating it as before, except that we now regard C as a variable; that is, forming, and then reducing, the equation

$$\frac{du}{dx}dx + \frac{du}{dy}dy + \frac{du}{dC}dC = 0,$$

we obtain

$$\sqrt{\mathfrak{C}Y}\,dx + \sqrt{\mathfrak{C}X}\,dy + \sqrt{XY}\,dC = 0,$$

or, what is the same thing,

$$\frac{dx}{\sqrt{X}} + \frac{dx}{\sqrt{Y}} + \frac{dC}{\sqrt{\mathfrak{C}}} = 0,$$

where $\quad \mathfrak{C} = ad^2 + b^2e - bcd + C[-4ae + bd + (C-c)^2]$,

a cubic function of C.

440. Write

$$C = \tfrac{2}{3}c - 2\omega;$$

then

$$\begin{aligned} \mathfrak{C} &= ad^2 + b^2e - bcd + [-4ae + bd + (\tfrac{1}{3}c + 2\omega)^2](\tfrac{2}{3}c - 2\omega) \\ &= -8\omega^3 + (8ae - 2bd + \tfrac{2}{3}c^2)\,\omega \\ &\qquad\qquad + (-\tfrac{8}{3}ace + ad^2 + b^2e - \tfrac{1}{3}bcd + \tfrac{2}{27}c^3). \end{aligned}$$

But the invariants of $a + bx + cx^2 + dx^3 + ex^4$ are

$$I = \tfrac{1}{12}(12ae - 3bd + c^2),$$

$$J = \tfrac{1}{432}(72ace - 27ad^2 - 27b^2e - 2c^3 + 9bcd);$$

whence

$$\mathfrak{C} = -8(\omega^3 - I\omega + 2J),$$

$$\sqrt{\mathfrak{C}} = 2i\sqrt{2}\sqrt{\omega^3 - I\omega + 2J}, = 2i\sqrt{2}\sqrt{\Omega}, \text{ suppose,}$$

$$dC = -2d\omega;$$

or the differential equation is

$$\frac{dx}{\sqrt{X}} + \frac{dy}{\sqrt{Y}} - \frac{d\omega}{i\sqrt{2}\sqrt{\Omega}} = 0;$$

viz. writing for C its value $\tfrac{2}{3}c - 2\omega$, the corresponding integral equation is

$$u = \lambda + 2\mu(\tfrac{2}{3}c - 2\omega) + \nu(\tfrac{2}{3}c - 2\omega)^2,$$

$$= \lambda + \tfrac{4}{3}c\mu + \tfrac{4}{9}c^2\nu + 2\omega(-2\mu - \tfrac{4}{3}c\nu) + \omega^2 . 4\nu, = 0;$$

or substituting for λ, μ, ν their values and reducing, this is

$$\left.\begin{array}{l} b^2 - \tfrac{8}{3}ac \\ + (\tfrac{4}{3}bc - 4ad)(x+y) \\ + (\tfrac{1}{9}c^2 - 4ae)(x^2+y^2) \\ + (-8ae - 2bd + \tfrac{16}{9}c^2)xy \\ + (\tfrac{4}{3}cd - 4be)xy(x+y) \\ + (d^2 - \tfrac{8}{3}ce)x^2y^2 \end{array}\right| + 2\omega \left|\begin{array}{l} + 4a \\ + 2b(x+y) \\ + \tfrac{2}{3}c(x^2+y^2) \\ + \tfrac{8}{3}c\,xy \\ + 2d\,xy(x+y) \\ + 4e\,x^2y^2 \end{array}\right. + \omega^2 . 4(x-y)^2 = 0$$

where the left-hand side is quadric in each of the variables x, y, ω: as there is no arbitrary constant, this is only a particular integral.

441. We may by a linear substitution performed on the ω bring the third radical $\sqrt{\Omega}$ to a like form with the other two radicals.

Write for convenience

$$a+bx+cx^2+dx^3+ex^4=e(x-\alpha)(x-\beta)(x-\gamma)(x-\delta);$$

then the substitution may be taken to be

$$2\omega=\tfrac{1}{3}e[-\beta\gamma-\gamma\alpha-\alpha\beta+2\delta(\alpha+\beta+\gamma)-3\delta^2]+\frac{e(\alpha-\delta)(\beta-\delta)(\gamma-\delta)}{z-\delta},$$

which, as will appear, makes the radical $\sqrt{\Omega}$ to depend on $\sqrt{Z}$, where $Z=a+bz+cz^2+dz^3+ez^4$. Some preliminary formulæ are required.

442. Reverting to the formulæ which contain $C(=\tfrac{2}{3}c-2\omega)$, assume

$$C_1=e(\beta+\gamma)(\alpha+\delta),$$
$$C_2=e(\gamma+\alpha)(\beta+\delta),$$
$$C_3=e(\alpha+\beta)(\gamma+\delta);$$

then we have

$$(C-C_1)(C-C_2)(C-C_3)=C[(C-c)^2-4ae+bd]+ad^2+b^2e-bcd,$$

viz. C_1, C_2, C_3 are the roots of the equation $\mathfrak{C}=0$.

Hence writing for C its value $=\tfrac{2}{3}c-2\omega$, we have

$$(\tfrac{2}{3}c-2\omega-C_1)(\tfrac{2}{3}c-2\omega-C_2)(\tfrac{2}{3}c-2\omega-C_3)$$
$$=-8(\omega^3-I\omega+2J)$$
$$=-8(\omega-\omega_1)(\omega-\omega_2)(\omega-\omega_3) \text{ suppose.}$$

We have $\omega_1=\tfrac{1}{3}c-\tfrac{1}{2}C_1, \;=\tfrac{1}{6}(2c-3C_1)$

$$=\tfrac{1}{6}e[2\beta\gamma+2\alpha\delta-\alpha\beta-\beta\delta-\alpha\gamma-\gamma\delta];$$

or putting

$$A=(\beta-\gamma)(\alpha-\delta)=\alpha\beta+\gamma\delta-\beta\delta-\alpha\gamma,$$
$$B=(\gamma-\alpha)(\beta-\delta)=\beta\gamma+\alpha\delta-\gamma\delta-\beta\alpha,$$
$$C=(\alpha-\beta)(\gamma-\delta)=\gamma\alpha+\beta\delta-\alpha\delta-\gamma\beta,$$

we have $\omega_1=\tfrac{1}{6}e(B-C)$; or forming the analogous equations

$$\omega_1=\tfrac{1}{6}e(B-C),$$
$$\omega_2=\tfrac{1}{6}e(C-A),$$
$$\omega_3=\tfrac{1}{6}e(A-B).$$

443. Now writing as above

$$2\omega = \tfrac{1}{3}e\left[-\beta\gamma-\gamma\alpha-\alpha\beta+2\delta(\alpha+\beta+\gamma)-3\delta^2\right]+\frac{e(\alpha-\delta)(\beta-\delta)(\gamma-\delta)}{z-\delta},$$

then if $z=\alpha, \beta, \gamma$ we have $\omega=\omega_1, \omega_2, \omega_3$ respectively: thus writing $z=\alpha$ we find

$$\begin{aligned}6\omega = \; & e\left[-\beta\gamma-\gamma\alpha-\alpha\beta+2\delta\alpha+2\delta\beta+2\delta\gamma-3\delta^2\right.\\ & \left.+3\beta\gamma \qquad\qquad -3\delta\beta-3\delta\gamma+3\delta^2\right]\\ = \; & e\left[2\alpha\delta+2\beta\gamma-(\alpha+\delta)(\beta+\gamma)\right], = 6\omega_1,\end{aligned}$$

and so for the others.

Hence

$$2(\omega-\omega_1) = -e(\beta-\delta)(\gamma-\delta)\frac{z-\alpha}{z-\delta},$$

$$2(\omega-\omega_2) = -e(\gamma-\delta)(\alpha-\delta)\frac{z-\beta}{z-\delta},$$

$$2(\omega-\omega_3) = -e(\alpha-\delta)(\beta-\delta)\frac{z-\gamma}{z-\delta};$$

and therefore

$$\begin{aligned}8(\omega-\omega_1)(\omega-\omega_2)(\omega-\omega_3), &= 8(\omega^3-I\omega+2J),\\ &= -\left[(\alpha-\delta)(\beta-\delta)(\gamma-\delta)\right]^2 e^3\frac{Z}{(z-\delta)^4};\end{aligned}$$

or say

$$2i\sqrt{2}\sqrt{\Omega} = (\alpha-\delta)(\beta-\delta)(\gamma-\delta)\,e\frac{\sqrt{Z}}{(z-\delta)^2}.$$

But from the expression for ω

$$2\delta\omega = -(\alpha-\delta)(\beta-\delta)(\gamma-\delta)\,e\frac{dz}{(z-\delta)^2};$$

whence $\qquad \dfrac{d\omega}{i\sqrt{2}\sqrt{\Omega}} = -\dfrac{dz}{\sqrt{Z}};$

or the equation

$$\frac{dx}{\sqrt{X}}+\frac{dy}{\sqrt{Y}}-\frac{d\omega}{i\sqrt{2}\sqrt{\Omega}}=0$$

is by the substitution

$$2\omega = \tfrac{1}{3}e\,[-\beta\gamma-\gamma\alpha-\alpha\beta+2\delta(\alpha+\beta+\gamma)-3\delta^2] + \frac{e(\alpha-\delta)(\beta-\delta)(\gamma-\delta)}{z-\delta}$$

transformed into

$$\frac{dx}{\sqrt{X}} + \frac{dy}{\sqrt{Y}} + \frac{dz}{\sqrt{Z}} = 0;$$

and if in the equation between x, y, ω we write for ω the above value, we have the corresponding integral equation between x, y, z. (This will be presently given in the particular case $a = 0$, No. 445.)

444. But we may in a different way make the transformation from $\dfrac{d\omega}{\sqrt{\Omega}}$ to $\dfrac{du}{\sqrt{U}}$, $U = a + bu + cu^2 + du^3 + eu^4$. Take as before I, J for the invariants, H for the Hessian, and Φ for the cubi-covariant,

$$\begin{aligned} H = \tfrac{1}{16}\,[(8ac-3b^2)u^0 &+ (24ad-4bc)\,u + (48ae+6bd-4c^2)\,u^2 \\ &+ (24be-4cd)\,u^3 + (8ce-3d^2)\,u^4], \end{aligned}$$

$$\begin{aligned} \Phi = \tfrac{1}{32}\,[\;&(-\ 8a^2d + 4abc - b^3)\,u^0 \\ &+(-32a^2e - 4abd + 8ac^2 - 2b^2c)\,u \\ &+(-40abe + 20acd + 5b^2d)\,u^2 \\ &+(\ 20ad^2 - 20b^2e)\,u^3 \\ &+(\ 40ade - 20bce + 5bd^2)\,u^4 \\ &+(\ 32ae^2 + 4bde - 8c^2e + 2cd^2)\,u^5 \\ &+(\ 8be^2 - 4cde + d^3)\,u^6]; \end{aligned}$$

then identically $\quad JU^3 - IU^2H + 4H^3 = -\Phi^2,$

whence assuming $\qquad \omega = \dfrac{2H}{U},$

we have

$$\omega^3 - I\omega + 2J = -\frac{2\Phi^2}{U^3}, \text{ or say } \sqrt{\Omega} = \frac{i\sqrt{2}\,\Phi\sqrt{U}}{U^2}.$$

From the expression of ω we find

$$d\omega = \frac{2(U\Pi' - U'\Pi)\,du}{U^2};$$

and hence $$\frac{d\omega}{\sqrt{\Omega}} = \frac{2(U\Pi' - U'\Pi)\,du}{i\sqrt{2}\,.\,\Phi\sqrt{U}},$$

where the multiplier of $\frac{du}{\sqrt{U}}$ is a constant. We in fact have

$$U\Pi' - U'\Pi = \tfrac{1}{16}(8a^2d - 4abc + b^3) + \&c.$$
$$= -2\Phi;$$

that is $$\frac{d\omega}{\sqrt{\Omega}} = 2i\sqrt{2}\,.\,\frac{du}{\sqrt{U}};$$

and the equation $\frac{dx}{\sqrt{X}} + \frac{dy}{\sqrt{Y}} - \frac{d\omega}{i\sqrt{2}\sqrt{\Omega}} = 0$ is thus converted into

$$\frac{dx}{\sqrt{X}} + \frac{dy}{\sqrt{Y}} - \frac{2du}{\sqrt{U}} = 0.$$

It is to be noticed that the above transformation $\omega = \frac{2\Pi}{U}$ leading to $\frac{d\omega}{\sqrt{\Omega}} = 2i\sqrt{2}\frac{du}{\sqrt{U}}$ is really a transformation of the order 4, degenerating into a multiplication by $(\sqrt{4}, =)\,2$. For it was shown above that $\frac{d\omega}{\sqrt{\Omega}}$ is by a linear substitution transformable into $\frac{dz}{\sqrt{Z}}$.

445. Reverting to the relation between (x, y, ω), leading to a relation between x, y, z, suppose in order to simplify that $a = 0$; that is, assume $X = bx + cx^2 + dx^3 + ex^4, = e\,x(x-\alpha)(x-\beta)(x-\gamma)$, the value of δ being thus zero.

Then the integral of

$$\frac{dx}{\sqrt{X}} + \frac{dy}{\sqrt{Y}} - \frac{d\omega}{i\sqrt{2}\sqrt{\Omega}} = 0$$

becomes

$$b^2 + 2bc(x+y) + c^2(x^2+y^2) + (2c^2 - 2bd)xy$$
$$\qquad + (2cd - 4be)xy(x+y) + d^2x^2y^2$$
$$+ 2\{-b(x+y) - c(x^2+y^2) - d\,xy(x+y) - 2e\,x^2y^2\}(\tfrac{2}{3}c - 2\omega)$$
$$+ \quad (x-y)^2.(\tfrac{2}{3}c - 2\omega)^2 = 0,$$

say this is

$$\lambda + 2\mu(\tfrac{2}{3}c - 2\omega) + \nu(\tfrac{2}{3}c - 2\omega)^2 = 0,$$

and writing herein

$$2\omega = \tfrac{1}{3}e(-\beta\gamma - \gamma\alpha - \alpha\beta) + \frac{e\,\alpha\beta\gamma}{z}, \; = -\tfrac{1}{3}c - \frac{b}{z},$$

that is $\qquad \tfrac{2}{3}c - 2\omega = c + \dfrac{b}{z},$

the equation becomes

$$\lambda z^2 + 2\mu(cz^2 + bz) + \nu(cz + b)^2 = 0\,;$$

or substituting for λ, μ, ν their values

$$b^2(x^2 + y^2 + z^2 - 2yz - 2zx - 2xy)$$
$$- 4bc\,xyz$$
$$- 2bd\,xyz(x + y + z)$$
$$- 4be\,xyz(yz + zx + xy)$$
$$+ (d^2 - 4ce)\,x^2y^2z^2 = 0\,;$$

viz. this is a particular integral of

$$\frac{dx}{\sqrt{X}} + \frac{dy}{\sqrt{Y}} + \frac{dz}{\sqrt{Z}} = 0,$$

where $X = bx + cx^2 + dx^3 + ex^4$, &c.

446. It would be easy to verify this by writing the integral equation successively in the forms

$$u = A + 2Bx + Cx^2 = A' + 2B'y + C'y^2 = A'' + 2B''z + C''z^2\,;$$

we then have B^2-AC, $B'^2-A'C'$, $B''^2-A''C''$ proportional to YZ, ZX, XY respectively.

Write $b, c, d, e = 1, 0, -I, 2J$; then X becomes $x-Ix^3+2Jx^4$, which putting therein $\frac{1}{x}$ instead of x is $\frac{1}{x^4}(x^3-Ix+2J)$; writing similarly $\frac{1}{y}$, $\frac{1}{z}$ for y, z, and putting finally

$$X=x^3-Ix+2J,\ Y=y^3-Iy+2J,\ Z=z^3-Iz+2J,$$

we have

$$\begin{aligned}&\Gamma\\ &-8J(x+y+z)\\ &+2I(yz+zx+xy)\\ &+y^2z^2+z^2x^2+x^2y^2-2xyz(x+y+z)=0,\end{aligned}$$

as a particular integral of

$$\frac{dx}{\sqrt{X}}+\frac{dy}{\sqrt{Y}}+\frac{dz}{\sqrt{Z}}=0;$$

this can of course be directly verified in the same manner.

CHAPTER XV.

ON THE DETERMINATION OF CERTAIN CURVES, THE ARC OF WHICH IS REPRESENTED BY AN ELLIPTIC INTEGRAL OF THE FIRST KIND.

Outline of the Solution. Art. Nos. 447 to 449.

447. In Chapter III. it was seen that the lemniscate was a curve such that its arc represented an elliptic integral of the first kind: but the problem of finding such a curve is obviously an indeterminate one; we have to find x, y functions of z, such that

$$dx^2 + dy^2 = \frac{dz^2}{\overline{1-z^2}\,.\,\overline{1-k^2z^2}};$$

for this being so then, writing $z = \sin\phi$, the arc of the curve, measured from the point for which $z = 0$, will be $s = F(k, \phi)$.

Similarly if a, α are conjugate imaginaries, and x, y are functions of z, such that

$$dx^2 + dy^2 = \frac{dz^2}{\overline{z^2-a^2}\,.\,\overline{z^2-\alpha^2}};$$

then the expression for the arc of the curve is

$$s = \int \frac{dz}{\sqrt{\overline{z^2-a^2}\,.\,\overline{z^2-\alpha^2}}},$$

a form in the nature of an elliptic integral of the first kind, and which can in fact be made to depend on elliptic integrals.

448. A very general mode of satisfying the equation is to assume

$$dx + i\,dy = \frac{(z-a)^m\,(z+a)^n\,dz}{(z-\alpha)^{m+1}\,(z+\alpha)^{n+1}};$$

for then x, y being real functions of s, we have also

$$dx - idy = \frac{(s-\alpha)^m (s+a)^n\, ds}{(s-a)^{m+1}(s+\alpha)^{n+1}};$$

and multiplying, we have the relation in question.

The above expression of $dx + idy$ as a multiple of ds is not in general integrable, but it is to be shown that if one of the indices m, n, say m, is a positive integer, and provided a single relation is satisfied between a, α (the form of this equation depending on m, n) then that the expression is integrable algebraically: viz. we obtain by means of it an algebraical (imaginary) value of $x + iy$; this of course gives x, y equal to real algebraical functions of (x, y), and thus determines a curve, the arc of which is expressed by the formula

$$s = \int \frac{dz}{\sqrt{z^2 - a^2\,.\,z^2 - \alpha^2}}.$$

449. The form of the relation between the (a, α) is a very remarkable one, viz. writing $\zeta = \dfrac{(a+\alpha)^2}{4a\alpha}$, then the relation is

$$\frac{1}{\zeta^{n-m}} \cdot \left(\frac{d}{d\zeta}\right)^m \zeta^n (\zeta - 1)^m = 0;$$

this is an equation of the order m in ζ giving for ζ m values which (n being within certain limits) are all or some of them real and less than unity, and the corresponding values of a, α are then conjugate imaginary values, in accordance with the original supposition.

Thus if $m = 1$, the equation is $\dfrac{1}{\zeta^{n-1}} \dfrac{d}{d\zeta} \cdot \zeta^n (\zeta - 1) = 0$, that is $(n+1)\zeta - n = 0$ or $\zeta = \dfrac{n}{n+1}$; which, n being positive, is positive and less than 1; if $m = 2$, it is $\dfrac{1}{\zeta^{n-2}} \left(\dfrac{d}{d\zeta}\right)^2 \zeta^n (\zeta - 1)^2 = 0$, viz. this is

$$(n+2)(n+1)\zeta^2 - 2(n+1)n\zeta + n(n-1) = 0,$$

or

$$(n+2)\zeta = n \pm \sqrt{\frac{2n}{n+1}}.$$

If n is positive and less than 1, one value of ζ; if n be greater than 1, each value of ζ; is positive and less than 1. It is to be observed that if n is integral, and less than m, the equation as above obtained contains the factor ζ^{m-n}, and throwing this out sinks to the degree n; the equation may in fact be written indifferently in the forms

$$\frac{1}{\zeta^{m-n}}\left(\frac{d}{d\zeta}\right)^m \zeta^m(\zeta-1)^n=0;\quad \frac{1}{(1-\zeta)^{n-m}}\left(\frac{d}{d\zeta}\right)^n \zeta^m(\zeta-1)^n=0,$$

the degree being m or n whichever is least. The values of ζ are in this case all of them positive and less than 1.

General Theorem of Integration. Art. Nos. 450 to 437.

450. The foregoing result depends on a general theorem of integration which is as follows: taking θ any positive integer, the integral

$$\int\frac{(u+p)^{m+n-\theta}(u+q)^{\theta}\,du}{u^{m+1}(u+p+q)^{n+1}}$$

has an algebraical value provided a single relation subsists between p, q, m, n: viz. writing

$$([m]\,p^{\prime}+[n]\,q^{\prime})^{\theta}$$

to denote

$$[m]^{\theta}p^{\theta}+\frac{\theta}{1}[m]^{\theta-1}[n]^{1}p^{\theta-1}q^{1}+\ldots+[n]^{\theta}q^{\theta},$$

where as usual $[m]^{\theta}$ represents the factorial

$$m(m-1)\ldots(m-\theta+1),$$

the required relation is

$$([m]\,p^{\prime}+[n]\,q^{\prime})^{\theta}=0.$$

451. If in this theorem, m being a positive integer, we take $\theta=m$, and writing $u=z-a$, take $p=a+\alpha$, $q=a-\alpha$, we have the integral

$$\int\frac{(z-\alpha)^{m}(z+\alpha)^{n}\,dz}{(z-a)^{m+1}(z+a)^{n+1}}$$

having an algebraical value, provided there is satisfied between a, α, m, n the relation

$$\{[m]\,(a+\alpha)^{\prime}+[n]\,(a-\alpha)^{\prime}\}^{m}=0.$$

Or taking as before $\zeta = \frac{(a+x)^2}{4ax}$, we have $\zeta - 1 = \frac{(a-x)^2}{4ax}$, and the equation may be written

$$\{[m]\zeta + [n](\zeta - 1)\}^m = 0,$$

which is the before-mentioned equation in ζ: thus, $m = 2$, the equation is

$$[2]^2 \zeta^2 + 2[2]^1 [n]^1 \zeta(\zeta - 1) + [n]^2 (\zeta - 1)^2 = 0,$$

that is, $\quad 2\zeta^2 + 4n\zeta(\zeta - 1) + (n^2 - n)(\zeta - 1)^2 = 0$;

or $\quad (n^2 - n)(\zeta^2 - 2\zeta + 1) + 4n(\zeta^2 - \zeta) + 2\zeta^2 = 0,$

which is $\quad (n+2)(n+1)\zeta^2 - 2(n+1)n\zeta + n(n-1) = 0,$

as above.

452. To prove the general theorem, write for shortness

$$U = (u+p)^{m+n-\theta+1}(u+p+q)^{-n}.$$

The integral then is

$$\int \frac{U(u+q)^\theta du}{u^{m+1}(u+p)(u+p+q)},$$

which we assume to be

$$= Uu^{-m}(A + Bu + Cu^2 \ldots + Ku^{\theta-1}),$$

say it is $\quad = UQ.$

This will be the case if

$$UQ' + U'Q = \frac{U(u+q)^\theta}{u^{m+1}(u+p)(u+p+q)},$$

or what is the same thing,

$$\frac{U'}{U}Q + Q' = \frac{(u+q)^\theta}{u^{m+1}(u+p)(u+p+q)};$$

viz. substituting for U' its value, this is

$$[(m+n-\theta+1)(u+p+q) - n(u+p)]\,Q$$
$$+ (u+p)(u+p+q)\,Q' = \frac{(u+q)^\theta}{u^{m+1}},$$

where Q' denotes $\frac{dQ}{du}$. The question therefore is to express that this differential equation has an integral

$$Q = u^{-m}(A + Bu + Cu^2 \ldots + Ku^{\theta-1}).$$

Substituting this value and equating coefficients, we have between the θ coefficients $A, B, C \ldots K$, a system of $\theta+1$ equations implying one relation between the quantities m, n, p, q: and this condition being satisfied, the coefficients $A, B \ldots K$ will be completely determined, or we have for Q an equation of the form in question.

453. For instance, if $\theta = 1$, the equation is

$$[mu + mp + \overline{m+n}\,q]\,Q + [u^2 + u(2p+q) + p^2 + pq]\,Q' = \frac{u+q}{u^{m+1}},$$

to be satisfied by $Q = Au^{-m}$: this gives

$$\begin{aligned}&[mp + (m+n)\,q + mu \quad] \quad Au^{-m}\\ &+ [p^2 + pq + (2p+q)\,u + u^2] \,.\, - mAu^{-m-1} = qu^{-m-1} + u^{-m},\end{aligned}$$

that is

$$\begin{array}{lllll} u^{-m-1} & [& & -m(p^2+pq)A - q &] \\ u^{-m} & [& [mp+(m+n)q]\,A & -m(2p+q)\,A - 1 &] \\ u^{-m+1} & [& +mA & -mA &] = 0, \end{array}$$

viz. the equations are

$$m(p^2 + pq)\,A + q = 0,$$

$$(mp - nq)\,A + 1 = 0,$$

whence eliminating A we have $m(p^2+pq) - q(mp - nq) = 0$, that is $mp^2 + nq^2 = 0$, as the required relation in the case in hand.

454. Similarly if $\theta = 2$, the differential equation is

$$\begin{aligned}&[\overline{m-1}\,p + \overline{m+n-1}\,q + \overline{m-1}\,u]\,Q\\ &\quad + [p^2 + pq + u(2p+q) + u^2]\,Q' = \frac{(u+q)^2}{u^{m+1}},\end{aligned}$$

satisfied by $Q = Au^{-m} + Bu^{-m+1}$. This gives

$$\begin{array}{c|c|c|c|l}
\overbrace{\qquad\qquad}^{u^{-m-1}} & \overbrace{\qquad\qquad}^{u^{-m}} & \overbrace{\qquad\qquad}^{u^{-m+1}} & \overbrace{\qquad\qquad}^{u^{-m+2}} & \\
 & (\overline{m-1}p+\overline{m+n-1}q)A & (\overline{m-1}p+\overline{m+n-1}q)B & & \\
\text{,,} & \text{,,} & \overline{m-1}\,A & (m-1)B & \\
-m(p^2+pq)A & -(m-1)(p^2+pq)B & \text{,,} & \text{,,} & =0, \\
\text{,,} & -m(2p+q)A & -(m-1)(2p+q)B & & \\
\text{,,} & & -mA & -(m-1)B & \\
-q^2 & -2q & -1 & &
\end{array}$$

that is

$$\begin{aligned}
&1 + [\overline{m-1}p - nq]\,B &&+ 1A = 0,\\
&2q + (m-1)(p^2+pq)\,B + (\overline{m+1}p - \overline{n-1}q)\,A &&= 0,\\
&q^2 \qquad\qquad + &&m(p^2+pq)\,A = 0,
\end{aligned}$$

and the elimination of A, B gives the required condition, for the case $\theta = 2$ now in question.

455. The series of equations are

$$\theta=1,\ 0=\begin{vmatrix} 1, & mp-nq \\ q, & m(p^2+pq) \end{vmatrix}$$

$$\theta=2,\ 0=\begin{vmatrix} 1, & \overline{m-1}p-nq, & 1 \\ 2q, & \overline{m-1}(p^2+pq), & \overline{m+1}p-\overline{n-1}q \\ q^2, & & m(p^2+pq) \end{vmatrix}$$

$$\theta=3,\ 0=\begin{vmatrix} 1, & \overline{m-2}p-nq, & 1 & \\ 3q, & \overline{m-2}(p^2+pq), & mp-\overline{n-1}q, & 2 \\ 3q^2, & . & \overline{m-1}(p^2+pq), & \overline{m+2}p-\overline{n-2}q \\ q^3, & & . & m(p^2+pq) \end{vmatrix}$$

$$\theta=4,\ 0=\begin{vmatrix} 1, & \overline{m-3}p-nq, & 1 & . & . \\ 4q, & \overline{m-3}(p^2+pq), & \overline{m-1}p-\overline{n-1}q, & 2 & . \\ 6q^2, & . & \overline{m-2}(p^2+pq), & \overline{m+1}p-\overline{n-2}q, & 3 \\ 4q^3, & & . & \overline{m-1}(p^2+pq), & \overline{m+3}p-\overline{n-3}q \\ q^4, & & . & . & m(p^2+pq) \end{vmatrix}$$

450. Expanding the several determinants the equations are, for the case $\theta = 1$,

$$\begin{array}{ll} [m]^1 p(p+q) & , = ([m]p^2+[n]q^2)^1, = 0 \\ -1(q[m]p-[n]q) & \end{array}$$

for the case $\theta = 2$

$$\begin{array}{lll} [m]^2 p^2 (p+q)^2 & & = ([m]p^2+[n]q^2)^2, = 0 \\ -2[m]^1 p(p+q) q([m-1]p-[n]q)^1 & & \\ +1 \qquad\qquad q^2([m]p-[n]q)^2 & & \end{array}$$

for the case $\theta = 3$

$$\begin{array}{ll} [m]^3 p^3 (p+q)^3 & = ([m]p^2+[n]q^2)^3, = 0 \\ -3[m]^2 p^2 (p+q)^2 q([m-2]p-[n]q)^1 & \\ +3[m]^1 p(p+q) q^2([m-1]p-[n]q)^2 & \\ -1 \qquad\qquad q^3([m]p-[n]q)^3 & \end{array}$$

for the case $\theta = 4$

$$\begin{array}{ll} [m]^4 p^4 (p+q)^4 & = ([m]p^2+[n]q^2)^4, \\ -4[m]^3 p^3 (p+q)^3 q([m-3]p-[n]q)^1 & \\ +6[m]^2 p^2 (p+q)^2 q^2([m-2]p-[n]q)^2 & \\ -4[m]^1 p(p+q) q^3([m-1]p-[n]q)^3 & \\ +1 \qquad\qquad q^4([m]p-[n]q) & \end{array}$$

and so on. The notations $([m]p-[n]q)^1$, $([m]p-[n]q)^2$ have a signification analogous to $([m]p^2+[n]q^2)^1$, $([m]p^2+[n]q^2)^2$, &c. already explained: for instance

$$([m]p-[n]q)^2 = [m]^2 p^2 - 2[m]^1 [n]^1 pq + [n]^2 q^2.$$

457. To show how the reduction is effected, consider for instance the second determinant; this contains terms multiplied by 1, $2q$, q^2 respectively:

the first is

$$1 \,.\, \overline{m-1}\,(p^2+pq) \,.\, m\,(p^2+pq), \;= 1 \,.\, [m]^2 p^2 (p+q)^2;$$

the second is

$$2q.-m(p^2+pq)\{(m-1)p-nq\}=-2[m]^2p(p+q)q([m-1]p-[n]q)^2;$$

the third is

$$q^2(\overline{m-1}\,p-nq)\{(\overline{m+1}\,p-\overline{n-1}\,q)-(m-1)(p^2+pq)\}$$
$$=q^2\{(m^2-m)p^2-2mnpq+(n^2-n)q^2\},\ =q^2([m]p-[n]q)^2.$$

And similarly the third determinant is composed of terms in 1, $3q$, $3q^2$, q^3, which are the four terms in the first reduced expression of the determinant: and so in other cases. These first reduced expressions give without difficulty the final forms

$$([m]p^2+[n]q^2)^2,\ ([m]p^3+[n]q^3)^3,\ \&c.$$

458. Writing $\theta=n$, and $z-a$, $\alpha-a$, $\alpha+a$ for u, p, q respectively, we have the originally mentioned theorem in regard to the integral

$$\int\frac{(z-\alpha)^m(z+\alpha)^n\,dz}{(z-a)^{m+1}(z+a)^{n+1}},$$

and thence, as already mentioned, the expressions of x, y as functions of a parameter z such that the arc of the curve is given by the formula

$$s=\int\frac{dz}{\sqrt{z^2-a^2\,.\,z^2-\alpha^2}},$$

viz. as an integral in the nature of an elliptic integral of the first kind.

CHAPTER XVI.

ON TWO INTEGRALS REDUCIBLE TO ELLIPTIC INTEGRALS.

459. An integral $\int \frac{dx}{\sqrt{P}}$, where P is a quintic function of x, is not in general reducible to elliptic integrals; but Jacobi has shown (*Crelle*, t. VIII. (1832) p. 416) that if P has the particular form

$$P = x(1-x)(1+\kappa x)(1+\lambda x)(1-\kappa\lambda x),$$

then the integrals $\int \frac{dx}{\sqrt{P}}$, $\int \frac{x\,dx}{\sqrt{P}}$, that is the two integrals

$$\int \frac{dx}{\sqrt{x\,.\,1-x\,.\,1+\kappa x\,.\,1+\lambda x\,.\,1-\kappa\lambda x}},\quad \int \frac{\sqrt{x}\,dx}{\sqrt{1-x\,.\,1+\kappa x\,.\,1+\lambda x\,.\,1-\kappa\lambda x}},$$

are reducible to elliptic integrals: and that by means of the theory an elliptic integral of the first kind $\int \frac{d\phi}{\sqrt{1-k^2\sin^2\phi}}$, where k is a complex imaginary quantity, say $k = \sin(\alpha+\beta i)$, can be reduced to the form $G + Hi$, where G and H are real integrals of the above-mentioned kind.

Investigation of the Formulæ. Art. Nos. 460 to 463.

460. Considering the integral

$$\int \frac{dx}{\sqrt{x}\sqrt{1-x\,.\,1+\kappa x\,.\,1+\lambda x\,.\,1-\kappa\lambda x}},\quad = \int \frac{dx}{\sqrt{x}X} \text{ for shortness,}$$

viz. X used to denote

$$1-x\,.\,1+\kappa x\,.\,1+\lambda x\,.\,1-\kappa\lambda x;$$

write

$$b = \sqrt{\kappa} + \sqrt{\lambda} \quad (+), \qquad c = \sqrt{\kappa} - \sqrt{\lambda} \quad (+),$$

$$b' = 1 - \sqrt{\kappa\lambda} \quad (+), \qquad c' = 1 + \sqrt{\kappa\lambda} \quad (+),$$

$$\text{denom.} = \sqrt{1+\kappa \,.\, 1+\lambda};$$

and therefore

$$b^2 + b'^2 = 1,$$

$$c^2 + c'^2 = 1.$$

Assume

$$\sqrt{x} = \frac{(b'+c')\sin\phi}{\sqrt{1-b^2\sin^2\phi} + \sqrt{1-c^2\sin^2\phi}},$$

$$= \frac{(b'+c')\sin\phi}{B+C} \text{ for shortness,}$$

we have

$$(1+\kappa x)(1+\lambda x) = (B+C)^4 + (\kappa+\lambda)(B+C)^2(b'+c')^2\sin^2\phi$$
$$+ \kappa\lambda(b'+c')^4\sin^4\phi \quad (+),$$

$$(1-x)(1-\kappa\lambda x) = (B+C)^4 - (1+\kappa\lambda)(B+C)^2(b'+c')^2\sin^2\phi$$
$$+ \kappa\lambda(b'+c')^4\sin^4\phi \quad (+),$$

$$\text{denom.} = (B+C)^4;$$

which after all reductions become

$$(1+\kappa x)(1+\lambda x) = \frac{1}{(B+C)^4} \,.\, 4(B+C)^2, \qquad = \frac{4}{(B+C)^2},$$

$$(1-x)(1-\kappa\lambda x) = \frac{1}{(B+C)^4} \,.\, 4(B+C)^2\cos^2\phi, \quad = \frac{4\cos^2\phi}{(B+C)^2}.$$

461. We have in fact

$$\frac{c'-b'}{c'+b'} = \sqrt{\kappa\lambda}, \quad \kappa\lambda = \left(\frac{c'-b'}{c'+b'}\right)^2,$$

$$\frac{b^2+c^2}{b'^2+c'^2} = \frac{\kappa+\lambda}{1+\kappa\lambda};$$

and thence

$$\kappa+\lambda = \left\{1 + \left(\frac{c'-b'}{c'+b'}\right)^2\right\}\frac{b^2+c^2}{b'^2+c'^2}, \quad = \frac{2(b^2+c^2)}{(b'+c')^2},$$

and $\quad (1+\kappa)(1+\lambda) = \dfrac{4}{(b'+c')^2}.$

Hence observing that

$$\kappa\lambda\,(b'+c')^4\sin^4\phi = (b'^2-c'^2)^2\sin^4\phi, \quad = (B^2-C^2)^2,$$

we have

$$(B+C)^4 + (\kappa+\lambda)\,(B+C)^2\,(b'+c')^2\sin^2\phi + \kappa\lambda\,(b'+c')^4\sin^4\phi$$
$$= (B+C)^4 + 2\,(b'^2+c'^2)\,(B+C)^2\sin^2\phi + (B^2-C^2)^2,$$
$$= (B+C)^2\,[(B+C)^2 + 2\,(b'^2+c'^2)\sin^2\phi + (B-C)^2]$$
$$= 2\,(B+C)^2\,\{B^2+C^2+(b'^2+c'^2)\sin^2\phi\}$$
$$= 4\,(B+C)^2.$$

Also

$$(B+C)^4 - (1+\kappa\lambda)\,(B+C)^2\,(b'+c')^2\sin^2\phi + \kappa\lambda\,(b'+c')^4\sin^4\phi$$
$$= (B+C)^4 - 2\,(b'^2+c'^2)\,(B+C)^2\sin^2\phi + (B^2-C^2)^2$$
$$= (B+C)^2\,[(B+C)^2 - 2\,(b'^2+c'^2)\sin^2\phi + (B-C)^2]$$
$$= 2\,(B+C)^2\,[B^2+C^2-(b'^2+c'^2)\sin^2\phi]$$
$$= 4\,(B+C)^2\cos^2\phi,$$

and we have thence the formulæ in question.

402. Moreover from the equation

$$\sqrt{x} = \frac{(b'+c')\sin\phi}{B+C},$$

we have

$$\frac{dx}{2\sqrt{x}} = \frac{(b'+c')\cos\phi}{(B+C)^2}\left\{B+C+\sin^2\phi\left(\frac{b'^2}{B}+\frac{c'^2}{C}\right)\right\}d\phi$$

$$= \frac{(b'+c')\cos\phi}{(B+C)^2\,BC}\left\{(B^2+b'^2\sin^2\phi)\,C + (C^2+c'^2\sin^2\phi)\,B\right\}d\phi$$

$$= \frac{(b'+c')\cos\phi\,d\phi}{(B+C)\,BC};$$

and combining herewith the foregoing equations

$$\sqrt{1-x\,.\,1-\kappa\lambda x} = \frac{2\cos\phi}{B+C},$$

$$\sqrt{1+\kappa x\,.\,1+\lambda x} = \frac{2}{B+C},$$

whence also

$$\sqrt{X} = \sqrt{1-x\,.\,1+\kappa x\,.\,1+\lambda x\,.\,1-\kappa\lambda x} = \frac{4\cos\phi}{(B+C)^2}\,;$$

we have therefore

$$\frac{dx}{\sqrt{xX}} = \tfrac{1}{2}(b'+c')\,d\phi\left(\frac{1}{B}+\frac{1}{C}\right).$$

463. Moreover

$$x = \frac{(b'+c')^2\sin^2\phi}{(B+C)^2},$$

and thence

$$\frac{\sqrt{x}\,dx}{\sqrt{X}} = \tfrac{1}{2}(b'+c')^3\sin^2\phi\,\frac{1}{BC(B+C)}\,d\phi$$

$$= \tfrac{1}{2}(b'+c')^3\sin^2\phi\,\frac{B-C}{BC(B^2-C^2)}\,d\phi.$$

Or since

$$B^2-C^2 = -(b'^2-c'^2)\sin^2\phi, \quad = (c'^2-c'^2)\sin^2\phi,$$

this is

$$\frac{\sqrt{x}\,dx}{\sqrt{X}} = \tfrac{1}{2}\,\frac{(b'+c')^2}{(b'-c')}\cdot\frac{B-C}{BC}\,d\phi$$

$$= \tfrac{1}{2}\,\frac{(c'+b')^2}{c'-b'}\left(\frac{1}{B}-\frac{1}{C}\right)d\phi,$$

viz. we have the two equations

$$\frac{dx}{\sqrt{xX}} = \tfrac{1}{2}(c'+b')\left(\frac{1}{B}+\frac{1}{C}\right)d\phi,$$

$$\frac{\sqrt{x}\,dx}{\sqrt{X}} = \tfrac{1}{2}\,\frac{(c'+b')^2}{c'-b'}\left(\frac{1}{B}-\frac{1}{C}\right)d\phi,$$

where

$$X = \sqrt{1-x\,.\,1+\kappa x\,.\,1+\lambda x\,.\,1-\kappa\lambda x},$$

$$B = \sqrt{1-b'^2\sin^2\phi},$$

$$C = \sqrt{1-c'^2\sin^2\phi},$$

and as above

$$\sqrt{x} = \frac{(b'+c')\sin\phi}{B+C}.$$

We thus see that the two integrals in question $\int\frac{dx}{\sqrt{xX}}$, $\int\frac{\sqrt{x}\,dx}{\sqrt{X}}$ depend upon the two elliptic integrals of the first kind,

$$\int\frac{d\phi}{\sqrt{1-b^2\sin^2\phi}},\quad \int\frac{d\phi}{\sqrt{1-c^2\sin^2\phi}},$$

which is the theorem in question.

Further Developments. Art. Nos. 404 to 408.

404. We may express ϕ as a function of x; viz. the last equation gives

$$B+C=\frac{(b'+c')\sin\phi}{\sqrt{x}},$$

and thence $\quad B^2+C^2-\frac{(b'+c')^2\sin^2\phi}{x}=-2BC,$

and $(B^2-C^2)^2-2(B^2+C^2)\frac{(b'+c')^2\sin^2\phi}{x}+\frac{(b'+c')^4\sin^4\phi}{x^2}=0$;

or since as before $B^2-C^2=(b'^2-c'^2)\sin^2\phi$, the whole equation divides by $(b'+c')^2\sin^2\phi$, and throwing out this factor it becomes

$$(b'-c')^2\sin^2\phi-2(B^2+C^2)\frac{1}{x}+\frac{(b'+c')^2\sin^2\phi}{x^2}=0,$$

that is

$$(b'-c')^2x^2\sin^2\phi-2x\left[2-(b'^2+c'^2)\sin^2\phi\right]+(b'+c')^2\sin^2\phi=0,$$

viz. $\quad \sin^2\phi\left[(b'+c')^2+2(b'^2+c'^2)x+(b'-c')^2x^2\right]=4x;$

that is

$$\sin^2\phi=\frac{4x}{(b'+c')^2+2(b'^2+c'^2)x+(b'-c')^2x^2}$$

$$=\frac{4x}{(b'+c')^2}\cdot\frac{1}{1+\frac{2(b'^2+c'^2)}{(b'+c')^2}x+\left(\frac{b'-c'}{b'+c'}\right)^2x^2}$$

$$=\frac{(1+\kappa)(1+\lambda)x}{1+\kappa x\,.\,1+\lambda x};$$

hence we may write

$$\sin^2\phi = 1+\kappa\,.\,1+\lambda\,.\,x \qquad (\div),$$

$$\cos^2\phi = 1-x\,.\,1-\kappa\lambda x \qquad (\div),$$

$$B^2 = (1-\sqrt{\kappa\lambda}\,x)^2 \qquad (\div),$$

$$C^2 = (1+\sqrt{\kappa\lambda}\,x)^2 \qquad (\div),$$

where $\qquad \text{denom.} = 1+\kappa x\,.\,1+\lambda x.$

465. It may be remarked that writing

$$1+\theta x = [(B+C)^2+\theta(b'+c')^2\sin^2\phi] \div (B+C)^2$$

$$= [2-(b^2+c^2)\sin^2\phi$$

$$+\theta(b'+c')^2\sin^2\phi+2BC] \div (B+C)^2,$$

and endeavouring to make the numerator a square, it will be the square of

$$\sqrt{1+b\sin\phi}\,\sqrt{1+c\sin\phi}+\sqrt{1-b\sin\phi}\,\sqrt{1-c\sin\phi},$$

or else of

$$\sqrt{1+b\sin\phi}\,\sqrt{1-c\sin\phi}+\sqrt{1-b\sin\phi}\,\sqrt{1+c\sin\phi};$$

viz. in the first case we must have

$$2-(b^2+c^2)\sin^2\phi+\theta(b'+c')^2\sin^2\phi = 2+2bc\sin^2\phi,$$

that is

$$-(b^2+c^2)+\theta(b'+c')^2 = 2bc, \text{ or } \theta = \frac{(b+c)^2}{(b'+c')^2}, = \kappa:$$

and in the second case

$$2-(b^2+c^2)\sin^2\phi+\theta(b'+c')^2\sin^2\phi = 2-2bc\sin^2\phi,$$

that is

$$-(b^2+c^2)+\theta(b'+c')^2 = -2bc, \text{ or } \theta = \frac{(b-c)^2}{(b'+c')^2}, = \lambda.$$

Hence the two equations are

$$1+\kappa x=[\sqrt{1+b\sin\phi}\sqrt{1+c\sin\phi}+\sqrt{1-b\sin\phi}\sqrt{1-c\sin\phi}]^2\div(B+C)^2,$$

$$1+\lambda x=[\sqrt{1+b\sin\phi}\sqrt{1-c\sin\phi}+\sqrt{1-b\sin\phi}\sqrt{1+c\sin\phi}]^2\div(B+C)^2,$$

leading to the before-mentioned equation

$$1+\kappa x\,.\,1+\lambda x=4\div(B+C)^2,$$

but there are no analogous values of $1-x$, $1-\kappa x$ to lead to

$$1-x\,.\,1-\kappa\lambda x=4\cos^2\phi\div(B+C)^2.$$

400. Write now

$$b=\sin(a+\beta),\quad c=\sin(a-\beta),$$

and therefore

$$b'=\cos(a+\beta),\quad c'=\cos(a-\beta);$$

we hence obtain

$$\sqrt{\kappa}=\frac{2\sin a\cos\beta}{2\cos a\cos\beta}=\tan a,\quad \sqrt{x}=\frac{2\cos a\cos\beta\sin\phi}{B+C},$$

$$\sqrt{\lambda}=\frac{2\sin\beta\cos a}{2\cos a\cos\beta}=\tan\beta,$$

$$B=\sqrt{1-\sin^2(a+\beta)\sin^2\phi},$$

$$C=\sqrt{1-\sin^2(a-\beta)\sin^2\phi},$$

$$X=(1-x)(1+x\tan^2 a)(1+x\tan^2\beta)(1-x\tan^2 a\tan^2\beta),$$

and therefore

$$\frac{dx}{\sqrt{xX}}=\cos a\cos\beta\left(\frac{1}{B}+\frac{1}{C}\right)d\phi,$$

$$\frac{\sqrt{x}\,dx}{\sqrt{X}}=\frac{\cos^2 a\cos^2\beta}{\sin a\sin\beta}\left(\frac{1}{B}-\frac{1}{C}\right)d\phi.$$

Writing this last equation in the form

$$\tan a\tan\beta\,\frac{\sqrt{x}\,dx}{\sqrt{X}}=\cos a\cos\beta\left(\frac{1}{B}-\frac{1}{C}\right)d\phi,$$

we have

$$2\cos\alpha\cos\beta\,\frac{d\phi}{B}=\frac{dx}{\sqrt{xX}}(1+x\tan\alpha\tan\beta),$$

$$2\cos\alpha\cos\beta\,\frac{d\phi}{C}=\frac{dx}{\sqrt{xX}}(1-x\tan\alpha\tan\beta).$$

If in these equations we write βi for β, X continues a real function, viz. we have

$$X=(1-x)(1+x\tan^2\alpha)(1+x\tan^2\beta i)(1-x\tan^2\alpha\tan^2\beta i);$$

and the formulæ are

$$\sqrt{x}=\frac{2\cos\alpha\cos\beta i\sin\phi}{B+C},\ \text{giving rise to}$$

$$2\cos\alpha\cos\beta i\,\frac{d\phi}{B}=\frac{dx}{\sqrt{xX}}(1+x\tan\alpha\tan\beta i),$$

$$2\cos\alpha\cos\beta i\,\frac{d\phi}{C}=\frac{dx}{\sqrt{xX}}(1-x\tan\alpha\tan\beta i),$$

where observe that $B+C$,

$$=\sqrt{1-\sin^2(\alpha+\beta i)\sin^2\phi}+\sqrt{1-\sin^2(\alpha-\beta i)\sin^2\phi}$$

is real; viz. these formulæ give the values of

$$\int\frac{d\phi}{\sqrt{1-\sin^2(\alpha+\beta i)\sin^2\phi}},\quad \int\frac{d\phi}{\sqrt{1-\sin^2(\alpha-\beta i)\sin^2\phi}},$$

in terms of the integrals

$$\int\frac{dx}{\sqrt{xX}}\ \text{and}\ \int\frac{\sqrt{x}\,dx}{\sqrt{X}}.$$

We may change the form by writing $\tan\beta i=i\sin\gamma$, whence

$$\cos\beta i=\frac{1}{\cos\gamma},\quad \sin\beta i=i\tan\gamma:$$

we thus have

$$\kappa=\tan^2\alpha,\quad \lambda=-\sin^2\gamma,$$

$$\sqrt{x} = \frac{2\cos\alpha}{\cos\gamma} \cdot \frac{\sin\phi}{B+C},$$

$$X = (1-x)(1+x\tan^2\alpha)(1-x\sin^2\gamma)(1+x\tan^2\alpha\sin^2\gamma),$$

$$\frac{2\cos\alpha}{\cos\gamma} \cdot \frac{d\phi}{B} = \frac{dx}{\sqrt{xX}}(1+ix\tan\alpha\sin\gamma),$$

$$\frac{2\cos\alpha}{\cos\gamma} \cdot \frac{d\phi}{C} = \frac{dx}{\sqrt{xX}}(1-ix\tan\alpha\sin\gamma),$$

and observing the equation

$$\sin^2\phi = \frac{(1+\kappa)(1+\lambda)x}{(1+\kappa x)(1+\lambda x)} = \frac{1}{(\cos^2\alpha + x\sin^2\alpha)} \frac{x\cos^2\gamma}{1-x\sin^2\gamma},$$

we see that to real values of ϕ there correspond values of x which are positive and less than 1, and that as x passes from 0 to 1, $\sin^2\phi$ passes from 0 to 1, or ϕ from 0 to 90°, X being thus always real and positive.

Writing $\sin\phi = y$, the relation between ϕ, x gives a relation between x, y: viz. this is

$$\sqrt{x} = \frac{(b'+c')y}{\sqrt{1-b'^2y^2}+\sqrt{1-c'^2y^2}},$$

or what is the same thing

$$y^2 = \frac{(1+\kappa)(1+\lambda)x}{(1+\kappa x)(1+\lambda x)},$$

viz. this is a quartic curve; and introducing z for homogeneity, or writing the equation in the form

$$y^2(z+\kappa x)(z+\lambda x) - (1+\kappa)(1+\lambda)xz^3 = 0,$$

we see that

$x=0$, $z=0$ is a flecnode, the tangents being $z+\kappa x=0$, $z+\lambda x=0$;

$y=0$, $z=0$ is a cusp, the tangent being $y=0$;

$x=0$, $y=0$ is an ordinary point, the tangent being $x=0$;

hence the curve, as having a node and cusp, is bicursal.

407. The transformation of a given imaginary modulus into the form $\sin(\alpha+\beta i)$ presents of course no difficulty: assuming that we have $k = e + fi$, then we have to find α, β such that $e + fi = \sin(\alpha+\beta i)$, or writing $\sin\alpha = \xi$, $\sin\beta i = i\eta$, to find ξ, η from the equations

$$e = \xi\sqrt{1+\eta^2}, \quad f = \eta\sqrt{1-\xi^2}:$$

these give $$e^2 = \xi^2 + \xi^2\eta^2, \quad f^2 = \eta^2 - \xi^2\eta^2,$$

whence $e^2 + f^2 = \xi^2 + \eta^2$, and thence easily

$$\xi^2 = \tfrac{1}{2}\{\ \ 1 + e^2 + f^2 - \sqrt{\nabla}\},$$
$$\eta^2 = \tfrac{1}{2}\{-1 + e^2 + f^2 + \sqrt{\nabla}\},$$

where $$\nabla = 1 + e^4 + f^4 - 2e^2 + 2f^2 + 2e^2f^2.$$

If as above $\sin\beta i = i\tan\gamma$, then $\tan\gamma = \eta$, or the equations give $\xi, = \sin\alpha$, and $\eta, = \tan\gamma$.

408. The integrals $\int\frac{dx}{\sqrt{P}}$, $\int\frac{xdx}{\sqrt{P}}$ are also reducible to elliptic integrals when the quintic function P has the form

$$P = x(1-x)(1+\kappa x)(1+\lambda x)(1+\overline{\kappa+\lambda+\kappa\lambda}\,x),$$

as shown by Prof. M. Roberts in his "Tract on the Addition of Elliptic and Hyper-elliptic Integrals," Dublin, 1871, p. 63; and in the Note, p. 82, to the same work, a simple demonstration is given of the theorem (due to Prof. Gordan) that the like integrals, wherein P denotes a sextic function the skew invariant of which vanishes, are reducible to elliptic integrals.

ADDITION. FURTHER THEORY OF THE LINEAR AND QUADRIC TRANSFORMATIONS.

The Linear Transformation. Art. Nos. 469 to 473.

469. We consider the transformation of the differential expression

$$\frac{dx}{\sqrt{x-\alpha\,.\,x-\beta\,.\,x-\gamma\,.\,x-\delta}},$$

where the new variable y is given by an equation of the form

$$xy + Bx + Cy + D = 0.$$

The coefficients B, C, D might be expressed in terms of any three pairs of corresponding values of the variables x, y, say the values α, β, γ of x, and the corresponding values α', β', γ' of y: but it is better to consider in a symmetrical manner four pairs of corresponding values, viz. the values α, β, γ, δ of x and the corresponding values α', β', γ', δ' of y. We have thus four equations from which B, C, D may be eliminated, and we obtain the relation

$$\begin{vmatrix} \alpha\alpha', & \alpha, & \alpha', & 1 \\ \beta\beta', & \beta, & \beta', & 1 \\ \gamma\gamma', & \gamma, & \gamma', & 1 \\ \delta\delta', & \delta, & \delta', & 1 \end{vmatrix} = 0,$$

which in fact expresses that the two sets of values $(\alpha, \beta, \gamma, \delta)$ and $(\alpha', \beta', \gamma', \delta')$ correspond homographically to each other.

470. Writing for convenience

$$a,\ b,\ c,\ f,\ g,\ h = \beta-\gamma,\ \ \gamma-\alpha,\ \ \alpha-\beta,\ \ \alpha-\delta,\ \ \beta-\delta,\ \ \gamma-\delta,$$

$$a',\ b',\ c',\ f',\ g',\ h' = \beta'-\gamma',\ \ \gamma'-\alpha',\ \ \alpha'-\beta',\ \ \alpha'-\delta',\ \ \beta'-\delta',\ \ \gamma'-\delta',$$

so that identically

$$af+bg+ch=0,\quad a'f'+b'g'+c'h'=0\,;$$

then, as is well known, the relation in question may be expressed in the several forms

$$af : bg : ch = a'f' : b'g' : c'h'\,;$$

or, what is the same thing, there exists a quantity N such that

$$\frac{a'f'}{af}=\frac{b'g'}{bg}=\frac{c'h'}{ch}=N^2.$$

471. The relation between (x, y) may now be expressed in the several forms,

$$\frac{y-\alpha'}{y-\delta'}=P\,\frac{x-\alpha}{x-\delta},\quad \frac{y-\beta'}{y-\delta'}=Q\,\frac{x-\beta}{x-\delta},\quad \frac{y-\gamma'}{y-\delta'}=R\,\frac{x-\gamma}{x-\delta},$$

and writing for (x, y) their corresponding values, the values of P, Q, R are found to be

$$P=\frac{b'h}{bh'}=\frac{c'g}{cg'}\,;\quad Q=\frac{c'f}{cf'}=\frac{a'h}{ah'}\,;\quad R=\frac{a'g}{ag'}=\frac{b'f}{bf'}\,;$$

and we thence obtain

$$f^2PN^2=f'^2QR,\ g^2QN^2=g'^2RP,\ h^2RN^2=h'^2PQ,\ \sqrt{PQR}=\frac{fgh}{f'g'h'}N^3.$$

472. Differentiating any one of the equations in (x, y), for instance the first of them, we find

$$\frac{f'dy}{(y-\delta')^2}=\frac{fP\,dx}{(x-\delta)^2},$$

and then forming the equation

$$\frac{\sqrt{y-\alpha'\,.\,y-\beta'\,.\,y-\gamma'}}{(y-\delta')\sqrt{y-\delta'}}=\frac{\sqrt{PQR}\,\sqrt{x-\alpha\,.\,x-\beta\,.\,x-\gamma}}{(x-\delta)\sqrt{x-\delta}},$$

or if we please

$$\frac{\sqrt{y-\alpha'\,.\,y-\beta'\,.\,y-\gamma'\,.\,y-\delta'}}{(y-\delta')^2}=\frac{\sqrt{PQR}\,\sqrt{x-\alpha\,.\,x-\beta\,.\,x-\gamma\,.\,x-\delta}}{(x-\delta)^2},$$

and attending to the relation $f^n PN' = f^n QR$, we obtain

$$\frac{N\,dy}{\sqrt{y-\alpha'\,.\,\overline{y-\beta'}\,.\,\overline{y-\gamma'}\,.\,\overline{y-\delta'}}} = \frac{dx}{\sqrt{x-\alpha\,.\,\overline{x-\beta}\,.\,\overline{x-\gamma}\,.\,\overline{x-\delta}}},$$

which is the required formula: $(\alpha, \beta, \gamma, \delta)$ and any three, say $(\alpha', \beta', \gamma')$, of the other set of quantities are arbitrary, and the values of δ', N in terms of these are given as above.

473. It is proper to remark that in this and similar formulæ the sign of the multiplier N may be assumed at pleasure: only, this being so, the radicals $\sqrt{X}$ and $\sqrt{Y}$ of the formulæ are not in general both positive; we have between the radicals a relation of the form $F\sqrt{X} = \pm\, G\sqrt{Y}$ (F, G rational functions) wherein the sign $\pm$ has a determinate signification; in fact the last-mentioned relation combined with the differential equation gives $\pm\, NG\,dy = F\,dx$, which equation substituting therein for $\frac{dy}{dx}$ its value, obtained by differentiation as a rational function of (x, y), is a rational equation equivalent, when the sign is taken properly, to the given rational equation between the variables (x, y). The sign $\pm$ of the equation $F\sqrt{X} = \pm\, G\sqrt{Y}$ might have been assumed at pleasure, and the sign of N would then have been determinate; but this is less convenient.

Transformation of a form into itself. Art. No. 474.

474. The homographic relation is satisfied by writing therein

$\alpha', \beta', \gamma', \delta' = (\alpha, \beta, \gamma, \delta), (\beta, \alpha, \delta, \gamma), (\gamma, \delta, \alpha, \beta)$, or $(\delta, \gamma, \beta, \alpha)$:

these values in fact give

a',	b',	c',	f',	g',	h', =
a,	b,	c,	f,	g,	h,
f,	$-g$,	$-c$,	a,	$-b$,	$-h$,
$-f$,	$-b$,	h,	$-a$,	$-g$,	c,
$-a$,	g,	$-h$,	$-f$,	b,	$-c$,

respectively, so that in each case

$$a'f' : b'g' : c'h' = af : bg : ch.$$

We have thus four solutions of the equation

$$\frac{dy}{\sqrt{y-\alpha\,.\,y-\beta\,.\,y-\gamma\,.\,y-\delta}}=\frac{dx}{\sqrt{x-\alpha\,.\,x-\beta\,.\,x-\gamma\,.\,x-\delta}};$$

viz. these are

$$\frac{y-\alpha}{y-\delta}=\frac{x-\alpha}{x-\delta},$$

$$\frac{y-\beta}{y-\gamma}=\frac{\beta-\delta}{\gamma-\alpha}\frac{x-\alpha}{x-\delta},$$

$$\frac{y-\gamma}{y-\beta}=-\frac{\gamma-\delta}{\alpha-\beta}\frac{x-\alpha}{x-\delta},$$

$$\frac{y-\delta}{y-\alpha}=-\frac{\beta-\delta\,.\,\gamma-\delta}{\gamma-\alpha\,.\,\alpha-\beta}\frac{x-\alpha}{x-\delta},$$

the first of them being the self-evident solution $y=x$.

In particular there are four solutions of

$$\frac{dy}{\sqrt{y^2-1\,.\,y^2-\frac{1}{k^2}}}=\frac{dx}{\sqrt{x^2-1\,.\,x^2-\frac{1}{k^2}}},$$

that is

$$\frac{dy}{\sqrt{1-y^2\,.\,1-k^2y^2}}=\frac{dx}{\sqrt{1-x^2\,.\,1-k^2x^2}};$$

viz. these are $y=x$, $y=-x$, $y=\frac{1}{kx}$, and $y=-\frac{1}{kx}$, respectively.

Application to the standard form. Art. Nos. 475 to 477.

475. Considering now the equation

$$\frac{N\,dy}{\sqrt{y^2-1\,.\,y^2-\frac{1}{\lambda^2}}}=\frac{dx}{\sqrt{x^2-1\,.\,x^2-\frac{1}{k^2}}},$$

or, writing $N=\frac{kM}{\lambda}$, say

$$\frac{M\,dy}{\sqrt{1-y^2\,.\,1-\lambda^2y^2}}=\frac{dx}{\sqrt{1-x^2\,.\,1-k^2x^2}},$$

if in the general form we assume

$$\alpha,\ \beta,\ \gamma,\ \delta = 1,\ -1,\ \frac{1}{k},\ -\frac{1}{k},$$

then we have in any one of the twenty-four orders

$$\alpha',\ \beta',\ \gamma',\ \delta' = 1,\ -1,\ \frac{1}{\lambda},\ -\frac{1}{\lambda};$$

and since, for any one of these orders, λ will be determined by a quadric equation, it would at first sight appear that there might be in all twenty-four pairs of solutions, belonging to forty-eight different values of λ, M. But the solutions corresponding to two orders in which $\frac{1}{\lambda}$, $-\frac{1}{\lambda}$ are interchanged, are equivalent; and moreover $y = \phi(x)$ being a solution belonging to determinate values of λ, M, then we have, belonging to the same values of λ, M, the four solutions $y = \phi(x)$, $y = \phi(-x)$, $y = \phi\left(\frac{1}{kx}\right)$ and $y = \phi\left(-\frac{1}{kx}\right)$: we have thus only three pairs of solutions, or say six solutions, belonging each to a different set of values of λ, M; and which correspond to the three orders

α',	β',	γ',	$\delta' =$
$1,$	$-1,$	$\frac{1}{\lambda},$	$-\frac{1}{\lambda},$
$1,$	$\frac{1}{\lambda},$	$-1,$	$-\frac{1}{\lambda},$
$1,$	$\frac{1}{\lambda},$	$-\frac{1}{\lambda},$	$-1.$

476. Forming for each of these the equation which determines λ, say in the form $\frac{a'f'}{af} = \frac{b'g'}{bg}$, we have successively the three equations

$$\left(\frac{1+\lambda}{1-\lambda}\right)^2 = \left(\frac{1+k}{1-k}\right)^2,\quad \frac{(1+\lambda)^2}{4\lambda} = \left(\frac{1+k}{1-k}\right)^2,\quad \frac{4\lambda}{(1+\lambda)^2} = \left(\frac{1+k}{1-k}\right)^2,$$

giving for λ the values

$$k,\ \frac{1}{k};\quad \left(\frac{1-\sqrt{k}}{1+\sqrt{k}}\right)^2,\ \left(\frac{1+\sqrt{k}}{1-\sqrt{k}}\right)^2;\quad \left(\frac{1-i\sqrt{k}}{1+i\sqrt{k}}\right)^2,\ \left(\frac{1+i\sqrt{k}}{1-i\sqrt{k}}\right)^2.$$

The corresponding values of N are derived from the equation $N^2 = \frac{a'f'}{af}$, viz. we thus obtain for $\frac{\lambda N}{k}$, that is for M, the values

$$\frac{1+\lambda}{1+k};\quad \frac{i(1+\lambda)}{1+k};\quad \frac{2i\sqrt{\lambda}}{1+k};$$

viz. substituting for λ its values, these are

$$1,\ \frac{1}{k};\quad \frac{2i}{(1+\sqrt{k})^2},\ \frac{2i}{(1-\sqrt{k})^2};\quad \frac{2i}{(1+i\sqrt{k})^2},\ \frac{2i}{(1-i\sqrt{k})^2}.$$

477. The six transformations

$$\frac{M\,dy}{\sqrt{1-y^2 \,.\, 1-\lambda^2 y^2}} = \frac{dx}{\sqrt{1-x^2 \,.\, 1-k^2 x^2}},$$

then are

$y =$	$\lambda =$	$M =$
$x,$	$k,$	$1,$
$\frac{1}{x},$	$\frac{1}{k},$	$\frac{1}{k},$
$\frac{1+\sqrt{k}}{1-\sqrt{k}}\,\frac{1-x\sqrt{k}}{1+x\sqrt{k}},$	$\left(\frac{1-\sqrt{k}}{1+\sqrt{k}}\right)^2,$	$\frac{2i}{(1+\sqrt{k})^2},$
$\frac{1-\sqrt{k}}{1+\sqrt{k}}\,\frac{1+x\sqrt{k}}{1-x\sqrt{k}},$	$\left(\frac{1+\sqrt{k}}{1-\sqrt{k}}\right)^2,$	$\frac{2i}{(1-\sqrt{k})^2},$
$\frac{1+i\sqrt{k}}{1-i\sqrt{k}}\,\frac{1-ix\sqrt{k}}{1+ix\sqrt{k}},$	$\left(\frac{1-i\sqrt{k}}{1+i\sqrt{k}}\right)^2,$	$\frac{2i}{(1+i\sqrt{k})^2},$
$\frac{1-i\sqrt{k}}{1+i\sqrt{k}}\,\frac{1+ix\sqrt{k}}{1-ix\sqrt{k}},$	$\left(\frac{1+i\sqrt{k}}{1-i\sqrt{k}}\right)^2,$	$\frac{2i}{(1-i\sqrt{k})^2},$

where it is to be remarked that the last four transformations are included under the form

$$y = \frac{1+\alpha}{1-\alpha}\,\frac{1-\alpha x}{1+\alpha x},\quad \lambda = \left(\frac{1-\alpha}{1+\alpha}\right)^2,\quad M = \frac{2i}{(1+\alpha)^2},$$

where α is a fourth root of k^2. These are in fact Abel's results referred to No. 416.

The Quadric Transformation for the standard form.

Art. Nos. 478 to 482.

478. Reckoning the number of linear transformations as six, that of the quadric transformations is reckoned as eighteen; viz. these are Abel's eighteen transformations referred to No. 418. Taking as before the differential relation to be

$$\frac{M\,dy}{\sqrt{1-y^2\,.\,1-\lambda^2y^2}}=\frac{dx}{\sqrt{1-x^2\,.\,1-k^2x^2}},$$

we have, Four transformations

$y=$	$\lambda=$	$M=$
$\dfrac{(1+k)\,x}{1+kx^2}$,	$\dfrac{2\sqrt{k}}{1+k}$,	$\dfrac{1}{1+k}$,
$\dfrac{(1-k)\,x}{1-kx^2}$,	$\dfrac{2i\sqrt{k}}{1-k}$,	$\dfrac{1}{1-k}$,
$\dfrac{2\sqrt{k}x}{1+kx^2}$,	$\dfrac{1+k}{2\sqrt{k}}$,	$\dfrac{1}{2\sqrt{k}}$,
$\dfrac{2i\sqrt{k}x}{1-kx^2}$,	$\dfrac{1-k}{2i\sqrt{k}}$,	$\dfrac{1}{2i\sqrt{k}}$;

479. Six transformations

$y=$	$\lambda=$	$M=$
$\dfrac{1+kx^2}{-1+kx^2}$,	$\dfrac{1-k}{1+k}$,	$\dfrac{i}{1+k}$,
$\dfrac{-1+kx^2}{1+kx^2}$,	$\dfrac{1+k}{1-k}$,	$\dfrac{i}{1-k}$,
$\dfrac{1-(1+k')\,x^2}{1-(1-k')\,x^2}$,	$\dfrac{1-k'}{1+k'}$,	$\dfrac{1}{1+k'}$,
$\dfrac{1-(1-k')\,x^2}{1-(1+k')\,x^2}$,	$\dfrac{1+k'}{1-k'}$,	$\dfrac{1}{1-k'}$,
$\dfrac{-(k'+ik)+ik\,x^2}{(k'-ik)+ik\,x^2}$,	$\dfrac{k'-ik}{k'+ik}$,	$k+ik'$,
$\dfrac{(k'-ik)+ik\,x^2}{-(k'+ik)+ik\,x^2}$,	$\dfrac{k'+ik}{k'-ik}$,	$-k+ik'$;

480. And lastly, Eight transformations

$$y = \frac{\sqrt{1+k} + \sqrt{2}\sqrt[4]{k}}{\sqrt{1+k} - \sqrt{2}\sqrt[4]{k}} \, \frac{1 + kx^2 + x\sqrt{2}\sqrt[4]{k}\sqrt{1+k}}{1 + kx^2 - x\sqrt{2}\sqrt[4]{k}\sqrt{1+k}},$$

$$\lambda = \left(\frac{\sqrt{1+k} - \sqrt{2}\sqrt[4]{k}}{\sqrt{1+k} + \sqrt{2}\sqrt[4]{k}}\right)^2, \text{ and } M = \frac{2i}{(\sqrt{1+k} + \sqrt{2}\sqrt[4]{k})^2}.$$

Do. with $-\sqrt[4]{k}$ for $\sqrt[4]{k}$,

„ „ $i\sqrt[4]{k}$ „

„ „ $-i\sqrt[4]{k}$ „

Do. with $\frac{1+i}{\sqrt{2}}\sqrt[4]{k}$ for $\sqrt[4]{k}$ and $-k$, $\sqrt{1-k}$ for k, $\sqrt{1+k}$,

„ „ $\frac{1-i}{\sqrt{2}}\sqrt[4]{k}$ „ „ „ „ „ „

„ „ $\frac{-1+i}{\sqrt{2}}\sqrt[4]{k}$ „ „ „ „

„ „ $\frac{-1-i}{\sqrt{2}}\sqrt[4]{k}$ „ „ „ „ „

481. The last formulæ, writing for shortness, β an eighth root of $16k^2$, and $\alpha = \sqrt{1 + \tfrac{1}{4}\beta^4}$, are included under the form

$$y = \frac{\alpha+\beta}{\alpha-\beta} \, \frac{1 + \alpha\beta x + \tfrac{1}{4}\beta^4 x^2}{1 - \alpha\beta x + \tfrac{1}{4}\beta^4 x^2}, \quad \lambda = \left(\frac{\alpha-\beta}{\alpha+\beta}\right)^2, \quad M = \frac{2i}{(\alpha+\beta)^2},$$

and the verification may be effected as follows: we have

$$1 + y = 1 - \alpha\beta x + \tfrac{1}{4}\beta^4 x^2 + \frac{\alpha+\beta}{\alpha-\beta}(1 + \alpha\beta x + \tfrac{1}{4}\beta^4 x^2) \qquad (\div),$$

$$= \frac{2\alpha}{\alpha-\beta}(1 + \tfrac{1}{2}\beta^2 x)^2 \qquad (\div),$$

$$1 - y = 1 - \alpha\beta x + \tfrac{1}{4}\beta^4 x^2 - \frac{\alpha+\beta}{\alpha-\beta}(1 + \alpha\beta x + \tfrac{1}{4}\beta^4 x^2) \qquad (\div),$$

$$= -\frac{2\beta}{\alpha-\beta}(1 + x)(1 + \tfrac{1}{4}\beta^4 x) \qquad (\div),$$

$$1+\lambda y = 1-\alpha\beta x+\tfrac{1}{4}\beta^4x^2+\frac{\alpha-\beta}{\alpha+\beta}(1+\alpha\beta x+\tfrac{1}{4}\beta^4x^2) \qquad (\div),$$

$$= \frac{2\alpha}{\alpha+\beta}(1-\tfrac{1}{2}\beta^2x)^2 \qquad (\div),$$

$$1-\lambda y = 1-\alpha\beta x+\tfrac{1}{4}\beta^4x^2-\frac{\alpha-\beta}{\alpha+\beta}(1+\alpha\beta x+\tfrac{1}{4}\beta^4x^2) \qquad (\div),$$

$$= \frac{2\beta}{\alpha+\beta}(1-x)(1-\tfrac{1}{4}\beta^4x) \qquad (\div),$$

where

$$\text{denom.} = 1-\alpha\beta x+\tfrac{1}{4}\beta^4x^2.$$

Hence

$$\sqrt{1-y^2.1-\lambda^2y^2} = \frac{4i\alpha\beta}{(\alpha+\beta)(\alpha-\beta)}(1-\tfrac{1}{4}\beta^4x^2)\sqrt{1-x^2.1-k^2x^2} \qquad (\div),$$

where k^2 is written instead of its value $\frac{1}{16}\beta^8$: and moreover

$$dy = \frac{2\alpha\beta(\alpha+\beta)}{\alpha-\beta}(1-\tfrac{1}{4}\beta^4x^2)dx \qquad (\div),$$

in which two formulæ the denominator is equal to the square of its above-mentioned value; we hence find the required formula,

$$\frac{M\,dy}{\sqrt{1-y^2.1-\lambda^2y^2}} = \frac{dx}{\sqrt{1-x^2.1-k^2x^2}},$$

where M has its proper value $= \frac{2i}{(\alpha+\beta)^2}$.

462. It is, as regards all the formulæ, convenient to remark that the value of M may be verified by taking x small; thus, if when x is small the equation for y becomes $y=\beta x$, then obviously $M=\frac{1}{\beta}$; if the equation becomes $y=\pm 1+\beta x^2$, then we have $M=\frac{\mp\sqrt{1-\lambda^2}}{\sqrt{\mp 2\beta}}$; and so in other cases.

Combined Transformations: Irrational Transformations.

Art. Nos. 483 to 487.

483. By combining two linear transformations, we obtain a transformation which is linear, and as such is a transformation belonging to the system; viz. it is either one of the six transformations, or it is at once reducible to one of these. Similarly, by combining a quadric transformation with a linear one, we obtain a transformation which is quadric, and as such is equivalent to one of the system. For instance, changing the letters, if with the quadric transformation

$$z = \frac{(1+k)x}{1+kx^2},$$

giving $$\frac{\frac{1}{1+k}dz}{\sqrt{1-z^2\,.\,1-\lambda^2z^2}} = \frac{dx}{\sqrt{1-x^2\,.\,1-k^2x^2}},\quad \lambda = \frac{2\sqrt{k}}{1+k},$$

we combine the linear transformation

$$y = \frac{1+\sqrt{k}}{1-\sqrt{k}}\,\frac{1-x\sqrt{k}}{1+x\sqrt{k}},$$

giving $$\frac{\frac{2i}{(1+\sqrt{k})^2}dy}{\sqrt{1-y^2\,.\,1-\gamma^2y^2}} = \frac{dx}{\sqrt{1-x^2\,.\,1-k^2x^2}},\quad \gamma = \left(\frac{1-\sqrt{k}}{1+\sqrt{k}}\right)^2,$$

we have z a quadric function such that

$$\frac{\frac{(1+\sqrt{k})^2}{2i(1+k)}dz}{\sqrt{1-z^2\,.\,1-\lambda^2z^2}} = \frac{dy}{\sqrt{1-y^2\,.\,1-\gamma^2y^2}};$$

and this must be one of the series of quadric transformations. We in fact find

$$\sqrt{k} = \frac{1-\sqrt{\gamma}}{1+\sqrt{\gamma}},\ \text{and thence}\ \lambda = \frac{1-\gamma}{1+\gamma},\quad \frac{(1+\sqrt{k})^2}{2i(1+k)} = \frac{-i}{1+\gamma},$$

$$x\sqrt{k} = \frac{1-y\sqrt{\gamma}}{1+y\sqrt{\gamma}},\quad \text{or}\ x = \frac{1+\sqrt{\gamma}}{1-\sqrt{\gamma}}\,\frac{1-y\sqrt{\gamma}}{1+y\sqrt{\gamma}};$$

and thence

$$z, = \frac{(1+k)x}{1+kx^2}, \quad = \frac{1+\gamma}{1-\gamma}\,\frac{1-\gamma y^2}{1+\gamma y^2};$$

or, what is the same thing,

$$-\frac{1}{\lambda z} = \frac{1+\gamma y^2}{-1+\gamma y^2},$$

giving $\dfrac{-\dfrac{i}{1+\gamma}dz}{\sqrt{1-z^2\,.\,1-\lambda^2 z^2}} = \dfrac{dy}{\sqrt{1-y^2\,.\,1-\gamma^2 y^2}}$, where $\lambda = \dfrac{1-\gamma}{1+\gamma}$,

which $\left(\text{with } z \text{ in place of } -\dfrac{1}{\lambda z}\right)$ is one of the series of quadric transformations.

484. If we combine two quadric transformations we obtain in general an irrational transformation: viz. neither of the two variables is a rational function of the other of them, but the two are connected by an equation: for instance, if the two transformations are

$$z = \frac{2\sqrt{k}x}{1+kx^2},$$

giving $\dfrac{\dfrac{1}{2\sqrt{k}}dz}{\sqrt{1-z^2\,.\,1-\lambda^2 z^2}} = \dfrac{dx}{\sqrt{1-x^2\,.\,1-k^2x^2}}$, $\lambda = \dfrac{1+k}{2\sqrt{k}}$; and

$$y = \frac{-1+kx^2}{1+k^2x^2},$$

giving $\dfrac{\dfrac{i}{1-k}dy}{\sqrt{1-y^2\,.\,1-\gamma^2 y^2}} = \dfrac{dx}{\sqrt{1-x^2\,.\,1-k^2x^2}}$, $\gamma = \dfrac{1+k}{1-k}$;

then we have here $y^2+z^2=1$; $\dfrac{1}{\gamma^2}+\dfrac{1}{\lambda^2}=1$, giving $\dfrac{1}{\lambda^2}=\dfrac{\gamma^2-1}{\gamma^2}$,

that is $\lambda^2 = -\dfrac{\gamma^2}{\gamma_1^{\,2}}$; or $\lambda = -\dfrac{i\gamma}{\gamma_1}$ if $\gamma_1, = \sqrt{1-\gamma}$, is the complementary modulus to γ; also $\frac{1-k}{2i\sqrt{k}} = \frac{\lambda}{i\gamma} = -\frac{1}{\gamma}$, and the relation is

$$\frac{-\frac{1}{\gamma}\,dz}{\sqrt{1-z^2 \,.\, 1+\frac{\gamma'^2}{\gamma^2}z^2}} = \frac{dy}{\sqrt{1-y^2 \,.\, 1-\gamma^2y^2}}:$$

or, what is the same thing, changing the letters, the transformation arrived at is $x^2+y^2=1$, giving

$$\frac{-\frac{1}{k'}\,dy}{\sqrt{1-y^2 \,.\, 1-\lambda^2y^2}} = \frac{dx}{\sqrt{1-x^2 \,.\, 1-k^2x^2}}, \quad \lambda = \frac{ik}{k'},$$

which is at once verified, since from the assumed relation $x^2+y^2=1$ we have

$$\frac{dx}{\sqrt{1-x^2}} = \frac{-dy}{\sqrt{1-y^2}}, \quad \sqrt{1-k^2x^2} = k'\sqrt{1-\lambda^2y^2}.$$

485. Observe that the equations which define the two new variables y, z in terms of x are in general of the form

$$y = \frac{A}{B}, \quad z = \frac{C}{D},$$

where A, B, C, D are quadric functions of x. Writing these equations in the form

$$y : z : 1 = AD : BC : BD,$$

then regarding (y, z) as the co-ordinates of a point of a plane curve, these expressions of y, z in terms of the arbitrary parameter x show that the curve in question is a unicursal curve, and, being of the order four, it is a trinodal quartic; viz. the equation $\phi(y, z)=0$, obtained as above by combining any two quadric transformations, being a solution of the equation

$$\frac{M\,dz}{\sqrt{1-z^2 \,.\, 1-\lambda^2z^2}} = \frac{dy}{\sqrt{1-y^2 \,.\, 1-\gamma^2y^2}},$$

we have the theorem that this equation $\phi(y, x) = 0$ represents a curve which is in general a trinodal quartic. It has been seen how in one case the curve is a circle.

486. It appears to be a conclusion of Abel's, that if for any given values of (λ, M) the equation

$$\frac{M\,dy}{\sqrt{1-y^2 \cdot 1-\lambda^2 y^2}} = \frac{dx}{\sqrt{1-x^2 \cdot 1-k^2x^2}}$$

admits of an irrational solution, then there is always an integer number n such that the equation

$$\frac{M\,dy}{\sqrt{1-y^2 \cdot 1-\lambda^2 y^2}} = \frac{n\,dx}{\sqrt{1-x^2 \cdot 1-k^2x^2}}$$

admits of a solution $y =$ rational function of x. So that, in fact, the general problem of transformation reduces itself to the problem of rational transformation. For instance, as just seen, the equation

$$\frac{-\frac{1}{k'}dy}{\sqrt{1-y^2 \cdot 1+\frac{k^2y^2}{k'^2}}} = \frac{dx}{\sqrt{1-x^2 \cdot 1-k^2x^2}}$$

has the irrational solution $y = \sqrt{1-x^2}$; the equation

$$\frac{-\frac{1}{k'}dy}{\sqrt{1-y^2 \cdot 1+\frac{k^2y^2}{k'^2}}} = \frac{2dx}{\sqrt{1-x^2 \cdot 1-k^2x^2}}$$

has a solution $y =$ rational function of x. To verify this, observe that the first equation is satisfied by $y = \text{cn}\, u$, $x = \text{sn}\, u$ (which are such that $y = \sqrt{1-x^2}$): hence the second equation is satisfied by the values $y = \text{cn}\, 2u$, $x = \text{sn}\, u$; we have $\text{cn}\, 2u$ a rational function of $\text{sn}\, u$, *ante* No. 100, and writing therein x for $\text{sn}\, u$ we obtain

$$y = \frac{1-2x^2+k^2x^4}{1-k^2x^4}$$

as a rational solution of the second equation: the solution can of course be at once verified.

487. It appears from the formulæ given No. 94, interchanging therein (z, x) and also k, k', that the equation

$$\frac{-i\,dz}{\sqrt{1-z^2 \,.\, 1-k'^2z^2}} = \frac{dx}{\sqrt{1-x^2 \,.\, 1-k^2x^2}},$$

has the irrational solution $z = \dfrac{1}{\sqrt{1-k^2x^2}}$; hence the equation

$$\frac{-i\,dz}{\sqrt{1-z^2 \,.\, 1-k'^2z^2}} = \frac{2dx}{\sqrt{1-x^2 \,.\, 1-k^2x^2}}$$

has a solution z = rational function of x; viz. the first equation being satisfied by $x = \operatorname{sn} u$, $z = \dfrac{1}{\operatorname{dn} u}$, the second equation is satisfied by $x = \operatorname{sn} u$, $z = \dfrac{1}{\operatorname{dn} 2u}$; or $\operatorname{dn} 2u$ being a rational function of $\operatorname{sn} u$, see No. 100, replacing $\operatorname{sn} u$ by x, we find

$$z = \frac{1-k^2x^4}{1-2k^2x^2+k^2x^4}$$

as a rational solution of the second equation.

INDEX.

The figures refer to the pages: where the reference extends to the whole or the bulk of a Chapter, the Chapter is also referred to. A heading "Function," with divisions (1), (2)...(13) has been introduced.

THE END.

June, 1876.

A CLASSIFIED LIST

OF

EDUCATIONAL WORKS

PUBLISHED BY

GEORGE BELL & SONS.

Full Catalogues will be sent post free on application.

BIBLIOTHECA CLASSICA.

A Series of Greek and Latin Authors, with English Notes; edited by eminent Scholars. 8vo.

Æschylus. By F. A. Paley, M.A. 18s.

Cicero's Orations. By G. Long, M.A. 4 vols. 16s., 14s., 16s., 18s.

Demosthenes. By R. Whiston, M.A. 2 vols. 16s. each.

Euripides. By F. A. Paley, M.A. 3 vols. 16s. each.

Homer. By F. A. Paley, M.A. Vol. I. 12s.; Vol. II. 14s.

Herodotus. By Rev. J. W. Blakesley, B.D. 2 vols. 32s.

Hesiod. By F. A. Paley, M.A. 10s. 6d.

Horace. By Rev. A. J. Macleane, M.A. 18s.

Juvenal and Persius. By Rev. A. J. Macleane, M.A. 12s.

Plato. By W. H. Thompson, D.D. 2 vols. 7s. 6d. each.

Sophocles. By Rev. F. H. Blaydes, M.A. Vol. I. 18s.

Tacitus: The Annals. By the Rev. P. Frost. 15s.

Terence. By E. St. J. Parry, M.A. 18s.

Virgil. By J. Conington, M.A. 3 vols. 12s., 14s., 14s.

An Atlas of Classical Geography; Twenty-four Maps. By W. Hughes and George Long, M.A. New edition, with coloured outlines. Imperial 8vo. 12s. 6d.

Uniform with above.

A Complete Latin Grammar. By J. W. Donaldson, D.D. 3rd edition. 14s.

A Complete Greek Grammar. By J. W. Donaldson, D.D. 3rd edition. 16s.

GRAMMAR-SCHOOL CLASSICS.

A Series of Greek and Latin Authors, with English Notes. Fcap 8vo.

Cæsar de Bello Gallico. By George Long, M.A. 5s. 6d.

—— Books I.–III. For Junior Classes. By G. Long, M.A. 2s. 6d.

Catullus, Tibullus, and Propertius. Selected Poems. With Life. By Rev. A. H. Wratislaw. 3s. 6d.

Cicero: De Senectute, De Amicitia, and Select Epistles. By George Long, M.A. 4s. 6d.

Cornelius Nepos. By Rev. J. F. Macmichael. 2s. 6d.

Homer: Iliad. Books I.–XII. By F. A. Paley, M.A. 6s. 6d.

Horace. With Life. By A. J. Macleane, M.A. 6s. 6d.

Juvenal: Sixteen Satires. By H. Prior, M.A. 4s. 6d.

Martial: Select Epigrams. With Life. By F. A. Paley, M.A. 6s. 6d.

Ovid: the Fasti. By F. A. Paley, M.A. 5s.

Sallust: Catilina and Jugurtha. With Life. By G. Long, M.A. 5s.

Tacitus: Germania and Agricola. By Rev. P. Frost. 3s. 6d.

Virgil: Bucolics, Georgics, and Æneid, Books I.–IV. Abridged from Professor Conington's edition. 5s. 6d.

(The Bucolics and Georgics, in one volume. 3s.)

—— Æneid, Bks. V.–XII. Abgd. from Prof. Conington's Ed. 5s. 6d.

Xenophon: the Anabasis. With Life. By Rev. J. F. Macmichael. 5s.

—— The Cyropædia. By G. M. Gorham, M.A. 6s.

—— Memorabilia. By Percival Frost, M.A. 4s. 6d.

A Grammar-School Atlas of Classical Geography. Containing Ten selected Maps. Imperial 8vo. 5s.

Uniform with the Series.

The New Testament, in Greek. With English Notes, &c. By Rev. J. F. Macmichael. 7s. 6d.

CAMBRIDGE GREEK AND LATIN TEXTS.

Æschylus. By F. A. Paley, M.A. 3s.

Cæsar de Bello Gallico. By G. Long, M.A. 2s.

Cicero de Senectute et de Amicitia, et Epistolæ Selectæ. By G. Long, M.A. 1s. 6d.

Ciceronis Orationes. Vol. I. (in Verrem). By G. Long, M.A. 3s. 6d.

Euripides. By F. A. Paley, M.A. 3 vols. 3s. 6d. each.

Herodotus. By J. G. Blakesley, B.D. 2 vols. 7s.

Homeri Ilias. I.–XII. By F. A. Paley, M.A. 2s. 6d.

Horatius. By A. J. Macleane, M.A. 2s. 6d.
Juvenal et Persius. By A. J. Macleane, M.A. 1s. 6d.
Lucretius. By H. A. J. Munro, M.A. 2s. 6d.
Sallusti Crispi Catilina et Jugurtha. By G. Long, M.A. 1s. 6d.
Terenti Comoediae. By W. Wagner, Ph.D. 3s.
Thucydides. By J. G. Donaldson, D.D. 2 vols. 7s.
Virgilius. By J. Conington, M.A. 3s. 6d.
Xenophontis Expeditio Cyri. By J. F. Macmichael, B.A. 2s. 6d.
Novum Testamentum Graecum. By F. H. Scrivener, M.A. 4s. 6d.
An edition with wide margin for notes, 7s. 6d.

CAMBRIDGE TEXTS WITH NOTES.

A Selection of the most usually read of the Greek and Latin Authors, Annotated for Schools. Fcap 8vo.

Euripides. Alcestis. By F. A. Paley, M.A. 1s. 6d.
—— Medea. By F. A. Paley, M.A. 1s. 6d.
Aeschylus. Prometheus Vinctus. By F. A. Paley, M.A. 1s. 6d.
Ovid. Selections. By A. J. Macleane, M.A. 1s. 6d.

PUBLIC SCHOOL SERIES.

A Series of Classical Texts, annotated by well-known Scholars. Crown 8vo.

Aristophanes. The Peace. By F. A. Paley, M.A. 4s. 6d.
—— The Acharnians. By F. A. Paley. [*Preparing.*
Cicero. The Letters to Atticus. Bk. I. By A. Pretor, M.A. 4s. 6d.
Demosthenes de Falsa Legatione. By R. Shilleto, M.A. 6s.
—— The Oration against the Law of Leptines. By B. W. Beatson, M.A. 6s.
Plato. The Apology of Socrates and Crito. By W. Wagner, Ph.D. 4s. 6d.
—— The Phaedo. By W. Wagner, Ph.D. 5s. 6d.
—— The Protagoras. By W. Wayte, M.A. 4s. 6d.
Plautus. The Aulularia. By W. Wagner, Ph.D. [*Preparing.*
—— Trinummus. By W. Wagner, Ph.D. 2nd Edition. 4s. 6d.
Sophoclis Trachiniae. By A. Pretor, M.A. [*Preparing.*
Terence. By W. Wagner, Ph.D. 10s. 6d.
Theocritus. By F. A. Paley, M.A. 4s. 6d.

Others in preparation.

CRITICAL AND ANNOTATED EDITIONS.

Ætna. By H. A. J. Munro, M.A. 3s. 6d.

Aristophanis Comœdiæ. By H. A. Holden, LL.D. 8vo. 2 vols. 23s. 6d. Plays sold separately.

—— Pax. By F. A. Paley, M.A. Fcap. 8vo. 4s. 6d.

Euripides. Fabulæ Quatuor. By J. H. Monk, S.T.P. Crown 8vo. 12s.

Separately—Hippolytus, cloth, 5s. Alcestis, sewed, 4s. 6d.

Horace. Quinti Horatii Flacci Opera. By H. A. J. Munro, M.A. Large 8vo. 1l. 1s.

Livy. The first five Books. By J. Prendeville. 12mo. roan, 5s. Or Books I.–III. 3s. 6d. IV. and V. 3s. 6d.

Lucretius. Titi Lucreti Cari de Rerum Natura Libri Sex. With a Translation and Notes. By H. A. J. Munro, M.A. 2 Vols. 8vo. Vol. I. Text, 16s. Vol. II. Translation, 6s. (Sold separately.)

Ovid. P. Ovidii Nasonis Heroides XIV. By A. Palmer, M.A. 8vo. 6s.

Propertius. Sex. Aurelii Propertii Carmina. By F. A. Paley, M.A. 8vo. Cloth, 9s.

Thucydides. The History of the Peloponnesian War. By Richard Shilleto, M.A. Book I. 8vo. 6s. 6d. (Book II. *in the press.*)

Greek Testament. By Henry Alford, D.D. 4 Vols. 8vo. (Sold separately.) Vol. I. 1l. 8s. Vol. II. 1l. 4s. Vol. III. 18s. Vol. IV. Part I. 18s.; Part II. 14s.; or in one Vol. 32s.

LATIN AND GREEK CLASS-BOOKS.

Auxilia Latina. A Series of Progressive Latin Exercises. By Rev. J. B. Baddeley, M.A. Fcap 8vo. 2s.

Latin Prose Lessons. By A. J. Church, M.A. 2nd Edit. Fcap. 8vo. 2s. 6d.

Latin Exercises and Grammar Papers. By T. Collins, M.A. Fcap. 8vo. 2s. 6d.

Analytical Latin Exercises. By C. P. Mason, B.A. Post 8vo. 3s. 6d.

Scala Græca: a Series of Elementary Greek Exercises. By Rev. J. W. Davis, M.A., and R. W. Baddeley, M.A. 3rd Edition. Fcap 8vo. 2s. 6d.

Greek Verse Composition. By G. Preston, M.A. Crown 8vo. 4s. 6d.

By the Rev. P. Frost, M.A., St. John's College, Cambridge.

Eclogæ Latinæ; or, First Latin Reading Book, with English Notes and a Dictionary. 13th Thousand. *Fcap 8vo.* 2s. 6d.

Materials for Latin Prose Composition. 5th Thousand. Fcap 8vo. 2s. 6d. Key, 4s.

A Latin Verse Book. An Introductory Work on Hexameters and Pentameters. 4th Thousand. Fcap 8vo. 3s. Key, 5s.

Analecta Græca Minora, with Introductory Sentences, English Notes, and a Dictionary. 10th Thousand. Fcap 8vo. 3s. 6d.

Materials for Greek Prose Composition. 2nd Edit. Fcap. 8vo. 3s. 6d. Key, 5s.

BY THE REV. F. E. GRETTON.

A First Cheque-Book for Latin Verse-makers. 1s. 6d.

A Latin Version for Masters. 2s. 6d.

Reddenda; or Passages with Parallel Hints for Translation into Latin Prose and Verse. Crown 8vo. 4s. 6d.

Reddenda Reddita (*see next page*).

BY H. A. HOLDEN, LL.D.

Foliorum Silvula. Part I. Passages for Translation into Latin Elegiac and Heroic Verse. 6th Edition. Post 8vo. 7s. 6d.

—— Part II. Select Passages for Translation into Latin Lyric and Comic Iambic Verse. 3rd Edition. Post 8vo. 5s.

—— Part III. Select Passages for Translation into Greek Verse. 3rd Edition. Post 8vo. 8s.

Folia Silvulæ, sive Eclogæ Poetarum Anglicorum in Latinum et Græcum conversæ. 8vo. Vol. I. 10s. 6d. Vol. II. 12s.

Foliorum Centuriæ. Select Passages for Translation into Latin and Greek Prose. Post 8vo. 8s.

TRANSLATIONS, SELECTIONS, &c.

*** Many of the following books are well adapted for school prizes.

Æschylus. Translated into English Prose by F. A. Paley, M.A. 2nd Edition. 8vo. 7s. 6d.

—— Translated by Anna Swanwick. Crown 8vo. 2 vols. 12s.

—— Folio Edition, with Thirty-three Illustrations from Flaxman's designs. 2l. 2s.

Anthologia Græca. A Selection of Choice Greek Poetry, with Notes. By Rev. F. St. John Thackeray. Fcap 8vo. 7s. 6d.

Anthologia Latina. A Selection of Choice Latin Poetry, from Nævius to Boëthius, with Notes. By Rev. F. St. John Thackeray. Fcap 8vo 6s. 6d.

Aristophanes: The Peace. Text and metrical translation. By B. B. Rogers, M.A. Fcap 4to. 7s. 6d.

—— The Wasps. Text and metrical translation. By B. B. Rogers, M.A. Fcap 4to. 7s. 6d.

Corpus Poetarum Latinorum. Edited by Walker. 1 vol. 8vo. 18s.

Horace. The Odes and Carmen Sæculare. In English verse by J. Conington, M.A. 8th edition. Fcap 8vo. 5s. 6d.

—— The Satires and Epistles. In English verse by J. Conington, M.A. 3rd edition. 6s. 6d.

—— Illustrated from Antique Gems by C. W. King, M.A. The text revised with Introduction by H. A. J. Munro, M.A. Large 8vo. 1l. 1s.

Musæ Etonenses sive Carminum Etonæ Conditorum Delectus. By Richard Okes. 2 vols 8vo. 15s.

Propertius. Verse translations from Book V., with revised Latin Text. By F. A. Paley, M.A. Fcap 8vo. 3s.

Plato. Gorgias. Translated by E. M. Cope, M.A. 8vo. 7s.

—— Philebus. Translated by F. A. Paley, M.A. Small 8vo. 4s.

—— Theætetus. Translated by F. A. Paley, M.A. Small 8vo. 4s.

—— Analysis and Index of the Dialogues. By Dr. Day. Post 8vo. 5s.

Reddenda Reddita: Passages from English Poetry, with a Latin Verse Translation. By F. E. Gretton. Crown 8vo. 6s.

Sabrinæ Corolla in hortulis Regiæ Scholæ Salopiensis contexuerunt tres viri floribus legendis. Editio tertia. 8vo. 8s. 6d.

Sertum Carthusianum Floribus trium Seculorum Contextum. By W. H. Brown. 8vo 14s.

Theocritus. In English Verse, by C. S. Calverley, M.A. Crown 8vo. 7s. 6d.

Translations into English and Latin. By C. S. Calverley, M.A. Post 8vo. 7s. 6d.

—— into Greek and Latin Verse. By R. C. Jebb. 4to. cloth gilt. 10s. 6d.

Virgil in English Rhythm. By Rev. R. C. Singleton. Large crown 8vo. 7s 6d

REFERENCE VOLUMES.

A Latin Grammar. By T. H. Key, M.A. 6th Thousand. Post 8vo. 8s.

A Short Latin Grammar for Schools. By T. H. Key, M.A., F.R.S. 6th Edition. Post 8vo. 3s. 6d.

A Guide to the Choice of Classical Books. By J. B. Mayor, M.A. Crown 8vo. 2s.

The Theatre of the Greeks. By J. W. Donaldson, D.D. Post 8vo. 5s.

A History of Roman Literature. By W. S. Teuffel, Professor at the University of Tübingen. By W. Wagner, Ph.D. 2 vols Demy 8vo. 21s.

Student's Guide to the University of Cambridge. Revised and corrected. 3rd Edition. Fcap. 8vo. 6s. 6d.

CLASSICAL TABLES.

Greek Verbs. A Catalogue of Verbs, Irregular and Defective; their leading formations, tenses, and inflexions, with Paradigms for conjugation, Rules for formation of tenses, &c. &c. By J. S. Baird, T.C.D. 2s. 6d.

Greek Accents (Notes on). On Card, 6d.

Homeric Dialect. Its Leading Forms and Peculiarities. By J. S. Baird, T.C.D. 1s. 6d.

Greek Accidence. By the Rev. P. Frost, M.A. 1s.

Latin Accidence. By the Rev. P. Frost, M.A. 1s.

Latin Versification. 1s.

Notabilia Quædam; or the Principal Tenses of most of the Irregular Greek Verbs and Elementary Greek, Latin, and French Constructions. New edition. 1s. 6d.

Richmond Rules for the Ovidian Distich, &c. By J. Tate, M.A. 1s. 6d.

The Principles of Latin Syntax. 1s.

CAMBRIDGE SCHOOL AND COLLEGE TEXT-BOOKS.

A Series of Elementary Treatises for the use of Students in the Universities, Schools, and Candidates for the Public Examinations. Fcap 8vo.

Arithmetic. By Rev. C. Elsee, M.A. Fcap. 8vo. 7th Edit. 3s. 6d.

Elements of Algebra. By the Rev. C. Elsee, M.A. 4th Edit. 4s.

Arithmetic. By A. Wrigley, M.A. 3s. 6d.

—— A Progressive Course of Examples. With Answers. By J. Watson, M.A. 3rd Edition. 2s. 6d.

An Introduction to Plane Astronomy. By P. T. Main, M.A. 2nd Edition. 4s.

Conic Sections treated Geometrically. By W. H. Besant, M.A. 2nd Edition. 4s. 6d.

Elementary Statics. By Rev. H. Goodwin, D.D. 2nd Edit. 3s.

Elementary Dynamics. By Rev. H. Goodwin, D.D. 2nd Edit. 3s.

Elementary Hydrostatics. By W. H. Besant, M.A. 7th Edit. 4s.

An Elementary Treatise on Mensuration. By B. T. Moore, M.A. 5s.

The First Three Sections of Newton's Principia, with an Appendix; and the Ninth and Eleventh Sections. By J. H. Evans, M.A. 4th Edition, by P. T. Main, M.A. 4s.

Elementary Trigonometry. By T. P. Hudson, M.A. 3s. 6d.

Geometrical Optics. By W. S. Aldis, M.A. 3s. 6d.

Analytical Geometry for Schools. By T. G. Vyvyan. 3rd Edit. 4s. 6d.

Companion to the Greek Testament. By A. C. Barrett, A.M. 3rd Edition. Fcap 8vo. 5s.

An Historical and Explanatory Treatise on the Book of Common Prayer. By W. G. Humphry, B.D. 5th Edition. Fcap. 8vo. 4s 6d.

Music. By H. C. Banister. 4th Edition. 5s.

Others in Preparation.

ARITHMETIC AND ALGEBRA.

Principles and Practice of Arithmetic. By J. Hind, M.A. 9th Edit. 4s. 6d.

Elements of Algebra. By J. Hind, M.A. 6th Edit. 8vo. 10s. 6d.

See also foregoing Series.

GEOMETRY AND EUCLID.

Text Book of Geometry. By T. S. Aldis, M.A. Small 8vo. 4s. 6d. Part I. 2s. 6d. Part II. 2s.

The Elements of Euclid. By H. J. Hose. Fcap. 8vo. 4s. 6d. Exercises separately, 1s.

—— **The First Six Books,** with Commentary by Dr. Lardner. 10th Edition. 8vo. 6s.

—— **The First Two Books explained to Beginners.** By C. P. Mason, B.A. 2nd Edition. Fcap 8vo. 2s. 6d.

The Enunciations and Figures to Euclid's Elements. By Rev. J. Brasse, D.D. 3rd Edition. Fcap. 8vo. 1s. On Cards, in case, 5s. 6d. Without the Figures, 6d.

Exercises on Euclid and in Modern Geometry. By J. McDowell, B.A. Crown 8vo. 6s. 6d.

Geometrical Conic Sections. By W. H. Besant, M.A. 2nd Edit. 4s. 6d.

The Geometry of Conics. By C. Taylor, M.A. 2nd Edit. 8vo. 4s. 6d.

Solutions of Geometrical Problems, proposed at St. John's College from 1830 to 1846. By T. Gaskin, M.A. 8vo. 12s.

TRIGONOMETRY.

The Shrewsbury Trigonometry. By J. C. P. Aldous. Crown 8vo. 2s.

Elementary Trigonometry. By T. P. Hudson, M.A. 3s. 6d.

Elements of Plane and Spherical Trigonometry. By J. Hind. M.A. 5th Edition. 12mo. 6s.

An Elementary Treatise on Mensuration. By B. T. Moore, M.A. 5s.

ANALYTICAL GEOMETRY AND DIFFERENTIAL CALCULUS.

An Introduction to Analytical Plane Geometry. By W. P. Turnbull, M.A. 8vo. 12s.

Treatise on Plane Co-ordinate Geometry. By M. O'Brien, M.A. 8vo. 9s.

Problems on the Principles of Plane Co-ordinate Geometry. By W. Walton, M.A. 8vo. 16s.

Trilinear Co-ordinates, and Modern Analytical Geometry of Two Dimensions. By W. A. Whitworth, M.A. 8vo. 16s.

Choice and Chance. By W. A. Whitworth. 2nd Edit. Cr. 8vo. 6s.

An Elementary Treatise on Solid Geometry. By W. S. Aldis, M.A. 2nd Edition, revised. 8vo. 8s.

Geometrical Illustrations of the Differential Calculus. By M. B. Pell. 8vo. 2s. 6d.

Elementary Treatise on the Differential Calculus. By M. O'Brien, M.A. 8vo. 10s. 6d.

Notes on Roulettes and Glissettes. By W. H. Besant, M.A. 8vo. 3s. 6d.

MECHANICS & NATURAL PHILOSOPHY.

Elementary Statics. By H. Goodwin, D.D. Fcap. 8vo. 2nd Edit. 3s.

Treatise on Statics. By S. Earnshaw, M.A. 4th Edit. 8vo. 10s. 6d.

A Treatise on Elementary Dynamics. By W. Garnett, B.A. Cr. 8vo. 6s.

Elementary Dynamics. By H. Goodwin, D.D. Fcap. 8vo. 2nd Edit. 3s.

Problems in Statics and Dynamics. By W. Walton, M.A. 8vo. 10s. 6d.

Problems in Theoretical Mechanics. By W. Walton. 2nd Edit. revised and enlarged. Demy 8vo. 16s.

An Elementary Treatise on Mechanics. By Prof. Potter. 4th Edit. revised. 8s. 6d.

Elementary Hydrostatics. By Prof. Potter. 7s. 6d.

——— By W. H. Besant, M.A. Fcap. 8vo. 7th Edition. 4s.

A Treatise on Hydromechanics. By W. H. Besant, M.A. 8vo. *New Edition in the press.*

A Treatise on the Dynamics of a Particle. *Preparing.*

Solutions of Examples on the Dynamics of a Rigid Body. By W. N. Griffin, M.A. 8vo. 6s. 6d.

Of Motion. An Elementary Treatise. By J. R. Lunn, M.A. 7s. 6d.

Geometrical Optics. By W. S. Aldis, M.A. Fcap. 8vo. 3s. 6d.

A Chapter on Fresnel's Theory of Double Refraction. By W. S. Aldis, M.A. 8vo. 2s.

An Elementary Treatise on Optics. By Prof. Potter. Part I. 3rd Edit. 9s. 6d. Part II. 12s. 6d.

Physical Optics; or the Nature and Properties of Light. By Prof. Potter, A.M. 6s. 6d. Part II. 7s. 6d.

Heat, An Elementary Treatise on. By W. Garnett, B.A. Crown 8vo. 2s. 6d.

Figures Illustrative of Geometrical Optics. From Schelbach. By W. B. Hopkins. Folio. Plates. 10s. 6d.

The First Three Sections of Newton's Principia, with an Appendix; and the Ninth and Eleventh Sections. By J. H. Evans, M.A. 5th Edit. Edited by P. T. Main, M.A. 4s.

An Introduction to Plane Astronomy. By P. T. Main, M.A. Fcap. 8vo. Cloth. 4s.

Practical and Spherical Astronomy. By R. Main, M.A. 8vo. 14s.

Elementary Chapters on Astronomy, from the "Astronomie Physique" of Biot. By H. Goodwin, D.D. 8vo. 3s. 6d.

A Compendium of Facts and Formulæ in Pure Mathematics and Natural Philosophy. By G. R. Smalley. Fcap 8vo. 3s. 6d.

Elementary Course of Mathematics. By H. Goodwin, D.D. 6th Edit. 8vo. 16s.

Problems and Examples, adapted to the "Elementary Course of Mathematics." 3rd Edition. 8vo. 5s.

Solutions of Goodwin's Collection of Problems and Examples. By W. W. Hutt, M.A. 3rd Edition, revised and enlarged. 8vo. 9s.

Elementary Examples in Pure Mathematics. By J. Taylor. 8vo. 7s. 6d.

Mechanical Euclid. By the late W. Whewell, D.D. 5th Edition. 5s.

Mechanics of Construction. With numerous Examples. By S. Fenwick, F.R.A.S. 8vo. 12s.

Table of Anti-Logarithms. By H. E. Filipowski. 3rd Edit. 8vo. 15s.

Mathematical and other Writings of R. L. Ellis, M.A. 8vo. 16s.

Notes on the Principles of Pure and Applied Calculation. By Rev. J. Challis, M.A. Demy 8vo. 15s.

The Mathematical Principle of Physics. By Rev. J. Challis, M.A. Demy 8vo. 5s.

HISTORY, TOPOGRAPHY, &c.

Rome and the Campagna. By R. Burn, M.A. With Eighty-five fine Engravings and Twenty-six Maps and Plans. 4to. 3l. 3s.

The History of the Kings of Rome. By Dr. T. H. Dyer. 8vo. 16s.

A Plea for Livy. By T. H. Dyer. 8vo. 1s.

Roma Regalis. By T. H. Dyer. 8vo. 2s. 6d.

The History of Pompeii; Its Buildings and Antiquities. By T. H. Dyer. 3rd Edition, brought down to 1874. Post 8vo. 7s. 6d.

Ancient Athens: Its History, Topography, and Remains. By T. H. Dyer. Super-royal 8vo. Cloth. 1l. 5s.

The Decline of the Roman Republic. By G. Long. 5 vols. 8vo. 14s. each.

A History of England during the Early and Middle Ages. By C. H. Pearson, M.A. 2nd Edit., revised and enlarged. 8vo. Vol. I. 16s. Vol. II. 14s.

Historical Maps of England. By C. H. Pearson. Folio. 2nd Edit. revised. 31s. 6d.

A Practical Synopsis of English History. By A. Bowes. 4th Edit. 8vo. 2s.

Student's Text-Book of English and General History. By D. Beale. Crown 8vo. 2s. 6d.

Lives of the Queens of England. By A. Strickland. 6 vols. post 8vo. 5s. each. Abridged edition. 1 vol. 6s. 6d.

Outlines of Indian History. By A. W. Hughes. Small post 8vo. 3s. 6d.

The Elements of General History. By Prof. Tytler. New Edition, brought down to 1874. Small post 8vo. 3s. 6d.

ATLASES.

An Atlas of Classical Geography. 24 Maps. By W. Hughes and G. Long, M.A. New Edition. Imperial 8vo. 12s. 6d.

A Grammar-School Atlas of Classical Geography. Ten Maps selected from the above. New Edition. Imperial 8vo. 5s.

First Classical Maps. By the Rev. J. Tate, M.A. 3rd Edition. Imperial 8vo. 7s. 6d.

Standard Library Atlas of Classical Geography. Imp. 8vo. 7s. 6d.

PHILOLOGY.

New Dictionary of the English Language. Combining Explanation with Etymology, and copiously illustrated by Quotations from the best Authorities. By Dr. Richardson. New Edition, with a Supplement. 2 vols. 4to. 4l. 14s. 6d.; half russia, 5l. 15s. 6d.; russia, 6l. 12s. Supplement separately. 4to. 12s.

An 8vo. Edition, without the Quotations, 15s.; half russia, 20s.; russia, 24s.

A Dictionary of the English Language. By Dr. Webster. Re-edited by N. Porter and C. A. Goodrich. With Dr. Mahn's Etymology. 1 vol. 21s. With Appendices and 70 additional pages of Illustrations, 31s 6d.

"THE BEST PRACTICAL ENGLISH DICTIONARY EXTANT."—*Quarterly Review.*

The Elements of the English Language. By E. Adams, Ph. D. 14th Edition. Post 8vo. 4s. 6d.

Philological Essays. By T. H. Key, M.A., F.R.S. 8vo. 10s. 6d.

Language, its Origin and Development. By T. H. Key, M.A., F.R.S. 8vo. 14s.

Varronianus. A Critical and Historical Introduction to the Ethnography of Ancient Italy and to the Philological Study of the Latin Language. By J. W. Donaldson, D.D. 3rd Edition. 8vo. 16s.

Synonyms and Antonyms of the English Language. By Archdeacon Smith. 2nd Edition. Post 8vo. 5s.

Synonyms Discriminated. By Archdeacon Smith. Demy 8vo. 16s.

A Syriac Grammar. By G. Phillips, D.D. 3rd Edit., enlarged. 8vo. 7s 6d.

A Grammar of the Arabic Language. By Rev. W. J. Beamont, M.A. 12mo. 7s

---

DIVINITY, MORAL PHILOSOPHY, &c.

Novum Testamentum Græcum, Textus Stephanici, 1550. Curante F. H. Scrivener, A.M., LL.D. 16mo. 4s. 6d.

By the same Author.

Codex Bezæ Cantabrigiensis. 4to. 26s.

A Full Collation of the Codex Sinaiticus with the Received Text of the New Testament, with Critical Introduction. 2nd Edition, revised. Fcap. 8vo. 5s.

A Plain Introduction to the Criticism of the New Testament. With Forty Facsimiles from Ancient Manuscripts. New Edition. 8vo. 16s.

Six Lectures on the Text of the New Testament. For English Readers. Crown 8vo. 6s.

The New Testament for English Readers. By the late H. Alford, D.D. Vol. I. Part I. 3rd Edit. 12s. Vol. I. Part II. 2nd Edit. 10s 6d. Vol. II. Part I. 2nd Edit. 16s. Vol. II. Part II. 2nd Edit. 16s.

The Greek Testament. By the late H. Alford, D.D. Vol. I. 6th Edit. 1*l*. 8*s*. Vol. II. 6th Edit. 1*l*. 4*s*. Vol. III. 5th Edit. 18*s*. Vol. IV. Part I. 4th Edit. 18*s*. Vol. IV. Part II. 4th Edit. 14*s*. Vol. IV., 1*l*. 12*s*.

Companion to the Greek Testament. By A. C. Barrett, M.A. 3rd Edition. Fcap. 8vo. 5*s*.

Hints for Improvement in the Authorised Version of the New Testament. By the late J. Scholefield, M.A. 4th Edit. Fcap. 8vo. 4*s*.

Liber Apologeticus. The Apology of Tertullian, with English Notes, by H. A. Woodham, LL.D. 2nd Edition. 8vo. 8*s*. 6*d*.

The Book of Psalms. A New Translation, with Introductions, &c. By Rev. J. J. Stewart Perowne, D.D. 8vo. Vol. I. 3rd Edition, 18*s*. Vol. II. 3rd Edit. 16*s*.

—— Abridged for Schools. Crown 8vo. 10*s*. 6*d*.

The Thirty-nine Articles of the Church of England. By the Ven. Archdeacon Welchman. New Edition. Fcap. 8vo. 2*s*. Interleaved, 3*s*.

Pearson on the Creed. Carefully printed from an early edition. With Analysis and Index by E. Walford, M.A. Post 8vo. 5*s*.

An Historical and Explanatory Treatise on the Book of Common Prayer. By Rev. W. G. Humphry, B.D. 5th Edition, enlarged. Small post 8vo. 4*s*. 6*d*.

The New Table of Lessons Explained. By Rev. W. G. Humphry, B.D. Fcap. 1*s*. 6*d*.

A Commentary on the Gospels for the Sundays and other Holy Days of the Christian Year. By Rev. W. Denton, A.M. New Edition. 3 vols. 8vo. 54*s*. Sold separately.

Commentary on the Epistles for the Sundays and other Holy Days of the Christian Year. 2 vols. 36*s*. Sold separately.

Commentary on the Acts. Vol. I. 8vo. 18*s*. *Vol. II. in preparation.*

Jewel's Apology for the Church of England, with a Memoir. 32mo. 2*s*.

Notes on the Catechism. By Rev. A. Barry, D.D. 2nd Edit. Fcap. 2*s*.

Catechetical Hints and Helps. By Rev. E. J. Boyce, M.A. 3rd Edition, revised. Fcap. 2*s*. 6*d*.

Examination Papers on Religious Instruction. By Rev. E. J. Boyce. Sewed. 1*s*. 6*d*.

The Winton Church Catechist. Questions and Answers on the Teaching of the Church Catechism. By the late Rev. J. S. B. Monsell, LL.D. 3rd Edition. Cloth, 3*s*.; or in Four Parts, sewed.

The Church Teacher's Manual of Christian Instruction. By Rev. M. F. Sadler. 3rd Edition. 2*s*. 6*d*.

Brief Words on School Life. By Rev. J. Kempthorne. Fcap. 3*s*. 6*d*.

Short Explanation of the Epistles and Gospels of the Christian Year, with Questions. Royal 32mo. 2*s*. 6*d*.; calf, 4*s*. 6*d*.

Butler's Analogy of Religion; with Introduction and Index by Rev. Dr. Steere. New Edition. Fcap. 3*s*. 6*d*.

Butler's Three Sermons on Human Nature, and Dissertation on Virtue. By W. Whewell, D.D. 4th Edition. Fcap 8vo. 2s. 6d.

Lectures on the History of Moral Philosophy in England. By W. Whewell, D.D. Crown 8vo. 8s.

Elements of Morality, including Polity. By W. Whewell, D.D. New Edition, in 8vo. 15s.

Astronomy and General Physics (Bridgewater Treatise). New Edition. 5s.

Kent's Commentary on International Law. By J. T. Abdy, LL.D. 8vo. 16s.

A Manual of the Roman Civil Law. By G. Leapingwell, LL.D. 8vo. 12s.

FOREIGN CLASSICS.

A series for use in Schools, with English Notes, grammatical and explanatory, and renderings of difficult idiomatic expressions. Fcap. 8vo.

Schiller's Wallenstein. By Dr. A. Buchheim. 2nd Edit. 6s. 6d. Or the Lager and Piccolomini, 3s. 6d. Wallenstein's Tod, 3s. 6d.

——— **Maid of Orleans.** By Dr. W. Wagner. 3s. 6d.

——— **Maria Stuart.** By V. Kastner. *In the press.*

Goethe's Hermann and Dorothea. By E. Bell, M.A., and E. Wölfel. 2s. 6d.

German Ballads, from Uhland, Goethe, and Schiller. By C. L. Bielefeld. 3s. 6d.

Charles XII., par Voltaire. By L. Direy. 3rd Edit. 3s. 6d.

Aventures de Télémaque, par Fénélon. By C. J. Delille. 2nd Edit. 4s. 6d.

Select Fables of La Fontaine. By F. E. A. Gasc. New Edition. 3s.

Picciola, by X. B. Saintine. By Dr. Dubuc. 4th Edit. 3s. 6d.

FRENCH CLASS-BOOKS.

Twenty Lessons in French. With Vocabulary, giving the Pronunciation. By W. Brebner. Post 8vo. 4s.

French Grammar for Public Schools. By Rev. A. C. Clapin, M.A. Fcap. 8vo. 2nd Edit. 2s. 6d. Separately, Part I. 2s.; Part II. 1s. 6d.

Le Nouveau Trésor; or, French Student's Companion. By M. E. S. 16th Edition. Fcap. 8vo. 3s. 6d.

F. E. A. Gasc's French Course.

First French Book. Fcap. 8vo. New Edition. 1s. 6d.

Second French Book. New Edition. Fcap. 8vo. 2s. 6d.

Key to First and Second French Books. Fcap. 8vo. 3s. 6d.

French Fables for Beginners, in Prose, with Index. New Edition. 16mo. 2s.

Select Fables of La Fontaine. New Edition. Fcap. 8vo. 3s.

Histoires Amusantes et Instructives. With Notes. New Edition. Fcap. 8vo. 2s. 6d.

Practical Guide to Modern French Conversation. Fcap. 8vo. 2s. 6d.

French Poetry for the Young. With Notes. Fcap. 8vo. 2s.

Materials for French Prose Composition; or, Selections from the best English Prose Writers. New Edition. Fcap. 8vo. 4s. 6d. Key, 6s.

Prosateurs Contemporains. With Notes. 8vo. New Edition, revised. 5s.

Le Petit Compagnon; a French Talk-Book for Little Children. 16mo. 2s. 6d.

An Improved Modern Pocket Dictionary of the French and English Languages. 20th Thousand, with additions. 16mo. Cloth. 4s.

Modern French and English Dictionary. Demy 8vo. In two vols. Vol. I. F. and E. 12s.; Vol. II. E. and F. 10s.

Gombert's French Drama.

Being a Selection of the best Tragedies and Comedies of Molière, Racine, Corneille, and Voltaire. With Arguments and Notes by A. Gombert. New Edition, revised by F. E. A. Gasc. Fcap. 8vo. 1s. each; sewed, 6d.

Contents.

Molière:—Le Misanthrope. L'Avare. Le Bourgeois Gentilhomme. Le Tartuffe. Le Malade Imaginaire. Les Femmes Savantes. Les Fourberies de Scapin. Les Précieuses Ridicules. L'Ecole des Femmes. L'Ecole des Maris. Le Médecin malgré Lui.

Racine:—Phèdre. Esther. Athalie.

P. Corneille:—Le Cid. Horace. Cinna.

Voltaire:—Zaïre.

Others in preparation.

GERMAN CLASS-BOOKS.

Materials for German Prose Composition. By Dr. Buchheim. 4th Edition, revised. Fcap. 4s. 6d.

A German Grammar for Public Schools. By the Rev. A. C. Clapin and F. Holl Muller. Fcap. 2s. 6d.

Kotzebue's Der Gefangene. With Notes, by Dr. W. Stromberg. 1s.

ENGLISH CLASS-BOOKS.

The Elements of the English Language. By E. Adams, Ph.D. 14th Edition. Post 8vo. 4s. 6d.

The Rudiments of English Grammar and Analysis. By E. Adams, Ph.D. New Edition. Fcap. 8vo. 2s.

By Rev. C. P. Mason, B.A. London.

First Notions of Grammar for Young Learners. Fcap. 8vo. Cloth. 8d.

First Steps in English Grammar, for Junior Classes. Demy 18mo. New Edition. 1s.

Outlines of English Grammar for the use of Junior Classes. Cloth. 1s 6d.

English Grammar, including the Principles of Grammatical Analysis. 16th Edition. Post 8vo. 3s. 6d.

The Analysis of Sentences applied to Latin. Post 8vo. 1s. 6d.

Analytical Latin Exercises: Accidence and Simple Sentences, &c. Post 8vo. 3s. 6d.

Edited for Middle-Class Examinations.

With Notes on the Analysis and Parsing, and Explanatory Remarks.

Milton's Paradise Lost, Book I. With Life. 3rd Edit. Post 8vo. 2s.

——— Book II. With Life. 2nd Edit. Post 8vo. 2s.

——— Book III. With Life. Post 8vo. 2s.

Goldsmith's Deserted Village. With Life. Post 8vo. 1s. 6d.

Cowper's Task, Book II. With Life. Post 8vo. 2s.

Thomson's Spring. With Life. Post 8vo. 2s.

——— **Winter.** With Life. Post 8vo. 2s.

Practical Hints on Teaching. By Rev. J. Menet, M.A. 4th Edit. Crown 8vo. Cloth, 2s. 6d.; paper, 2s.

Test Lessons in Dictation. Paper cover, 1s. 6d.

Questions for Examinations in English Literature. By Rev. W. W. Skeat. 2s. 6d.

Drawing Copies. By P. H. Delamotte. Oblong 8vo. 12s. Sold also in parts at 1s. each.

Poetry for the School-room. New Edition. Fcap. 8vo. 1s. 6d.

Select Parables from Nature, for Use in Schools. By Mrs. A. Gatty. Fcap 8vo. Cloth. 1s.

School Record for Young Ladies' Schools. 6d.

Geographical Text-Book; a Practical Geography. By M. E. S. 12mo. 2s.

The Blank Maps done up separately. 4to. 2s. coloured.

A First Book of Geography. By Rev. C. A. Johns, B.A., F.L.S. &c. Illustrated. 12mo. 2s. 6d.

Loudon's (Mrs.) Entertaining Naturalist. New Edition. Revised by W. S. Dallas, F.L.S. 5s.

——— **Handbook of Botany.** New Edition, greatly enlarged by D. Wooster. Fcap. 2s. 6d.

The Botanist's Pocket-Book. With a copious Index. By W. R. Hayward. Crown 8vo. Cloth limp, 4s. 6d.

Experimental Chemistry, founded on the Work of Dr. Stöckhardt. By C. W. Heaton. Post 8vo. 5s.

Cambridgeshire Geology. By T. G. Bonney, F.G.S. &c. 8vo. 3s.

Double Entry Elucidated. By B. W. Foster. 7th Edit. 4to. 3s. 6d.

A New Manual of Book-keeping. By P. Crellin, Accountant. Crown 8vo. 3s. 6d.

Picture School-Books. In simple Language, with numerous Illustrations. Royal 16mo.

School Primer. 6d.—School Reader. By J. Tilleard. 1s.—Poetry Book for Schools. 1s.—The Life of Joseph. 1s.—The Scripture Parables. By the Rev. J. E. Clarke. 1s.—The Scripture Miracles. By the Rev. J. E. Clarke. 1s.—The New Testament History. By the Rev. J. G. Wood, M.A. 1s.—The Old Testament History. By the Rev. J. G. Wood, M.A. 1s.—The Story of Bunyan's Pilgrim's Progress. 1s.—The Life of Christopher Columbus. By Sarah Crompton. 1s.—The Life of Martin Luther. By Sarah Crompton. 1s.

BY THE LATE HORACE GRANT.

Arithmetic for Young Children. 1s. 6d.

——— **Second Stage.** 18mo. 3s.

Exercises for the Improvement of the Senses. 18mo. 1s.

Geography for Young Children. 18mo. 2s.

Books for Young Readers. In Eight Parts. Limp cloth, 8d. each; or extra binding, 1s. each.

Part I. contains simple stories told in monosyllables of not more than four letters, which are at the same time sufficiently interesting to preserve the attention of a child. Part II. exercises the pupil by a similar method in slightly longer easy words; and the remaining parts consist of stories graduated in difficulty, until the learner is taught to read with ordinary facility.

BELL'S READING-BOOKS.

FOR SCHOOLS AND PAROCHIAL LIBRARIES.

The popularity which the Series of Reading-books, known as "Books for Young Readers," has attained is a sufficient proof that teachers and pupils alike approve of the use of interesting stories, with a simple plot in place of the dry combination of letters and syllables, making no impression on the mind, of which elementary reading-books generally consist.

The publishers have therefore thought it advisable to extend the application of this principle to books adapted for more advanced readers.

Now Ready.

Masterman Ready. By Captain Marryat. 1s. 6d.

Parables from Nature (selected). By Mrs. Gatty. Fcap. 8vo. 1s.

Friends in Fur and Feathers. By Gwynfryn. 1s.

Robinson Crusoe. 1s. 6d.

Andersen's Danish Tales. By [illegible]ell, M.A. 1s.

In [illegible]:—

Our [illegible]. By Miss Mitford (selections).

Grimm's [illegible] Tales. ([illegible]ctions.)

London: Printed by JOHN STRANGEWAYS, Castle St. Leicester Sq.

www.ingramcontent.com/pod-product-compliance
Lightning Source LLC
Chambersburg PA
CBHW021350210326
41599CB00011B/818
9783742807397